Interdisciplinary Research into Cultural and Biological Transformations in the Paleolithic Period

Interdisciplinary Research into Cultural and Biological Transformations in the Paleolithic Period

Guest Editor

Ran Barkai

MDPI

Basel • Beijing • Wuhan • Barcelona • Belgrade • Novi Sad • Cluj • Manchester

Guest Editor
Ran Barkai
Archaeology and Near
Eastern Cultures
Tel-Aviv University
Tel Aviv
Israel

Editorial Office
MDPI AG
Grosspeteranlage 5
4052 Basel, Switzerland

This is a reprint of the Special Issue, published open access by the journal *Quaternary* (ISSN 2571-550X), freely accessible at: www.mdpi.com/journal/quaternary/special_issues/VKTTAHHFKA.

For citation purposes, cite each article independently as indicated on the article page online and using the guide below:

Lastname, A.A.; Lastname, B.B. Article Title. *Journal Name* **Year**, *Volume Number*, Page Range.

ISBN 978-3-7258-3162-3 (Hbk)
ISBN 978-3-7258-3161-6 (PDF)
https://doi.org/10.3390/books978-3-7258-3161-6

Cover image courtesy of Ran Barkai

Contents

About the Editor

Ran Barkai

Ran Barkai is a professor of prehistoric archaeology and former chair of the Department of Archaeology and Ancient Near EasternCultures at Tel Aviv University. His research focuses on the technological, cultural and social evolution during the Paleolithic and Neolithic periods. For the past twenty years, he has co-directed excavations and research at the Middle Pleistocen Qesem Cave. His research also addresses the study of stone tools at the Revadim and Jaljulia sites, human–elephnat interactions, cosmological and ontological perceptions in the paleolithic and the significance of caves in early humans' ways of life and perception.

Preface

This collection of papers represents the collective effort of the authors to demonstrate innovative ways of analysing and presenting new data regarding transformations in human culture and biology throughout the Paleolithic period. All the papers address key issues in recent research into the ways of life, adaptation, technology and behaviour of our early ancestors and present a plethora of new and interesting data that are aimed at diverse audiences in the fields of Archaeology, Human evolution, Palaeontology, Anthropology, Geology and more.

Ran Barkai
Guest Editor

quaternary

Article

The Evolution of Paleolithic Hunting Weapons: A Response to Declining Prey Size

Miki Ben-Dor * and Ran Barkai *

Department of Archaeology, Tel Aviv University, P.O. Box 39040, Tel Aviv 6997801, Israel
* Correspondence: bendor.michael@gmail.com (M.B.-D.); barkaran@tauex.tau.ac.il (R.B.)

Abstract: This paper examines the hypothesis that changes in hunting weapons during the Paleolithic were a direct response to a progressive decline in prey size. The study builds upon a unified hypothesis that explains Paleolithic human evolutionary and behavioral/cultural phenomena, including improved cognitive capabilities, as adaptations to mitigate declined energetic returns due to a decline in prey size. Five selected case studies in Africa and Europe were analyzed to test this hypothesis, focusing on the relative presence of megaherbivores (>1000 kg) in the transition between the Acheulean/Early Stone Age and the Middle Paleolithic/Middle Stone Age. The findings indicate a decline in megaherbivores' presence and biomass contribution in the studied transition period associated with the introduction of Levallois technology. We review the evolution of hunting weapons, including wooden-tipped and stone-tipped spears and bows and arrows. Analysis of tip size and breakage patterns indicate a reduction in point size over time, aligning with the declining prey size. We propose that changes in hunting weapons and strategies were driven by the practical and ontological incentives presented by the availability and size of prey. Developing smaller, more precise weapons required increased cognitive capacities, leading to the parallel evolution of human cognitive abilities.

Keywords: human evolution; hunting weapons; Levallois; prey size; Paleolithic

check for updates

Citation: Ben-Dor, M.; Barkai, R. The Evolution of Paleolithic Hunting Weapons: A Response to Declining Prey Size. *Quaternary* **2023**, *6*, 46. https://doi.org/10.3390/quat6030046

Academic Editor: Juan Manuel López García

Received: 16 July 2023
Revised: 31 July 2023
Accepted: 8 August 2023
Published: 24 August 2023

1. Introduction

In our previous paper in this journal [1], we proposed a unified hypothesis that explains key evolutionary and behavioral/cultural phenomena and changes in prehistory as adaptations to mitigate declining energetic returns due to prey size decline. Additionally, as research has linked weapon systems complexity to increased cognitive abilities [2–4], we proposed that prey size decline is causally related to brain size growth. This hypothesis also implies a connection between humans' evolutionary pressures and their environment. The decline in prey size would have represented a significant challenge for early humans, pushing them towards increasing adaptability and innovation in their hunting practices. These changes, in turn, would have affected their social structures, cultural practices, and overall survival strategies.

In this paper, we test a sub-hypothesis that the evolution of hunting weapons during the Paleolithic era was a direct response to a progressive decline in prey size. Smaller, precision-focused weapons were required to replace large, strength-based weapons.

The unifying hypothesis assumes there were ecological changes, possibly coupled with human predation, that led to a reduction in the size of commonly hunted animals during the Paleolithic era. To continue their subsistence lifestyle, humans had to adapt their hunting strategies and tools. The sub-hypothesis proposes that larger, heavier weapons, such as thrusting spears, which would have been effective for large prey like megaherbivores, became less useful as their abundance decreased. As a result, there was a shift towards more lethal and precise hunting tools.

These changes in weapon technology also reflect the evolution of human cognitive abilities, including problem-solving, creativity, and the ability to adapt to changing circumstances. Developing smaller, more precise weapons may have required greater skill and planning, leading to parallel cognitive development.

To test the above-mentioned hypothesis, we need to establish the basic premise that prey size declined throughout prehistory in the various regions of the Paleolithic world. We also need to present key prehistoric phenomena that changed in temporal association with the alleged decrease in prey size. We started, with partners, mapping prey size dynamics in the Levant. We used data from 133 stratigraphic layers from 58 archaeological sites throughout the Pleistocene and early Holocene. We identified a statistically significant decline in prey size between each successive cultural entity. It began with an average prey weight of three tons in the Lower Paleolithic Acheulean and ended with an average of 50 kg in the Epi-Paleolithic Natufian [5].

Other studies also found evidence of prey size decline in other areas of the world at different times during the Pleistocene. Faith et al. [6] documented a continuous decline in large and medium-sized species and an increase in small-size herbivores in East Africa throughout the Pleistocene. Yravedra [7] identified a decline in prey size in Spain between successive cultures throughout the Paleoarchaeological record. A decline in prey size megafauna richness between the Middle Paleolithic (M.P.)/Middle Stone Age (MSA) and the Upper Paleolithic (UP)/Later Stone Age (LSA) throughout the world is well documented as The Late Quaternary Megafaunal Extinction, e.g., [8,9]. Thus, scholars generally agree with the decline in prey size found in archaeological sites throughout the Pleistocene. The decline might be differently represented at different localities due to variability in animal taxa and specific environmental and topographic conditions. Thus, changes in animal taxa are to be expected.

While larger taxa were plentiful at the beginning of the Pleistocene, mostly small taxa survived into the Holocene. We will not investigate the causes of this decline. Several studies link it exclusively to stochastic ecological changes [10–12], whereas others argue that over-predation by humans was a major contributor [13–15]. In the Levant, humans were either directly responsible for the decline or made larger species more vulnerable to climate change [5]. However, our hypothesis attempts to explain the implications of size decline, not its causes. It is not dependent on the reasons for this decline.

This paper analyzes selected case studies of several key sites in South Africa, East Africa, Spain, and France. Archaeological records at these sites and groups of sites provide information about trends in prey size over time. We concentrated data collection and analysis on the transition between the Acheulean/Early Stone Age (ESA) and the MP/MSA as prey size declines in the later transition from the MP/MSA to the UP/LSA are better documented in The Late Quaternary Megafaunal Extinction framework [8,9,16,17].

Stone tools are the most conspicuous phenomenon in the prehistoric archaeological record. However, unequivocal evidence for using stone-tipped hunting weapons is found only as early as the MP [18–22], although some claims were put forward for making stone projectiles as early as half a million years ago [23]. We have solid evidence for using stone tools in animal carcass processing as early as stone tool production began three million years ago [24]. Human dependence on stone tools for meat, fat, and marrow characterizes human existence and adaptation throughout the Pleistocene and even later [25–31].

Stone tools were used to shape wood, bone, and antler into tools. However, we believe their primary role was to provide early humans access to animal carcasses' calories. It might be argued that changes in animal availability and prey size were both practical and ontological incentives for developing stone tool technologies throughout the Pleistocene. However, this paper will focus only on hunting tools rather than the entire technological repertoire. We focused on evidence of evolved morphology and complexity in hunting weapons production and use as a key phenomenon in human cultural and biological evolution. Due to the considerable time and cognitive demands of producing and maintaining complex hunting tools [32,33], it is safe to assume that they would not have evolved merely through 'cultural drift' sensu Koerper and Stickel [34].

No archaeological evidence exists for stone-tipped hunting weapons for the first two million years of human evolution. However, there is compelling evidence for hunting large prey, beginning in the Acheulean and possibly earlier, e.g., [35–37].

A wooden spear from Clacton, dated to MIS 11 (427 thousand years ago (ka) [38], and a wooden spear and hunting sticks from Schoningen, dated to MIS 9 (330 ka) [39–42], are the earliest archaeological evidence for hunting weapons. Schoningen spears' diameter suggests thrusting spears [43]. Nonetheless, they were also interpreted as throwing spears, except for one long spear, which was unanimously considered a thrusting spear. A double-ended short 64.5 cm spear was interpreted as a throwing stick [44]. Persistent hunting was considered to have been the method of choice in the Lower Paleolithic [45]. However, other hunting strategies were proposed for big game hunting in the past and present [46,47]. Recent indigenous hunters sometimes use clubs or spears to kill animals after their collapse at the end of persistent hunting [48]. Wooden spears were probably used for hunting before 400 ka. This was done by other methods [49,50] that may not be employed today due to alternative weapons. Based on the ethnographic record, Churchill [47] associated wooden thrusting spears with disadvantaging (limiting the movement) as a hunting method for very large animals. Persistence hunting can also be considered disadvantaging as the hunt aims to bring the prey to a static state of exhaustion. Milks [51], however, showed that wooden spears were used recently to hunt various animals sizes, both as throwing and thrusting spears.

The undisputed second stage of weapon evolution was wooden spear stone-tipping. The stone-tipping of spears may have begun as early as 500 ka [52]. However, stone tipping became prevalent in archaeological sites during the MP/MSA, beginning at 300 ka [21,53–56].

The third wave of hunting weapons inventions was mechanically projected weaponry systems, such as bows and arrows and spear throwers, and darts, which may have originated 100–64 ka in Africa [57–59] but were only widely used during the UP/Late Stone Age (LSA) [56,60] in Africa and the Old World. Other inventions include the domestication of dogs for use as hunter assistants in the middle to late UP [61–64] and the use of traps to capture small game [65–67].

Unlike stone, antler, and bone points, hunting weapons' other components are seldom preserved. Therefore, point size and breakage patterns are the most common indicators of hunting tool use. Reduction of point size has been a prominent trend since their appearance in the MP/MSA. Tip cross-sectional area (TCSA) is an accepted measure of tip size [68] though tip cross-sectional perimeter (TCSP) may be a more accurate measure [69]. A reasonable approximation of the TCSA can be obtained by multiplying half the tip's maximum width by its maximum thickness [43]. American ethnographic data on the average TCSA of tips attached to their shafts show that thrusting spear tips, dart (spear thrower) tips, and arrow tips have declining mean TCSAs of 168, 58, and 33, respectively [43]. Reviewing the South African record [53] found that the TCSA of points declined, beginning at 300 ka until reaching arrow tip size during the MIS 4 glacial (~71 ka). A further decline in TCSA points occurred in Africa in the LSA [70]. In the Levant, TCSA values were stable during the M.P. and declined during the UP, ending at the lowest values in the Neolithic [43] (Tables 2, 4 and 5). In Europe, there was a substantial decline in TCSA values of points between the M.P. and the UP, with no clear trend during the UP [43] (Figure 10).

Shea [43] proposed that projectile weapons were developed to provide an advantage in intraspecific human quarrels over diminishing resources in the UP/LSA. A case of interpersonal violence using a weapon is known from the Late Paleolithic [71], but ethnographic and archaeological evidence suggests prey-directed reasoning. "Each species… has its weapon", writes Bruce Winterhalder on the Cree of northern Ontario [72] (p. 85), and Churchill [47] shows that different hunting weapons were used with different hunting methods to capture different kinds of prey. In many cases, weapons and methods were associated with prey size though bows and arrows were used for hunting a wide range of prey sizes [47,73]. To our knowledge, besides Shea's [43] hypothesis, no hypothesis that associates hunting weapons evolution with an ecological driver has ever been proposed.

The chronological evolution of hunting weapons is widely accepted, namely wooden tools in the Lower Paleolithic, stone-tipped spears in the MP/MSA, and bows and atlatls mainly in the UP/LSA. As noted, a decline in megafauna presence in the Late Pleistocene is also accepted. Our testing was influenced by previous studies showing that megaherbivore biomass was high in Acheulean assemblages [5,74,75]. Consequently, we tested whether megaherbivore representation in archaeological sites declined between the LP/ESA/Acheulean and the MP/MSA in tandem with changes in hunting gear and proposed a causal explanation.

2. Material and Methods

We selected five multilevel sites with faunal data and long chronological records in the transition between the Lower Paleolithic (LP)/ESA and the MP/MSA. We compared the relative presence of megaherbivores between the two periods. Herbivores weighing more than 1000 kg are considered megaherbivores [76]. Our definition of megaherbivores is slightly different: herbivores weighing more than 1000 kg that do not escape. Owen-Smith [76] (p. 1) names megaherbivores as elephants, rhinos, and hippos "while giraffe slips marginally into the category". We did not include giraffes or bos/bison, which may sometimes have had bodyweights exceeding 1000 kg, because they tend to escape, though, for *bos/bison*, this is not always true, especially in the herd's safety. According to Churchill [47], elephants and hippos are hunted by disadvantaging and dispatching with thrusting spears, presumably the oldest hunting weapon, when untipped with stone points. A reduction in the number of megaherbivores in the assemblages suggests the need for new hunting weapons to hunt smaller, fleeing species. It was, therefore, imperative to compare the relative abundance of megaherbivores that did not escape. It should be noted that although megaherbivores usually do not comprise a high percentage of the NISP or the MNI, they may provide a dominant portion of bioenergy. For example, a six to twelve tons elephant could have provided over a hundred times more calories than smaller animals like Impala at 50 kg, so the energetic impact of their decline could have had dire consequences [74,75,77]. To study the significance of the change in megaherbivores' presence, we compared their relative contribution to total biomass in each case.

NISP (number of identified specimens) has been proposed to be superior to MNI (minimum number of individuals) in studies of taxa relative abundance [78,79]. In recent historical reviews of these indexes [80], there was no clear consensus on whether MNI or NISP indexes should be employed. There is scanty literature on using NISP or MNI to compare prey abundance by size instead of taxa. Usually, the prey abundance for a specific taxon is expressed as the relative NISP or MNI of that taxon. Our work shows that for abundance by prey size, MNI is superior to NISP because a markedly higher percentage of bones of smaller animals are brought to a central site [75]. It should be noted that only pre-depositional biases were considered in our actualistic work, so large bones, post-depositional preservation bias, and identification bias can change the conclusion. NISP, however, can be biased to a greater extent by taphonomy than MNI [81].

We used two comparison methods to elucidate megaherbivores' presence trend between the LP/ESA or Acheulean and the MP/MSA. In one method, we used only MNI. In the second, we added a consideration of the relative prey mass to sample the bioenergetic consequences of the change better.

1. **Relative abundance of megaherbivores**—MNI (megaherbivores)/total MNI—This is close to the common abundance index where, instead of one taxon, we include megaherbivores as a group of taxa.
2. **MNI-based relative biomass contribution of megaherbivores**. We multiplied the mass of each megaherbivore species by its MNI, summed it across all megaherbivore species, and divided it by the total biomass. In the case of the Hadza [75], introducing

biomass to the relative MNI index resulted in the closest estimate to the actual relative biomass contribution of the taxon to the total biomass of the assemblage.

In the data, we included only large animals known as prey, i.e., mainly herbivores >20 kg. We excluded avians and small animals such as Testudines, Leporidae, and Carnivora since they would have diluted the megaherbivore percentages and made the differences appear smaller. Animal weights were taken from Smith et al. ([82] in Supplementary Materials).

Many, if not most, Pleistocene zooarchaeological assemblages contain anthropogenic and non-anthropogenic bone accumulations that can rarely be reliably and completely separated. We analyzed the complete assemblages since the availability of prey to humans represented by non-anthropogenic accumulations also provides relevant information, even if biased against megaherbivores if the carnivorous contribution to the assemblage is significant.

Figure 1 shows the location of the studied sites in Africa and Europe and Table 1 list the sites, the reference for the data and the total MNI in each site.

Figure 1. Location of sites: 1—Elandsfontein, 2—Duinefontain, 3—Bundu Farm, 4—Pniel 6, 5—Florisbad, 6—Olorgesailie, 7—Bolomor Cave, 8—Sima del Elefante and Gran Dolina, 9—Orgnac 3.

Table 1. List of sites and references.

Country	Site	Period	Total MNI	Reference
South Africa	Elandsfontein	ESA	1140	[83–86]
South Africa	Duinefontain	ESA	225	[87,88]
South Africa	Bundu Farm	Early MSA	29	[89,90]
South Africa	Pniel 6	Early MSA	48	[89,90]
South Africa	Florisbad	Early MSA	29	[91]
East Africa	Olorgesailie Member 10	Acheulean	31	[92,93] (Supplementary Materials)
East Africa	Olorgesailie BOK layers	Early MSA	45	[92,93] (Supplementary Materials)
Europe, Spain	Bolomor Cave	Ancient MP	92	[94,95]
Europe, Spain	Sima del Elefante	Early Pleistocene	46	[96]
Europe, Spain	Gran Dolina	Early Pleistocene, Late Acheulean	57	[96]
Europe, France	Orgnac 3	Upper Acheulean, Early MP	410	[97,98]

3. Results

3.1. Africa

3.1.1. South Africa

Smith et al. [99] published a pan-African meta-study comparing Early Middle Pleistocene (EMP) to Late Middle Pleistocene (LMP) faunal assemblages to determine human behavioral changes in the period when modern humans and the MSA appeared. They concluded that adding cave assemblages to the LMP record is responsible for the apparent decline in prey size between the EMP and LMP.

We used Smith, Ruebens, Gaudzinski-Windheuser, and Steele's [99] site list, which compared EMP to LMP layers in open-air sites. We, however, divided the sites into Acheulean or MSA sites regardless of their temporal position in the EMP or LMP. The Acheulean open-air sites contain six EMP layers from Elandsfontein and two LMP/Acheulean layers from Duinefontein. Two of the MSA layers are from the LMP sites, Pniel 6 Level 3 and Bundu Farm Groups 4–6, and one is from the EMP site of Florisbad. There is doubt about the attribution of Florisbad to the MSA because the lithics were assigned to the MSA as "None of the assemblages contradicts an MSA characterization" [100] (p. 1414). Excluding Florisbad from the comparison would have produced a higher degree of megaherbivores decline in the MSA as the site contains five hippos of a total MNI of 29. It should be noted that the Acheulean Elandsfontein and the early MSA sites are situated in two distinct South African regions, the coastline and the interior, respectively. The environment of Elandsfontein, the main site in the Acheulean group, had an open landscape with a significant woody or bushy component [84] (p. 15). The early MSA sites in the interior were close to a water source in what is described as a drier environment with a more variable climate [89,100]. Based on ecological structure analysis, Haradon [92] (p. 251) attributes grassland continuity to the interior region during the Acheulean–MSA transition.

The MNI/NISP data in the open-air sites show a marked decline in megaherbivores in the MSA from 22% to 10% (Figure 2). Considering biomass, the high biomass contribution of megaherbivores is maintained despite a decline from 58% in the Acheulean to 47% of the total biomass in the Early MSA (Figure 2).

Figure 2. Percentage of MNI and Biomass of megaherbivores in the studied South African layers. Megaherbivores abundance declined between the Acheulean and the early MSA. In all the proceeding site charts, the light patterned columns represent the periods when Levallois became a significant technology.

3.1.2. East Africa, Olorgesailie

Olorgesailie is the only area with reports of NISP and MNI in East Africa's Acheulean and MSA layers [92]. Faith et al. [101] identified a reduction in prey size as part of a species turnover in East Africa between 500 and 400 ka. Potts et al. [102] temporally associated the species turnover with the appearance of the MSA and H. sapiens. Potts indicate that the fauna of the Olorgesailie BOK MSA layers matches the fauna from open, arid settings.

The decline in the relative presence of megaherbivores is very clear in the two methods. It was very high in the Acheulean at 90% of the biomass, which was like levels that we identified in the Levant Acheulean [5,74], yet, in contrast to the Levant, even after the decline in the MSA, megaherbivores still composed 60% of the biomass (Figure 3).

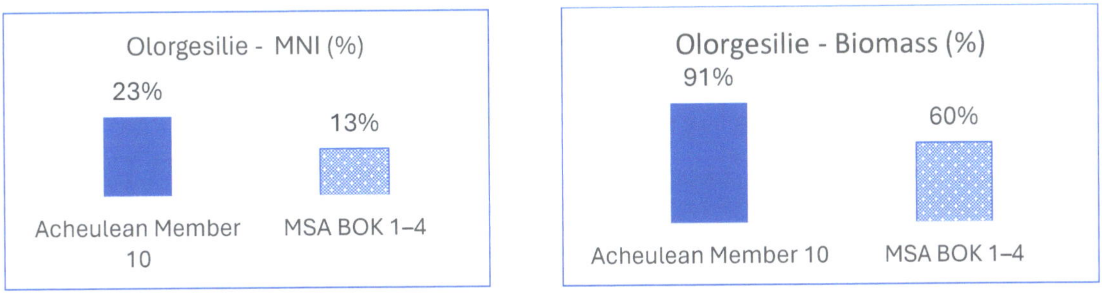

Figure 3. Percentage of MNI and Biomass of megaherbivores in Olorgesailie. Prey size declined between the Acheulean and the early MSA.

3.2. Europe

3.2.1. Spain, Bolomor Cave

Bolomor Cave, near Valencia, Spain, presents a late Middle Pleistocene MIS 9 to MIS 5e record. The stratigraphy of the cave consists of 17 levels. We compared Levels XVIIa (dated to MIS 9, 350 ka), XII, XI (dated to MIS 6), and IV (dated to MIS 5e), sampling the full length of the record. Controlled use of fire is evident, beginning in level XIII in MIS 7. The lithic assemblage shows progressing usage of Levallois techniques, although the tools' morphology was not classic. Pseudo-Levallois points appear in level XII, MIS 6 [103]. Animals associated with open space dominate the cave's assemblages, indicating a stable open landscape throughout [95].

The decline in megaherbivores is manifested in the two methods and is concurrent with the appearance of points in level XII. There is an increase in megaherbivores' presence between levels XI and IV. In any event, MIS 6 and MIS 5e layers (XII, XI, and IV) exhibit a marked decline in megaherbivores' biomass contribution compared to the MIS 9 layer (XVIIa) (Figure 4).

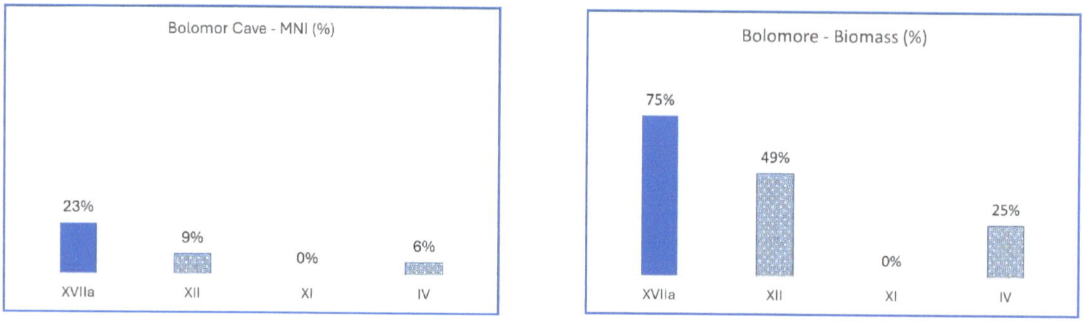

Figure 4. Percentage of MNI and Biomass of megaherbivores in Bolomor Cave. Prey size declined through the MIS 9 level (XVIIa) to the MIS 6 level (XII and XI) and MIS 5e level (IV).

3.2.2. Spain, Atapuerca

The Atapuerca archaeological complex contains several caves and galleries that provide the deepest perspective on European cultural and faunal trends. We compared layers from three assemblages hundreds of thousands of years apart. Layers TE9 to TE14 of Sima del Elefante cave are dated to the Early Pleistocene, before 1220 ka [104] (Table 1). Remains of *Homo antecessor* were found in TE9. Mode 1 lithic assemblage recovered from TE9, made on raw materials close to the cave. The environment of TE9 to TE14 was defined as "Temperate open woodlands and meadows with lagoons". Gran Dolina cave is also part of the Atapuerca archaeological complex. Layer TD6.2, dated to 731 ± 63 ka, and layer TD10.1, dated to 337 ± 29, present significant faunal assemblages [104]. Both levels were interpreted as base camps. The environment of layer 6.2 was interpreted as "open woodland and steppe" and layer 10.1 as "Mediterranean open woodland; thus, the later Gran Dolina layers were not significantly different from the TE9-TE14 layers' environment. Regarding lithic technology, layer TD6.2 presents Mode 1 technology, like the TE9-14 layers but with a high diversity of knapping strategies and raw material sourcing. Layer TD10.1 presents a transition from mode 2 to mode 3 in centripetal cores tending to Levallois [105].

The two comparison methods show the same pattern. There is a modest difference in megaherbivores presence between the Early Pleistocene TE-TE14 and the early Middle Pleistocene TD6.2, but a significant decline in megaherbivores' presence and biomass contribution from 44–54% during the Early Pleistocene to 15% during the Middle Pleistocene (Figure 5).

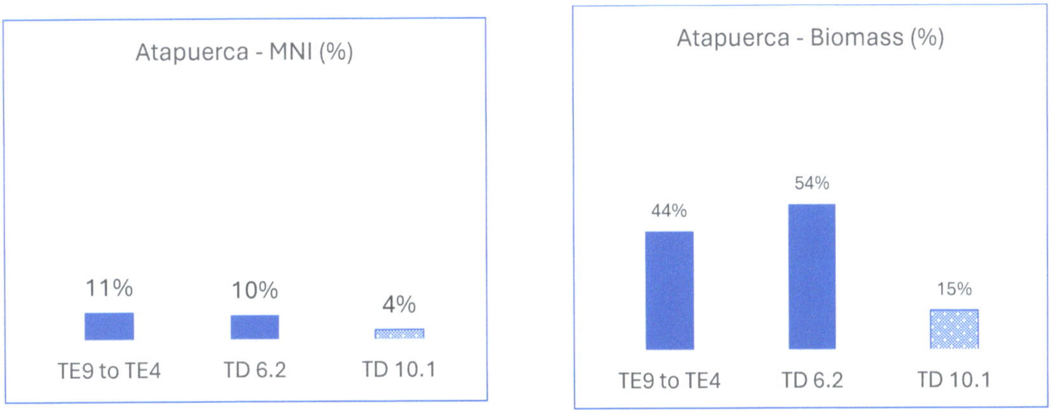

Figure 5. Percentage of MNI and Biomass of megaherbivores in Atapuerca. There is not much change in the MNI ratio of megaherbivores decline from the TE9 to TE 14 layers of Sima del Elefante dated to before 1220 ka, to layer TD 6.2 of Grand Dolina, dated to 771 ka. However, there is a marked decline in megaherbivores presence (MNI% and biomass%) between layer TD 6.2 and layer TD 10.1 of Grad Dolina dated to 337 ka.

3.2.3. France, Orgnac 3

The Orgnac 3 site is on a plateau near the Rhone River valley (France), on the right-hand bank of the Ardeche River gorges. The site layers cover MIS 10 to MIS 8 occupations, during which the European transition to Levallois-MP culture occurred [97,98]. The earlier 5–7 layers and the 4a and 4b layers were formed in a cave, but the later layers were formed in an open site after the cave roof receded. Levallois core technologies appeared in the middle levels 4a and 4b but became dominant in the MIS 8 layers 3 to 1. The climate in the period of layers 5–7 was humid and temperate at the beginning and became gradually cooler but remained humid with a forested landscape. There was no evidence of Levallois technology in layers 5–7.

The presence and biomass contribution of megaherbivores declined substantially in the MIS 8 levels 1–3, side by side with the emerging dominance of Levallois technology [98]. The climate became cold and dry, and the landscape became open, normally relatively favorable conditions for megaherbivores such as mammoths [106], so climate change cannot explain the decline (Figure 6).

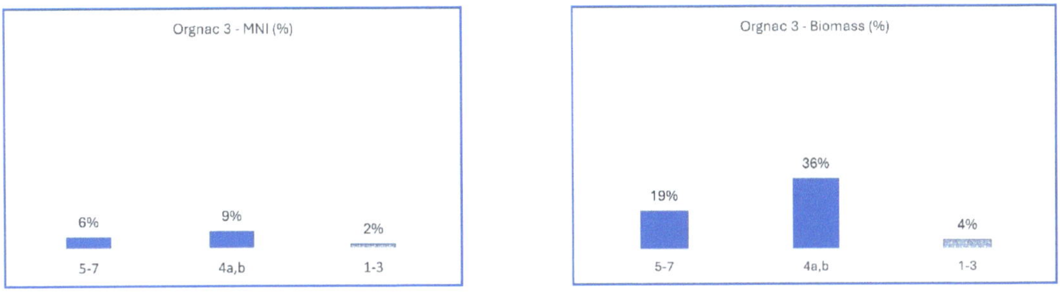

Figure 6. Percentage of MNI and Biomass of megaherbivores in Ornac. There is an increase in the presence of megaherbivores between MIS 9 layers 5–7 and layers 4A and B but a marked decline in the ratio of megaherbivores in the MIS 8 layers 1–3, compared to both 5–7 and 4A and B layers.

4. Discussion

The results should be considered exploratory because of the small sample size, excluding the Levant [5]. Both methods show that megaherbivores' presence and biomass contribution declined over time in all five cases. There seems to be a clear association between megaherbivore decline and the transition between the Acheulean and the MP/MSA in the cases where the data spanned the Acheulean and the MP/MSA. While the data do not allow examination of the transition between modes of adaptation in the three other cases, Atapuerca, Bolomor Cave, and Orgnac 3, the decline of megaherbivores was associated with an increase in Levallois-like technologies. In the Levant, Levallois technology at the end of the Acheulean was associated with elephant disappearance [107]. In all but one case (Olorgesailie), megaherbivore declines were not directly associated with climate change.

Concerning Africa in general, our findings do not corroborate Smith et al.'s (2019b) [99] assertion that the prey size decline observed during the Middle Pleistocene is an artifact of the increased number of caves used by humans between the early and late Middle Pleistocene. The increase in caves in Smith et al.'s data relates to South Africa and North Africa, so apriori could not apply to open-air sites in East Africa, where in addition to our analysis, prey size decline was evident by faunal turnover [93,101]. Additionally, we showed, via separate analysis of open-air sites in Africa, that a decline occurred in South and East Africa, which cannot be an artifact of a change in site type. It is also noteworthy that Bolomor is a cave, and at certain times, Atapuerca and Orgnac 3 were caves. They all show megaherbivores' presence decline in the Middle Pleistocene. It may well be that the causation in Smith, Ruebens, Gaudzinski-Windheuser, and Steele [99] is reversed. It is not the increased dwelling in caves that presents a faulty picture of prey size decline. It was the decline in megaherbivore reliance in the LMP that allowed humans to dwell in caves more often.

Figure 7 demonstrates the temporal association between the dynamics of prey size and hunting weapons' change, using data from the Levant.

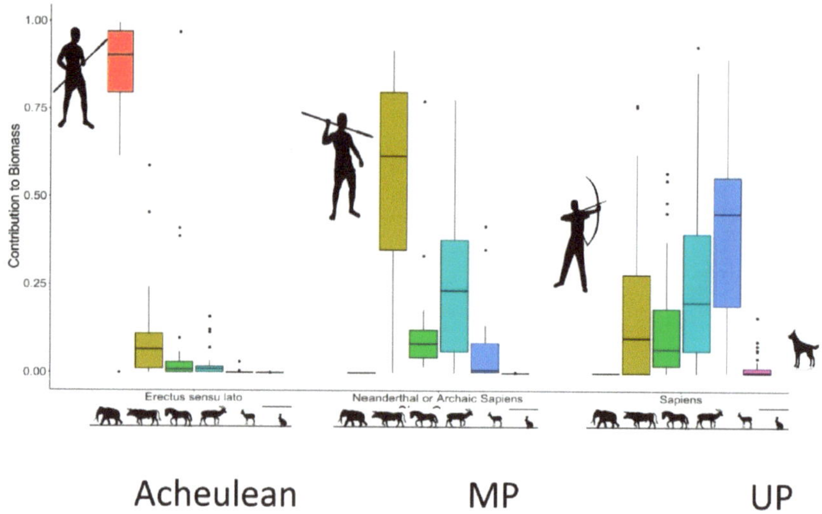

Figure 7. The graph is taken from [5]. The prey size decline is made more apparent by considering the prey biomass as documented for the Paleolithic Levant. Megaherbivores dominated the Acheulean prey biomass in the Levant, whereas in the MP, *Bos/bison* size prey (700–1000 kg) dominated the assemblages' biomass. Later, smaller prey like gazelles provided most of the biomass in the UP. Placing the figure of the typical hunting weapon of the period next to the prey size biomass distribution demonstrates the temporal association.

4.1. Energetic Return and Prey Size

Prey size is not the only factor affecting weapon choice [108]. Landscape factors such as hiding opportunities may facilitate a closer approach and improve weapon effectiveness. Small prey was also hunted throughout the Paleolithic, necessitating multiple hunting gears and strategies. Despite this acknowledged variability, this paper discusses the most appropriate hunting weapons for each phase's major dietary source. Furthermore, we follow the logic that the disappearance/decline of larger taxa led to the adoption of more appropriate weapons for smaller animals. For simplicity, the argument is presented dichotomously, although, in real life, nuanced approaches prevail.

Due to the wider range of weapons available to hunters at the end of the Pleistocene, the hunter could better match prey size and other circumstances of the hunt to maximize energy returns [72,85].

Prior to the MP/MSA, early humans used mostly wooden spears and throwing sticks for hunting. What would be the impact of hunting smaller prey with wooden-tipped spears on total energetic return? The same question about the transition from MP/MSA to UP/LSA and stone-tipped spears must be asked. What would have been the energetic returns of hunting small prey like gazelles with a stone-tipped spear had the bow and arrow not been invented? Studies in the Levant (Figure 7) and in Tanzania [75], as well as other studies [77], indicate that large prey provides most of the animal-sourced subsistence energy, so a decline in its abundance should have presented a significant loss of resources to the hunters. Existing hunting tools could not bridge the gap since they were no longer suitable for smaller and faster game. The energetic return on hunting megaherbivores with wooden-tipped spears should be compared to the return on hunting many more fleeing smaller prey animals to fill the biomass gap with the same wooden-tipped spears. There are no data on net energetic returns for such a comparison. It may have been possible to use wooden-tipped spears to hunt an increased number of smaller prey, but the energetic return is likely to have been insufficient.

To maximize their return under varying landscapes and prey sizes, recent hunter-gatherers had a full arsenal of hunting weapons at their disposal [72]. San hunters in the Kalahari sometimes carried bows, arrows, and spears in their hunting kits, and they used both thrusting and throwing spears [109]. Morin et al. [110]'s conclusion that ethnography-based hunter-gatherers' returns are not associated with prey size is thus not relevant to explaining prehistoric behaviors when hunting technologies were much more limited. After all, bows and arrows were unavailable for over 95% of the Pleistocene, and dogs, pervading Morin et al.'s data, were unavailable for 99% of it.

If anything, Morin, Bird, Winterhalder, and Bird's [110] conclusion of the non-association of net energetic returns with prey sizes confirms our assertion that the innovation of new hunting weapons and new technologies was aimed at mitigating the decline in net energetic returns due to prey size decline [1]. Their conclusion confirms that the mitigation was successful.

Figure 8 provides a graphic description of the main premise of the paper—In each period, the weapons were adapted to energetically efficiently hunting the dominant size prey of that period.

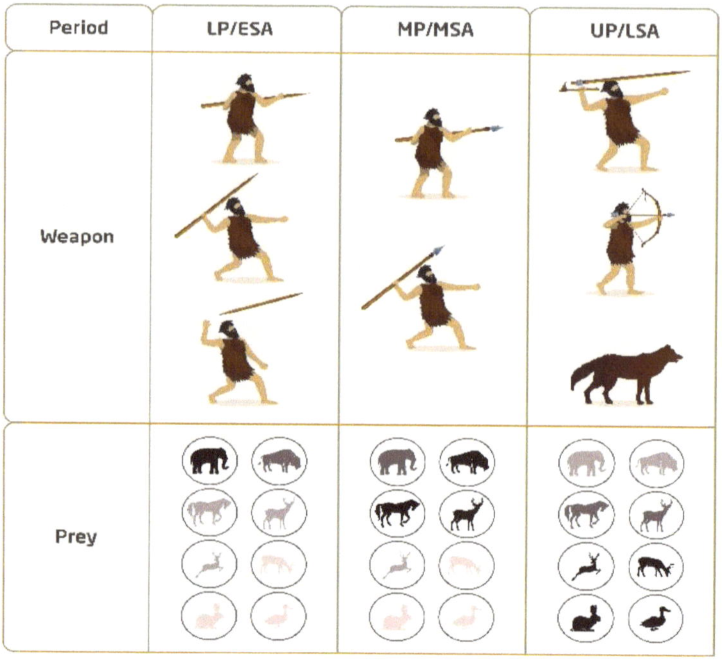

Figure 8. Each period is depicted with its typical hunting weapons and dominant prey in terms of potential caloric contribution (in darker colors). All sizes of prey were acquired in every period.

4.2. Tipping Spears with Stone Points

At geographically widespread archaeological sites, this paper shows a decline in megaherbivore abundance between the Acheulean and MP/MSA periods. There is clear evidence for the MP/MSA appearance of hafted stone points (see references in [23]). Therefore, we can determine a temporal correlation between the decline in megaherbivore presence and the appearance of hafted stone points. There is convincing evidence that hafted points were used in hunting proboscideans in North America and Europe [46,111,112], though their use has not been unanimously accepted [113]. This phenomenon may be explained by the global availability of stone points and hafting technology during the Late Pleistocene and the high relative abundance of proboscideans in new territories. Stone-tipped spears may be more effective than wooden-tipped spears in hunting proboscideans under certain circumstances. Still, in America, stone point size declined with prey size [114,115]. In the transition between MP/MSA and UP, we find a similar association. There was a decline in the size of prey in archaeological sites during this transition [5,8,9], and smaller stone tips appeared to be interpreted as dart tips and arrowheads [56,60]. The association between prey size decline and stone-tipped spears and later spear thrower-darts and bow-and-arrow hunting weapons systems must be explained.

So why would it be advantageous to tip wooden spears with stone points in the MP/MSA when megaherbivores' share of humans' food resources declined? Many changes in prehistory could be explained by adaptations to hunting smaller prey animals without exceeding a limited energetic budget [1]. Using a stone point instead of a wooden tip should have produced an increased energetic return in MP/MSA hunts of non-megaherbivores if the hypothesis holds true.

In comparing the energetic cost of hunting megaherbivores like elephants and large herbivores like bison, it is not clear if, apriori, a search for a megaherbivore is more time-consuming than a search for a large herbivore. Megaherbivores may be scarce in numbers, but they are conspicuous in the landscape because they are bigger. They leave noticeable traces in large spoons, dung piles, and broken branches [76]. Elephants are relatively easy to find, following expected paths to water sources. Many large herbivore species like *bos/bison* and equids are also conspicuous as they are large and tend to live in herds. The ethnographic record shows that stone tipping was used mainly in throwing spears [108], although stone points were used in thrusting spears. However, stone tipping of thrusting spears is not applicable to the Lower Paleolithic. Using a boat or a horse often accompanies ethnographic stone point usage in thrusting spears. This allows the stocking of several spears as the stone points tend to break and thus prevent repeated usage of the thrusting spear. In throwing spears, the stone tip allows skin penetration with a lighter spear. A lighter spear could be thrown from a longer distance, thereby increasing the chances of maintaining surprise and increasing the success rate of hitting the prey. In addition, it would reduce the energetic cost of the first stage of the hunt [108].

The next energy-consuming stage of the hunt is the pursuit, which in the hunt of large herbivores replaces the stage of disadvantaging the megaherbivores [47]. The pursuit stage can be energetically costly [110], and we can assume that the severity of the hunting weapon wound will be inversely associated with the time and distance of pursuit. Testing replicas of projectile points on gelatin, Anderson [116] found that stone points created a wound from 183 percent to 311 percent wider than that produced by a pointed wooden shaft. The same effect could be assumed in stone-tipped spears. The primacy of increased lethality in hunter consideration is supported by an ethnographic review of stone-tipped spears, spear throwers (atlatl), and arrows [108,117]. Tyua hunters of the northern Kalahari state that spears are superior to bows and arrows for killing large animals. They state that spear impact results in immediate death or sufficient blood loss to weaken the animal relatively quickly [109]. Thus, we can infer that spear stone tipping provides a larger interior wound area than wooden spear tips. Weapons kill by bleeding and damaging vital organs [68,118,119]. Thus, increased bleeding from the larger wound should shorten the escape/pursuit stage. Additionally, hyenas and other carnivores are less likely to steal the escaping prey because the pursuit is shorter. Although the breakability of the stone tip is often presented as a negative attribute, it should be considered that leaving the whole tip or a broken part of the tip in the animal tissue may cause greater blood flow as the prey escapes and thus shorten the escape/pursuit. Another consideration for MP/MSA hunters in the attempt to increase the lethality of the spears may have been safety, as the large prey they targeted, like *bos/bison*, could also attack the hunter [108,109,120].

4.3. Developing Complex Weapon Systems

There is strong evidence that prey size declined during UP/LSA as part of the Late Quaternary Megafaunal Extinction, e.g., [5,8]. The widespread usage of bows and arrows as hunting tools likely began in the Upper Paleolithic period [43].

The earliest known evidence of bow and arrow use comes from the South African site of Sibudu Cave, where researchers discovered small, pointed pieces of stone that could have been used as arrowheads, dating back approximately 64,000 years ago [58,121].

The next question is whether there is a causal association between the switch to complex projectile weapons (arrows, darts) and the further decline in prey size between the MP/MSA and the UP/LSA. A review of experimental and ethnographic literature concludes that humans traded wound areas for accuracy by switching to bows and arrows from spears. Small animals are as fast or faster than large escaping animals (Figure 9). For example, the African elephant has a maximum speed of 35 km/h, whereas the Zebra is 70 km/h, and the Impala is 90 km/h [122]. At the same time, the target organs (mostly the heart and lungs [119]) are smaller than those of large animals. At the same time, the depth of penetration and the wound size required to subdue the animal also decline. Arrows travel almost three times faster than spears [68]. The increased arrow speed enables better aiming as the trajectory is flattened (more accurate placement) [68,123]. Thus, the bow provides better accuracy at the cost of a smaller wound area. Hughes [68] mentions two other attributes of bows that increase the probability of hitting prey: the ability to produce repeated fire and the relative quietness of the arrow delivery, which maintains the element of surprise. Both attributes seem effective in hunting the fastest-flying small prey. To summarize, the main energetic savings from the transition to spear-throwers and bows and arrows may have come, as is the case with the transition from wooden to stone-tipped spears, from the increased success rate and the reduced pursuit cost when hunting small animals.

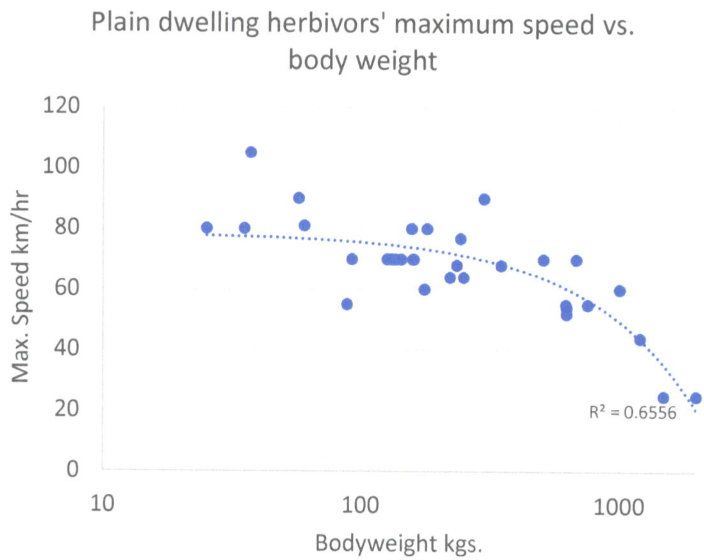

Figure 9. Speed vs. Weight in Plane-dwelling herbivores. Plane-dwelling herbivore becomes quicker as their size decrease but the association levels off at about 100 kg. Based on data from [122].

The domestication of dogs is another technology that can reduce the costs of pursuing small prey. Dog domestication occurred during the UP [124] when prey size continued to decline as megafauna became extinct in the LQME. Ethnographic evidence shows that dogs are predominantly used for hunting small prey [63,109]. During small game hunting, dogs do part of the pursuit and sometimes assist in the search, saving energy.

Studies of the association between prey size and weapons are particularly interesting when projectile points appear briefly. One such case is Sibudu Cave in South Africa. Bows and arrows are interpreted to have appeared in the Howiesons Poort and likely disappeared in the post-Howiesons Poort layers [125], when points were interpreted as spear points [21]. Explanations for the 'regression' from the use of bow and arrow are centered on demographic developments (see references in [125]). However, the decline in prey size in the Howiesons Poort, coupled with the reversal of the decline in post-Howiesons Poort assemblages [126], may also explain the appearance and disappearance of the bow. The bow introduction followed a decline in prey size in North America after 2000 B.P. [127–129]. In spite of the fact that bows were used in the Old World before humans migrated to America, they first appeared here more than ten thousand years after humans arrived. It is likely that the early Paleoindians were exposed to bow technology but did not use it until the size of prey decreased.

The model predicts that the gradual introduction of the light javelin in South Africa during the MSA, as is evident by the decline in TSCA values of the points [53], was associated with a decline in the density of large prey during the MSA. This prediction will have to be tested in the future.

4.4. Prey Size Decline and the Evolution of Cognition

The emergence of a causal brain size—prey size (reversed) association is a key implication of the causal association between prey size decline and the development of novel hunting weapons. Human biological and cultural evolution is closely linked to technological innovations [33,130]. As weapons become more complex, they require more cognitive ability [131,132]. Additionally, tracking prey may also require enhanced cognitive abilities [133,134]. As noted, a key difference in energy expenditure between disadvantaging megaherbivores and ambush hunting smaller prey is that disadvantaging saves pursuit. In contrast, ambush hunting with a spear or bow and arrow is associated with a longer pursuit of the smaller and faster prey (Figure 9). As far as we could ascertain, none of the past literature discusses a unifying ecological driver for weapon technology evolution or the need for enhanced cognitive capability for the tracking stage of smaller prey. We argue that the production of complex weapons and the employment of gradually more advanced tracking behaviors, at the cost of increasing cognitive resources, contributed to energetic savings in the increasingly longer pursuit stages of the hunt. We can thus infer that prey size decline and the resulting need to mitigate the additional energetic expenses imposed by the decline were, at least partly, driving human cognition evolution.

5. Conclusions

The findings in this paper contribute to the growing body of evidence that suggests prey sizes declined during the transition between the LP, ESA, and Acheulean. The decline was first associated with the appearance of spear-sized stone points, mostly made by the Levallois method. Later, the prey size decline known as The Late Quaternary Megafaunal Extinction was associated with the invention and adoption of complex projectile systems, trapping devices, and dog domestication. We hypothesized that the employment of the new weapon technology resulted in the mitigation of potential energetic cost increases with the decline in prey size. We expanded on the mechanism for the weapons to improve the energetic return from hunting smaller prey. Additionally, we discussed the emerging causal relationship between prey size decline and cognitive abilities extension. Wider geographical and temporal research is needed to increase the resolution of the association between prey size trends and the evolution of hunting weapons, technologies, and methods.

Supplementary Materials: The following supporting information can be downloaded at: https://www.mdpi.com/article/10.3390/quat6030046/s1. The data that support the findings are available from the corresponding author upon request.

Author Contributions: Conceptualization, M.B.-D. and R.B.; methodology, M.B.-D. and R.B.; data curation, M.B.-D.; writing—original draft preparation, M.B.-D. and R.B.; writing—review and editing, M.B.-D. and R.B. All authors have read and agreed to the published version of the manuscript.

Funding: This research received no external funding.

Data Availability Statement: The data is available as Supplementary Materials on this paper.

Acknowledgments: We would like to thank the three reviewers for their fruitful comments.

Conflicts of Interest: The authors declare no conflict of interest.

References

1. Ben-Dor, M.; Barkai, R. Prey Size Decline as a Unifying Ecological Selecting Agent in Pleistocene Human Evolution. *Quaternary* **2021**, *4*, 7. [CrossRef]
2. Lombard, M.; Gardenfors, P. Tracking the evolution of causal cognition in humans. *J. Anthropol. Sci.* **2017**, *95*, 219–234. [CrossRef]
3. Pargeter, J.; Khreisheh, N.; Stout, D. Understanding stone tool-making skill acquisition: Experimental methods and evolutionary implications. *J. Hum. Evol.* **2019**, *133*, 146–166. [CrossRef]
4. Wadley, L. Compound-adhesive manufacture as a behavioral proxy for complex cognition in the Middle Stone Age. *Curr. Anthropol.* **2010**, *51*, S111–S119. [CrossRef]
5. Dembitzer, J.; Barkai, R.; Ben-Dor, M.; Meiri, S. Levantine overkill: 1.5 Million Years of Hunting Down the Body Size Distribution in the Paleolithic Southern Levant. *Quat. Sci. Rev.* **2021**, *276*, 107316. [CrossRef]

6. Faith, J.T.; Rowan, J.; Du, A. Early hominins evolved within non-analog ecosystems. *Proc. Natl. Acad. Sci. USA* **2019**, *116*, 21478–21483. [CrossRef] [PubMed]
7. Yravedra, J. Zooarqueología de la Península Ibérica: Implicaciones tafonómicas y paleoecológicas en el debate de los homínidos del Pleistoceno Medio-Superior. *BAR Int. Ser.* **2001**, *979*. [CrossRef]
8. Koch, P.L.; Barnosky, A.D. Late Quaternary Extinctions: State of the Debate. *Annu. Rev. Ecol. Evol. Syst.* **2006**, *37*, 215–252. [CrossRef]
9. Smith, F.A.; Smith, R.E.E.; Lyons, S.K.; Payne, J.L.; Villaseñor, A. The accelerating influence of humans on mammalian macroecological patterns over the late Quaternary. *Quat. Sci. Rev.* **2019**, *211*, 1–16. [CrossRef]
10. Meltzer, D.J. Overkill, glacial history, and the extinction of North America's Ice Age megafauna. *Proc. Natl. Acad. Sci. USA* **2020**, *117*, 28555–28563. [CrossRef]
11. Faith, J.T.; Rowan, J.; Du, A.; Barr, W.A. The uncertain case for human-driven extinctions prior to Homo sapiens. *Quatern. Res.* **2020**, *96*, 88–104. [CrossRef]
12. Wroe, S.; Field, J.H.; Archer, M.; Grayson, D.K.; Price, G.J.; Louys, J.; Faith, J.T.; Webb, G.E.; Davidson, I.; Mooney, S.D. Climate change frames debate over the extinction of megafauna in Sahul (Pleistocene Australia-New Guinea). *Proc. Natl. Acad. Sci. USA* **2013**, *110*, 8777–8781. [CrossRef] [PubMed]
13. Faurby, S.; Svenning, J.C. Historic and prehistoric human-driven extinctions have reshaped global mammal diversity patterns. *Divers. Distrib.* **2015**, *21*, 1155–1166. [CrossRef]
14. Johnson, C.N.; Alroy, J.; Beeton, N.; Bird, M.I.; Brook, B.W.; Cooper, A.; Gillespie, R.; Herrando-Pérez, S.; Jacobs, Z.; Miller, G.H. What caused extinction of the Pleistocene megafauna of Sahul? *Proc. R. Soc. Lond. B Biol. Sci.* **2016**, *283*, 20152399. [CrossRef]
15. Saltré, F.; Rodríguez-Rey, M.; Brook, B.W.; Johnson, C.N.; Turney, C.S.; Alroy, J.; Cooper, A.; Beeton, N.; Bird, M.I.; Fordham, D.A.; et al. Climate change not to blame for late Quaternary megafauna extinctions in Australia. *Nat. Commun.* **2016**, *7*, 10511. [CrossRef]
16. Faith, J.T. Late Pleistocene and Holocene mammal extinctions on continental Africa. *Earth-Sci. Rev.* **2014**, *128*, 105–121. [CrossRef]
17. Barnosky, A.D.; Lindsey, E.L. Timing of Quaternary megafaunal extinction in South America in relation to human arrival and climate change. *Quat. Int.* **2010**, *217*, 10–29. [CrossRef]
18. Yaroshevich, A.; Oron, M.; Sharon, G. Big-game hunting during the late Middle Paleolithic in the Levant: Insights into technology and behavior from Nahal Mahanayeem Outlet, Upper Jordan River, israel. *J. Archaeol. Sci. Rep.* **2023**, *47*, 103777. [CrossRef]
19. Boëda, E.; Geneste, J.-M.; Griggo, C.; Mercier, N.; Muhesen, S.; Reyss, J.L.; Taha, A.; Valladas, H. A Levallois point embedded in the vertebra of a wild ass (*Equus africanus*): Hafting, projectiles and Mousterian hunting weapons. *Antiquity* **1999**, *73*, 394–402. [CrossRef]
20. Lazuén, T. European Neanderthal stone hunting weapons reveal complex behaviour long before the appearance of modern humans. *J. Archaeol. Sci.* **2012**, *39*, 2304–2311. [CrossRef]
21. Villa, P.; Lenoir, M. Hunting and hunting weapons of the Lower and Middle Paleolithic of Europe. In *The Evolution of Hominin Diets*; Springer: Berlin/Heidelberg, Germany, 2009; pp. 59–85.
22. Lombard, M.; Moncel, M.-H. Neanderthal Hunting Weapons Re-Assessed: A Tip Cross-Sectional Area Analysis of Middle Palaeolithic Point Assemblages from South Eastern France. *Quaternary* **2023**, *6*, 17. [CrossRef]
23. Wilkins, J.; Schoville, B.J.; Brown, K.S.; Chazan, M. Evidence for early hafted hunting technology. *Science* **2012**, *338*, 942–946. [CrossRef]
24. Plummer, T.W.; Oliver, J.S.; Finestone, E.M.; Ditchfield, P.W.; Bishop, L.C.; Blumenthal, S.A.; Lemorini, C.; Caricola, I.; Bailey, S.E.; Herries, A.I.; et al. Expanded geographic distribution and dietary strategies of the earliest Oldowan hominins and Paranthropus. *Science* **2023**, *379*, 561–566. [CrossRef] [PubMed]
25. Venditti, F.; Cristiani, E.; Nunziante-Cesaro, S.; Agam, A.; Lemorini, C.; Barkai, R. Animal residues found on tiny Lower Paleolithic tools reveal their use in butchery. *Sci. Rep.* **2019**, *9*, 13031. [CrossRef] [PubMed]
26. Venditti, F.; Agam, A.; Tirillo, J.; Nunziante-Cesaro, S.; Barkai, R. An integrated study discloses chopping tools use from Late Acheulean Revadim (Israel). *PLoS ONE* **2021**, *16*, e0245595. [CrossRef] [PubMed]
27. Assaf, E.; Caricola, I.; Gopher, A.; Rosell, J.; Blasco, R.; Bar, O.; Zilberman, E.; Lemorini, C.; Baena, J.; Barkai, R.; et al. Shaped stone balls were used for bone marrow extraction at Lower Paleolithic Qesem Cave, Israel. *PLoS ONE* **2020**, *15*, e0230972. [CrossRef] [PubMed]
28. Zupancich, A.; Shemer, M.; Barkai, R. Biface use in the Lower Paleolithic Levant: First insights from late acheulean Revadim and jaljulia (Israel). *J. Archaeol. Sci. Rep.* **2021**, *36*, 102877. [CrossRef]
29. Marinelli, F.; Lemorini, C.; Barkai, R. Lower Palaeolithic small flakes and megafauna: The contribution of experimental approach and use-wear analysis to reveal the link. In *Human-Elephant Interactions: From Past to Present*; Tübingen University Press: Tübingen, Germany, 2021.
30. Solodenko, N.; Zupancich, A.; Cesaro, S.N.; Marder, O.; Lemorini, C.; Barkai, R. Fat residue and use-wear found on Acheulian biface and scraper associated with butchered elephant remains at the site of Revadim, Israel. *PLoS ONE* **2015**, *10*, e0118572. [CrossRef]
31. Finkel, M.; Barkai, R. Technological persistency following faunal stability during the Pleistocene: A model for reconstructing Paleolithic adaptation strategies based on mosaic evolution. *L'Anthropologie* **2021**, *125*, 102839. [CrossRef]

32. Lombard, M. Mountaineering or ratcheting? Stone Age hunting weapons as proxy for the evolution of human technological, behavioral and cognitive flexibility. In *The Nature of Culture*; Springer: Berlin/Heidelberg, Germany, 2016; pp. 135–146.
33. Toth, N.; Schick, K. An overview of the cognitive implications of the Oldowan Industrial Complex. *Azania Archaeol. Res. Afr.* **2018**, *53*, 3–39. [CrossRef]
34. Koerper, H.C.; Stickel, E.G. Cultural drift: A primary process of culture change. *J. Anthropol. Res.* **1980**, *36*, 463–469. [CrossRef]
35. Domínguez-Rodrigo, M.; Pickering, T.R. The meat of the matter: An evolutionary perspective on human carnivory. *Azania Archaeol. Res. Afr.* **2017**, *52*, 4–32. [CrossRef]
36. Linares Matás, G.J.; Yravedra, J. 'We hunt to share': Social dynamics and very large mammal butchery during the Oldowan–Acheulean transition. *World Archaeol.* **2022**, *53*, 224–254. [CrossRef]
37. Oliver, J.S.; Plummer, T.W.; Hertel, F.; Bishop, L.C. Bovid mortality patterns from Kanjera South, Homa Peninsula, Kenya and FLK-Zinj, Olduvai Gorge, Tanzania: Evidence for habitat mediated variability in Oldowan hominin hunting and scavenging behavior. *J. Hum. Evol.* **2019**, *131*, 61–75. [CrossRef]
38. Bridgland, D.R.; Field, M.H.; Holmes, J.A.; McNabb, J.; Preece, R.C.; Selby, I.; Wymer, J.J.; Boreham, S.; Irving, B.G.; Parfitt, S.A.; et al. Middle Pleistocene interglacial Thames–Medway deposits at Clacton-on-Sea, England: Reconsideration of the biostratigraphical and environmental context of the type Clactonian Palaeolithic industry. *Quat. Sci. Rev.* **1999**, *18*, 109–146. [CrossRef]
39. Milks, A.G.; Lehmann, J.; Leder, D.; Sietz, M.; Koddenberg, T.; Böhner, U.; Wachtendorf, V.; Terberger, T. A double-pointed wooden throwing stick from Schöningen, Germany: Results and new insights from a multianalytical study. *PLoS ONE* **2023**, *18*, e0287719. [CrossRef]
40. Conard, N.J.; Serangeli, J.; Bigga, G.; Rots, V. A 300,000-year-old throwing stick from Schöningen, northern Germany, documents the evolution of human hunting. *Nat. Ecol. Evol.* **2020**, *4*, 690–693. [CrossRef]
41. Thieme, H. Lower Palaeolithic hunting spears from Germany. *Nature* **1997**, *385*, 807–810. [CrossRef]
42. Richter, D.; Krbetschek, M. The age of the Lower Paleolithic occupation at Schöningen. *J. Hum. Evol.* **2015**, *89*, 46–56. [CrossRef]
43. Shea, J.J. The origins of lithic projectile point technology: Evidence from Africa, the Levant, and Europe. *J. Archaeol. Sci.* **2006**, *33*, 823–846. [CrossRef]
44. Conard, N.J.; Serangeli, J.; Böhner, U.; Starkovich, B.M.; Miller, C.E.; Urban, B.; Van Kolfschoten, T. Excavations at Schöningen and paradigm shifts in human evolution. *J. Hum. Evol.* **2015**, *89*, 1–17. [CrossRef]
45. Bramble, D.M.; Lieberman, D.E. Endurance running and the evolution of Homo. *Nature* **2004**, *432*, 345–352. [CrossRef] [PubMed]
46. Agam, A.; Barkai, R. Elephant and mammoth hunting during the Paleolithic: A Review of the relevant archaeological, ethnographic and ethno-historical records. *Quaternary* **2018**, *1*, 3. [CrossRef]
47. Churchill, S.E. Weapon technology, prey size selection, and hunting methods in modern hunter-gatherers: Implications for hunting in the Palaeolithic and Mesolithic. *Archeol. Pap. Am. Anthropol. Assoc.* **1993**, *4*, 11–24. [CrossRef]
48. Liebenberg, L. Persistence hunting by modern hunter-gatherers. *Curr. Anthropol.* **2006**, *47*, 1017–1026. [CrossRef]
49. Bunn, H.T. Large ungulate mortality profiles and ambush hunting by Acheulean-age hominins at Elandsfontein, Western Cape Province, South Africa. *J. Archaeol. Sci.* **2019**, *107*, 40–49. [CrossRef]
50. Kübler, S.; Owenga, P.; Reynolds, S.C.; Rucina, S.M.; King, G.C. Animal movements in the Kenya Rift and evidence for the earliest ambush hunting by hominins. *Sci. Rep.* **2015**, *5*, 14011. [CrossRef]
51. Milks, A. A review of ethnographic use of wooden spears and implications for pleistocene hominin hunting. *Open Quat.* **2020**, *6*, 12. [CrossRef]
52. Wilkins, J.; Schoville, B.J. Edge damage on 500-thousand-year-old spear tips from Kathu Pan 1, South Africa: The combined effects of spear use and taphonomic processes. In *Multidisciplinary Approaches to the Study of Stone Age Weaponry*; Springer: Berlin/Heidelberg, Germany, 2016; pp. 101–117.
53. Lombard, M. Variation in hunting weaponry for more than 300,000 years: A tip cross-sectional area study of Middle Stone Age points from southern Africa. *Quat. Sci. Rev.* **2021**, *264*, 107021. [CrossRef]
54. McBrearty, S.; Tryon, C. From Acheulean to middle stone age in the Kapthurin formation, Kenya. In *Transitions before the Transition*; Springer: Berlin/Heidelberg, Germany, 2006; pp. 257–277.
55. Mercier, N.; Valladas, H. Reassessment of TL age estimates of burnt flints from the Paleolithic site of Tabun Cave, Israel. *J. Hum. Evol.* **2003**, *45*, 401–409. [CrossRef]
56. Villa, P.; Soriano, S. Hunting weapons of Neanderthals and early modern humans in South Africa: Similarities and differences. *J. Anthropol. Res.* **2010**, *66*, 5–38. [CrossRef]
57. Brooks, A.S.; Nevell, L.; Yellen, J.E.; Hartman, G. Projectile technologies of the African MSA. In *Transitions Before the Transition*; Springer: Boston, MA, USA, 2006; pp. 233–255.
58. Backwell, L.; Bradfield, J.; Carlson, K.J.; Jashashvili, T.; Wadley, L.; d'Errico, F. The antiquity of bow-and-arrow technology: Evidence from Middle Stone Age layers at Sibudu Cave. *Antiquity* **2018**, *92*, 289–303. [CrossRef]
59. Lombard, M.; Phillipson, L. Indications of bow and stone-tipped arrow use 64,000 years ago in KwaZulu-Natal, South Africa. *Antiquity* **2010**, *84*, 635–648. [CrossRef]
60. Sano, K.; Arrighi, S.; Stani, C.; Aureli, D.; Boschin, F.; Fiore, I.; Spagnolo, V.; Ricci, S.; Crezzini, J.; Boscato, P.; et al. The earliest evidence for mechanically delivered projectile weapons in Europe. *Nat. Ecol. Evol.* **2019**, *3*, 1409–1414. [CrossRef]

61. Germonpré, M.; Lázničková-Galetová, M.; Sablin, M.V.; Bocherens, H. *Could Incipient Dogs Have Enhanced Differential Access to Resources among Upper Palaeolithic Hunter-Gatherers in Europe?* McDonald Institute for Archaeological Research: Cambridge, UK, 2020.
62. Guagnin, M.; Perri, A.R.; Petraglia, M.D. Pre-Neolithic evidence for dog-assisted hunting strategies in Arabia. *J. Anthropol. Archaeol.* **2018**, *49*, 225–236. [CrossRef]
63. Lupo, K.D. When and where do dogs improve hunting productivity? The empirical record and some implications for early Upper Paleolithic prey acquisition. *J. Anthropol. Archaeol.* **2017**, *47*, 139–151. [CrossRef]
64. Yeomans, L.; Martin, L.; Richter, T. Close companions: Early evidence for dogs in northeast Jordan and the potential impact of new hunting methods. *J. Anthropol. Archaeol.* **2019**, *53*, 161–173. [CrossRef]
65. Wadley, L. Were snares and traps used in the Middle Stone Age and does it matter? A review and a case study from Sibudu, South Africa. *J. Hum. Evol.* **2010**, *58*, 179–192. [CrossRef]
66. Sinitsyn, A.; Stepanova, K.; Petrova, E. New direct evidence of mammoth hunting from Kostenki. *Prehist. Archaeol. J. Interdiscip. Stud.* **2019**, *1*, 149–158. [CrossRef]
67. Bergman, C.A. Hafting and use of bone and antler points from Ksar Akil, Lebanon. *MOM Ed.* **1987**, *15*, 117–126.
68. Hughes, S.S. Getting to the point: Evolutionary change in prehistoric weaponry. *J. Archaeol. Method Theory* **1998**, *5*, 345–408. [CrossRef]
69. Sisk, M.L.; Shea, J.J. Experimental use and quantitative performance analysis of triangular flakes (Levallois points) used as arrowheads. *J. Archaeol. Sci.* **2009**, *36*, 2039–2047. [CrossRef]
70. Bradfield, J. The Evolution of Bone Points as Hunting Weapons in South Africa. Ph.D. Thesis, University of the Witwatersrand, Johannesburg, South Africa, 2010.
71. Bocquentin, F.; Bar-Yosef, O. Early Natufian remains: Evidence for physical conflict from Mt. Carmel, Israel. *J. Hum. Evol.* **2004**, *47*, 19–23. [CrossRef]
72. Winterhalder, B. Foraging Strategies in the Boreal Forest: An Analysis of Cree Hunting and Gathering. In *Hunter-Gathere Foraging Startegies*; Winterhalder, B., Smith, E.A., Eds.; University of Chicago Press: Chicago, IL, USA, 1981.
73. Villa, P.; D'errico, F. Bone and ivory points in the Lower and Middle Paleolithic of Europe. *J. Hum. Evol.* **2001**, *41*, 69–112. [CrossRef] [PubMed]
74. Ben-Dor, M.; Gopher, A.; Hershkovitz, I.; Barkai, R. Man the fat hunter: The demise of Homo erectus and the emergence of a new hominin lineage in the Middle Pleistocene (ca. 400 kyr) Levant. *PLoS ONE* **2011**, *6*, e28689. [CrossRef]
75. Ben-Dor, M.; Barkai, R. Supersize does matter: The importance of large prey in Paleolithic subsistence and a method for measurement of its significance in zooarchaeological assemblages. In *Human-Elephant Interactions: From Past to Present*; Konidaris, G., Barkai, R., Tourloukis, V., Harvati, K., Eds.; Tübingen University Press: Tübingen, Germany, 2021. [CrossRef]
76. Owen-Smith, R.N. *Megaherbivores: The Influence of Very Large Body Size on Ecology;* Cambridge University Press: Cambridge, UK, 1988.
77. Brammer, J.R.; Menzies, A.K.; Carter, L.S.; Giroux-Bougard, X.; Landry-Cuerrier, M.; Leblanc, M.-L.; Neelin, M.N.; Studd, E.K.; Humphries, M.M. Weighing the importance of animal body size in traditional food systems. *FACETS* **2022**, *7*, 286–318. [CrossRef]
78. Grayson, D.K. *Quantitative Zooarchaeology: Topics in the Analysis of Archaelogical Faunas*; Elsevier: Amsterdam, The Netherlands, 2014.
79. Lyman, R.L. *Quantitative Paleozoology;* Cambridge University Press: Cambridge, UK, 2008.
80. Lyman, R.L. Observations on the history of zooarchaeological quantitative units: Why NISP, then MNI, then NISP again? *J. Archaeol. Sci. Rep.* **2018**, *18*, 43–50. [CrossRef]
81. Domínguez-Rodrigo, M. Critical review of the MNI (minimum number of individuals) as a zooarchaeological unit of quantification. *Archaeol. Anthropol. Sci.* **2012**, *4*, 47–59. [CrossRef]
82. Smith, F.A.; Smith, R.E.E.; Lyons, S.K.; Payne, J.L. Body size downgrading of mammals over the late Quaternary. *Science* **2018**, *360*, 310–313. [CrossRef]
83. Braun, D.R.; Levin, N.E.; Stynder, D.; Herries, A.I.; Archer, W.; Forrest, F.; Roberts, D.L.; Bishop, L.C.; Matthews, T.; Lehmann, S.B.; et al. Mid-Pleistocene hominin occupation at Elandsfontein, Western Cape, South Africa. *Quat. Sci. Rev.* **2013**, *82*, 145–166. [CrossRef]
84. Forrest, F.L. *Zooarchaeological and Palaeoenvironmental Reconstruction of Newly Excavated Middle Pleistocene Deposits from Elandsfontein, South Africa;* City University of New York: New York, NY, USA, 2017.
85. Forrest, F.L.; Stynder, D.D.; Bishop, L.C.; Levin, N.E.; Lehmann, S.B.; Patterson, D.B.; Matthews, T.; Braun, D.R. Zooarchaeological reconstruction of newly excavated Middle Pleistocene deposits from Elandsfontein, South Africa. *J. Archaeol. Sci. Rep.* **2018**, *17*, 19–29. [CrossRef]
86. Klein, R.G. Patterns of ungulate mortality and ungulate mortality profiles from Langebaanweg(Early Pliocene) and Elandsfontein(Middle Pleistocene), south-western Cape Province, South Africa. *Ann. S. Afr. Mus.* **1982**, *90*, 49–64.
87. Klein, R.G.; Avery, G.; Cruz-Uribe, K.; Halkett, D.; Hart, T.; Milo, R.G.; Volman, T.P. Duinefontein 2: An Acheulean site in the western Cape province of South Africa. *J. Hum. Evol.* **1999**, *37*, 153–190. [CrossRef] [PubMed]
88. Cruz-Uribe, K.; Klein, R.G.; Avery, G.; Avery, M.; Halkett, D.; Hart, T.; Milo, R.G.; Sampson, C.G.; Volman, T.P. Excavation of buried late Acheulean (mid-quaternary) land surfaces at Duinefontein 2, Western Cape Province, South Africa. *J. Archaeol. Sci.* **2003**, *30*, 559–575. [CrossRef]

89. Hutson, J.M. The faunal remains from Bundu Farm and Pniel 6: Examining the problematic Middle Stone Age archaeological record within the southern African interior. *Quat. Int.* **2018**, *466*, 178–193. [CrossRef]
90. Hutson, J.M. *A Comparative Study of Life and Death at Middle Stone Age Open-Air Sites within the Southern African Interior*; University of Nevada, Reno: Reno, NV, USA, 2012.
91. Brink, J.S. *The Archaeozoology of Florisbad, Orange Free State*; Stellenbosch University: Stellenbosch, South Africa, 1987.
92. Haradon, C.M. The Ecological Context of the Acheulean to Middle Stone Age Transition in Africa. Ph.D. Thesis, The George Washington University, Washington, DC, USA, 2010.
93. Potts, R.; Behrensmeyer, A.K.; Faith, J.T.; Tryon, C.A.; Brooks, A.S.; Yellen, J.E.; Deino, A.L.; Kinyanjui, R.; Clark, J.B.; Haradon, C.M.; et al. Environmental dynamics during the onset of the Middle Stone Age in eastern Africa. *Science* **2018**, *360*, 86–90. [CrossRef]
94. Blasco, R.; Fernández Peris, J. A uniquely broad spectrum diet during the Middle Pleistocene at Bolomor Cave (Valencia, Spain). *Quat. Int.* **2012**, *252*, 16–31. [CrossRef]
95. Blasco, R.; Rosell, J.; Fernandez Peris, J.; Luis Arsuaga, J.; Maria Bermudez de Castro, J.; Carbonell, E. Environmental availability, behavioural diversity and diet: A zooarchaeological approach from the TD10-1 sublevel of Gran Dolina (Sierra de Atapuerca, Burgos, Spain) and Bolomor Cave (Valencia, Spain). *Quat. Sci. Rev.* **2013**, *70*, 124–144. [CrossRef]
96. Huguet, R.; Saladié, P.; Cáceres, I.; Díez, C.; Rosell, J.; Bennàsar, M.; Blasco, R.; Esteban-Nadal, M.; Gabucio, M.J.; Rodríguez-Hidalgo, A.; et al. Successful subsistence strategies of the first humans in south-western Europe. *Quat. Int.* **2013**, *295*, 168–182. [CrossRef]
97. Moncel, M.-H.; Moigne, A.-M.; Combier, J. Towards the Middle Palaeolithic in western Europe: The case of Orgnac 3 (southeastern France). *J. Hum. Evol.* **2012**, *63*, 653–666. [CrossRef]
98. Bahain, J.-J.; Mercier, N.; Valladas, H.; Falguères, C.; Masaoudi, H.; Joron, J.-L.; Froget, L.; Moigne, A.-M.; Combier, J.; Moncel, M.-H. Reappraisal of the chronology of Orgnac 3 Lower-to-Middle Paleolithic site (Ardèche, France), a regional key sequence for the Middle Pleistocene of southern France. *J. Hum. Evol.* **2022**, *162*, 103092. [CrossRef] [PubMed]
99. Smith, G.M.; Ruebens, K.; Gaudzinski-Windheuser, S.; Steele, T.E. Subsistence strategies throughout the African Middle Pleistocene: Faunal evidence for behavioral change and continuity across the Earlier to Middle Stone Age transition. *J. Hum. Evol.* **2019**, *127*, 1–20. [CrossRef] [PubMed]
100. Kuman, K.; Inbar, M.; Clarke, R. Palaeoenvironments and cultural sequence of the Florisbad Middle Stone Age hominid site, South Africa. *J. Archaeol. Sci.* **1999**, *26*, 1409–1425. [CrossRef]
101. Faith, J.T.; Potts, R.; Plummer, T.W.; Bishop, L.C.; Marean, C.W.; Tryon, C.A. New perspectives on middle Pleistocene change in the large mammal faunas of East Africa: Damaliscus hypsodon sp. nov.(Mammalia, Artiodactyla) from Lainyamok, Kenya. *Palaeogeogr. Palaeoclimatol. Palaeoecol.* **2012**, *361*, 84–93. [CrossRef]
102. Potts, R.; Dommain, R.; Moerman, J.W.; Behrensmeyer, A.K.; Deino, A.L.; Riedl, S.; Beverly, E.J.; Brown, E.T.; Deocampo, D.; Kinyanjui, R.; et al. Increased ecological resource variability during a critical transition in hominin evolution. *Sci. Adv.* **2020**, *6*, eabc8975. [CrossRef]
103. Fernández Peris, J.; Barciela, V.; Blasco, R.; Cuartero Monteagudo, F.; Sañudo Die, P. El Paleolítico Medio en el territorio valenciano y la variabilidad tecno-económica de la Cova del Bolomor. *Treballs d'Arqueologia* **2008**, *14*, 141–169.
104. Rodríguez, J.; Burjachs, F.; Cuenca-Bescós, G.; García, N.; Van der Made, J.; González, A.P.; Blain, H.-A.; Expósito, I.; López-García, J.M.; Antón, M.G. One million years of cultural evolution in a stable environment at Atapuerca (Burgos, Spain). *Quat. Sci. Rev.* **2011**, *30*, 1396–1412. [CrossRef]
105. García-Medrano, P.; Ollé, A.; Mosquera, M.; Cáceres, I.; Carbonell, E. The nature of technological changes: The Middle Pleistocene stone tool assemblages from Galería and Gran Dolina-subunit TD10. 1 (Atapuerca, Spain). *Quat. Int.* **2015**, *368*, 92–111. [CrossRef]
106. Lindstedt, S.L.; Boyce, M.S. Seasonality, fasting, endurance, and body size in mammals. *Am. Nat.* **1985**, *125*, 873–878. [CrossRef]
107. Rosenberg-Yefet, T.; Shemer, M.; Barkai, R. Lower Paleolithic winds of change: Prepared core technologies and the onset of the Levallois method in the Levantine late Acheulian. *Front. Earth Sci.* **2022**, *10*, 847358. [CrossRef]
108. Ellis, C.J. Factors influencing the use of stone projectile tips. In *Projectile Technology*; Springer: New York, NY, USA, 1997; pp. 37–74.
109. Hitchcock, R.; Bleed, P. Each according to need and fashion. In *Projectile Technology*; Springer: New York, NY, USA, 1997; pp. 345–368.
110. Morin, E.; Bird, D.; Winterhalder, B.; Bird, R.B. Deconstructing Hunting Returns: Can We Reconstruct and Predict Payoffs from Pursuing Prey? *J. Archaeol. Method Theory* **2022**, *29*, 561–623. [CrossRef]
111. Kilby, J.D.; Surovell, T.A.; Huckell, B.B.; Ringstaff, C.W.; Hamilton, M.J.; Haynes, C.V., Jr. Evidence supports the efficacy of Clovis points for hunting proboscideans. *J. Archaeol. Sci. Rep.* **2022**, *45*, 103600. [CrossRef]
112. Moore, C.R.; Kimball, L.R.; Goodyear, A.C.; Brooks, M.J.; Daniel, I.R., Jr.; West, A.; Taylor, S.G.; Weber, K.J.; Fagan, J.L.; Walker, C.M. Paleoamerican exploitation of extinct megafauna revealed through immunological blood residue and microwear analysis, North and South Carolina, USA. *Sci. Rep.* **2023**, *13*, 9464. [CrossRef] [PubMed]
113. Eren, M.I.; Story, B.; Perrone, A.; Bebber, M.; Hamilton, M.; Walker, R.; Buchanan, B. North American Clovis point form and performance: An experimental assessment of penetration depth. *Lithic Technol.* **2020**, *45*, 263–282. [CrossRef]
114. Prates, L.; Rivero, D.; Perez, S.I. Changes in Projectile design and size of prey reveals the role of Fishtail points in megafauna hunting in South America. *Sci. Rep.* **2022**, *12*, 16964. [CrossRef]

115. Buchanan, B.; Collard, M.; Hamilton, M.J.; O'Brien, M.J. Points and prey: A quantitative test of the hypothesis that prey size influences early Paleoindian projectile point form. *J. Archaeol. Sci.* **2011**, *38*, 852–864. [CrossRef]
116. Anderson, D. How to make a bigger hole: An experimental analysis of projectile point morphology in wound creation. In *Furthering Perspectives: Anthropological Views of the World Volume 4: 2010*; Colorado State University: Fort Collins, CO, USA, 2010.
117. Davidson, D.S. Australian spear-traits and their derivations. *J. Polyn. Soc.* **1934**, *43*, 41–72.
118. Wood, J.; Fitzhugh, B. Wound ballistics: The prey specific implications of penetrating trauma injuries from osseous, flaked stone, and composite inset microblade projectiles during the Pleistocene/Holocene transition, Alaska U.S.A. *J. Archaeol. Sci.* **2018**, *91*, 104–117. [CrossRef]
119. Frison, G.C. *Prehistoric Hunters of the High Plains*; Academic Press: New York, NY, USA, 1978; Volume 1.
120. Shea, J.J. Middle Paleolithic spear point technology. In *Projectile Technology*; Knecht, H., Ed.; Springer: New York, NY, USA, 1997; pp. 79–106.
121. Lombard, M. Quartz-tipped arrows older than 60 ka: Further use-trace evidence from Sibudu, KwaZulu-Natal, South Africa. *J. Archaeol. Sci.* **2011**, *38*, 1918–1930. [CrossRef]
122. Hirt, M.R.; Jetz, W.; Rall, B.C.; Brose, U. A general scaling law reveals why the largest animals are not the fastest. *Nat. Ecol. Evol.* **2017**, *1*, 1116. [CrossRef]
123. Odell, G.H.; Cowan, F. Experiments with spears and arrows on animal targets. *J. Field Archaeol.* **1986**, *13*, 195–212.
124. Germonpré, M.; Sablin, M.V.; Lázničková-Galetová, M.; Després, V.; Stevens, R.E.; Stiller, M.; Hofreiter, M. Palaeolithic dogs and Pleistocene wolves revisited: A reply to Morey (2014). *J. Archaeol. Sci.* **2015**, *54*, 210–216. [CrossRef]
125. Lombard, M.; Parsons, I. What happened to the human mind after the Howiesons Poort? *Antiquity* **2011**, *85*, 1433–1443. [CrossRef]
126. Clark, J.L. The Howieson's poort fauna from Sibudu cave: Documenting continuity and change within Middle Stone Age industries. *J. Hum. Evol.* **2017**, *107*, 49–70. [CrossRef] [PubMed]
127. Janetski, J.C. Fremont hunting and resource intensification in the eastern Great Basin. *J. Archaeol. Sci.* **1997**, *24*, 1075–1088. [CrossRef]
128. Smith, C.S. The bow and arrow, population, environment, and seeds: Intensification in southwest Wyoming. *J. Anthropol. Archaeol.* **2021**, *62*, 101300. [CrossRef]
129. Bettinger, R.L.; Eerkens, J. Point typologies, cultural transmission, and the spread of bow-and-arrow technology in the prehistoric Great Basin. *Am. Antiq.* **1999**, *64*, 231–242. [CrossRef]
130. Ambrose, S.H. Paleolithic technology and human evolution. *Science* **2001**, *291*, 1748–1753. [CrossRef] [PubMed]
131. Wynn, T. Hafted spears and the archaeology of mind. *Proc. Natl. Acad. Sci. USA* **2009**, *106*, 9544–9545. [CrossRef]
132. Coolidge, F.L.; Haidle, M.N.; Lombard, M.; Wynn, T. Bridging theory and bow hunting: Human cognitive evolution and archaeology. *Antiquity* **2016**, *90*, 219–228. [CrossRef]
133. Gärdenfors, P.; Lombard, M. Causal cognition, force dynamics and early hunting technologies. *Front. Psychol.* **2018**, *9*, 87. [CrossRef] [PubMed]
134. Liebenberg, L. *The Origin of Science-The Evolutionary Roots of Scientific Reasoning and Its Implications for Citizen Science*; CyberTracker: Cape Town, South Africa, 2013.

quaternary

MDPI

Article

Lower Paleolithic Shaped Stone Balls—What Is Next? Some Cultural–Cognitive Questions

Ella Assaf [1,*], Javier Baena Preysler [2] and Emiliano Bruner [3,*]

1 The Sonia & Marco Nadler Institute of Archaeology, Tel Aviv University, Tel Aviv 6997801, Israel
2 Departamento de Prehistoria y Arqueología, Facultad de Filosofía y Letras, Universidad Autónoma de Madrid, 28049 Madrid, Spain; javier.baena@uam.es
3 Centro Nacional de Investigación Sobre la Evolución Humana, 09002 Burgos, Spain
* Correspondence: ellaassa@tauex.tau.ac.il (E.A.); emiliano.bruner@cenieh.es (E.B.)

Abstract: Shaped stone balls (SSBs) were an integral part of human culture across the Old World for nearly 2 million years. They are one of the oldest implements made and used by humans. In this significant era, which was characterised by biological and cultural transformations, these round implements were a stable hallmark throughout the Lower Paleolithic period and beyond. However, while much research progress has been made in other stone tool categories, and despite the increased research efforts in recent years, attempts to define SSB function and typology have remained inconclusive, and broader cultural and cognitive aspects related to their production techniques, use and dispersal have yet to be explored in depth. What is the significance of their continuous presence and wide geographical distribution? What do these imply regarding the cognitive abilities of Oldowan and Acheulian humans? In this paper, we turn the spotlight on these enigmatic items. We address unresolved issues and explore the ergonomic and perceptual properties stimulated by the geometry of these items. We show that stone ball variability matches (modern) human hand palm variability. Moreover, when using SSBs in percussion activities, they are handled with the entire palm. Following, we discuss the role of SSBs in the context of socio-cultural processes. We suggest that SSBs reflect some of the earliest evidence in human history of a specific form that was conceptualised and recurrently acted upon.

Keywords: shaped stone balls; Paleolithic; lithic technology; cognitive archaeology; haptics

check for updates

Citation: Assaf, E.; Baena Preysler, J.; Bruner, E. Lower Paleolithic Shaped Stone Balls—What Is Next? Some Cultural–Cognitive Questions. *Quaternary* **2023**, *6*, 51. https://doi.org/10.3390/quat6040051

Academic Editors: Anna Maria Mercuri and Ran Barkai

Received: 2 August 2023
Revised: 4 September 2023
Accepted: 12 September 2023
Published: 1 October 2023

1. The Shaped Stone Ball Phenomenon—What Do We Know So Far?

Shaped stone balls are common in Oldowan and Acheulian sites in Africa [1–9], Asia [10–14], and Europe [15,16]; although in northwest Europe they are scarce, see [1], as well as at Middle Stone-Age African sites [17].

These are spherical items that show signs of intentional shaping via faceted intersecting negative flake scars over some or their entire surface. While these general features are common to all SSBs, these implements vary in size and regularity, as some are more rounded, angular, intensively shaped, rounded, or symmetrical than others. Some SSBs are pecked and/or battered to create a nearly smooth surface, while others have projecting ridges (Figure 1). Over the years, several typological classifications have been proposed based on these different morphologies. Clark [18] 2 defined shaped stone balls as lithic artifacts of a spherical shape achieved via the knapping of facets. Kleindienst [19] further classified them into missiles (manuports or slightly shaped implements used as projectiles), polyhedrons (facetted implements shaped by intersecting flake scars), and bolas (almost completely pecked and battered implements with a smooth surface). Leakey [20] defined polyhedrons as having an angular form displaying at least three worked edges, spheroids as items whose entire surface has been facetted or smoothed, and sub-spheroids that are less symmetrical and more angular. Sahnouni et al. [21] differentiated between polyhedrons, items flaked on at least three different faces with some relatively acute edges but a fairly obtuse average core angle, and spheroids, items heavily flaked over much or all of the exterior with very obtuse angles. Each of these categories is considered to reflect a specific technological/functional process [22].

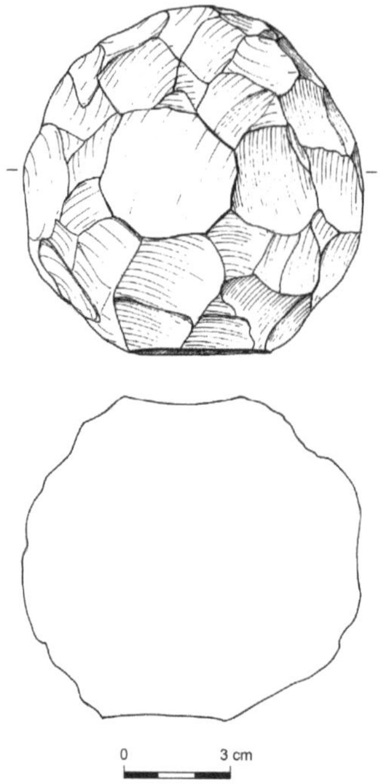

Figure 1. Shaped stone ball made of carbonatic rock from Qesem Cave, Israel, dated back to earlier than 300,000 BP (by Rodika Pinhas).

Some interpreted SSBs as end products of a preconceived shaping process [22,23], used as bolas or throwing stones for capturing animals [24–26] and as food-pounding tools [27], while others considered them not as predetermined tools but as by-products of specific technological or functional trajectories: exhausted cores [21,28], hammerstones [5,9,29], or battering tools for processing vegetal material or tendering meat [6].

In the Levant, a notable assemblage of 50 SSBs was found at the late LP site of Qesem Cave (dated to ca. 300 ka ago, see [10], Figure 2). In a previous techno-functional study, we showed that ten items bear traces of use and organic residues on the high, obtuse-angled ridges, which indicate bone-breaking activities (as percussive tools) for extracting fat/marrow [30].

Our experiments indicated that the spherical shape promotes some useful outline properties; it is particularly suitable for intensive percussion activities such as bone breaking or processing other hard materials, as it concentrates the force and prevents the breakage of the tool itself, while allowing a comfortable grip. The intersecting high, obtuse-angled ridges represent multiple working edges that are sharp enough to effectively and quickly break the bone without small pieces splintering and contaminating the marrow. We suggest, therefore, that the different morphologies of SSBs reflect a variable number of working areas and/or different degrees of wear resulting from varied intensities of use; the smooth outline of the spheroids/bolas, for example, is the result of their highly intensive percussive use, leading to smoothed ridges and surfaces (see [31] for further details and Figure 3).

Figure 2. A sample of shaped stone balls from Qesem Cave, dated to earlier than 300,000 BP. As is evident, some items have smoother faces compared to others with more angular morphologies.

SSBs were usually made of carefully selected rocks. Cabanes et al. [1] noted that hard stones (with high resistance to a physical constraint) available locally were commonly selected to produce SSBs, as well as soft sedimentary rocks, whereas siliceous materials were left aside. The notable specific selection of homogeneous carbonatic rocks such as limestone and dolomite, which were conspicuously different to other lithic artifacts in the assemblages, was detected at dozens of sites worldwide, and was further investigated in a series of experiments in which stone ball replicas were shaped from different materials and used to break bones [31]. The results show that limestone/dolomite (that are softer and more elastic rocks compared to flint and quartz) is preferable for the peripheral knapping of stone balls because it allows better control of the knapping process, which prevents the balls from splitting.

Figure 3. An experimental SSB replica made of limestone used for bone breaking. It can be seen that while some areas are more angular, showing high ridges and detachments, other areas (showing white dots) are smoother. The smooth areas were used for bone breaking.

Cabanes et al. [1] attribute this selectivity to the possible function of these items, pointing out that raw materials of SSBs are often similar to those of heavy-duty tools in assemblages, which could provide clues about their functions for tasks requiring a resistance to shock. However, our own experiments showed no difference with regard to the efficiency of flint versus limestone SSBs as percussion tools; therefore, we attribute this selectivity to the shaping process itself. We suggest that the distinct spherical morphology of SSBs is a result of a complex knapping trajectory, which required careful planning and necessitated a high degree of knowhow and precision. Titton et al. [16] carried out a study of SSBs from the Oldowan site of Barranco León (Spain), combining diacritical and 3D geometric morphometric analysis with an evaluation of the raw material (limestone) and of percussive use-wear. The authors suggest that subspheroids are the result of a systematic management of rounded cobbles. According to their analysis, while the organisation of the removal negatives reflects the same type of management as that observed on multidirectional cores, they differ in that the last phase of reduction was aimed at creating more open facet angles, following the morphology of the initial rounded cobbles and giving the final subspheroid form. De Weyer [22] demonstrated that polyhedrons, spheroids and bolas from Ounjougou (Mali) were shaped through independent chaînes opératoires for specific tasks. Opportunistic knapping was ruled out by the author, while the evidence of shaping was highlighted by diacritic schemes showing that the flakes detached are not controlled for their usability, but for shaping the morphology of the spheroid. Our experiments further support these assumptions, indicating that these items are likely to have been desired tools rather than by-products of functional activity (as suggested by Sahnouni et al. [21].

According to our reconstruction, the shaping process of SSBs began with the selection of a suitable blank in terms of the type, morphology and quality of material, followed by the creation of a flat, right-angled surface, which would have been the main challenge for the knapper, and lastly, the creation of a symmetrical, spherical morphology. In this process, specific technical and technological procedures were applied sequentially. These appear separately in other knapping processes but altogether, they are endemic to SSB production. In addition, it seems that some of these items underwent several "life cycles," indicating a very long history of use, and that their production process was not always continuous. In Ubeidiya (layer I-15, personal observation) and Qesem Cave (see Assaf et al. [30] and personal observation), for example, few SSBs show phases of the re-shaping of (sometimes patinated) surfaces reflecting various knapping stages and a possible time gap between them.

2. Interpretations and Unresolved Issues

What are the insights obtained from the studies carried out so far? Technology-wise, we suggest that this is some of the earliest evidence in human history of a specific form that was conceptualised and recurrently acted upon. For this kind of operational complexity, long-term planning, decision making, control, and precise intentionality are needed [32,33]. In our view, imposing form through multi-stage, multi-directional techniques at such an early stage of human history set the stage for the development of later technologies. Furthermore, the peculiar (and regular) geometry of stone balls necessarily stimulates a debate on its ergonomic and perceptual properties.

2.1. Stone Balls, Haptics, and Technological Extension

Humans were occasional tool users probably before 2 million years ago, then became habitual tool users after 1.7 my, and finally became obligatory toolusers in the last 300,000 years [34]. This suggests an increasing "prosthetic capacity", in which tools are progressively integrated into the body structure and into the neural schemes of the body [35]. In archaeology and evolutionary anthropology, tooluse and toolmaking have been the main topic of technological evolution, in this sense. Nonetheless, before *using* and *making*, we should consider possible changes in the process of *sensing*, which is the very foundation of body–tool structural and functional integration [36]. Sensing and perceptual integration are the processes through which the body, as the interface, assimilates the tool into the cognitive system. In this context, haptics is of crucial importance, intended as the ability of the body to sense and perceive spatial, gravitational, and material properties of external elements, involving both proprioceptive and exteroceptive feedbacks [37,38]. Tactile feedback is, in this sense, a crucial element to remap the sensorimotor process into spatial and chronological frames [39], and it is based on a comprehensive integration between action and perception, in which somatosensorial and motor adjustments are mutually dependent [40]. During tool sensing and integration, the whole body is used to achieve a new dynamic balance that is able to include the tool within the structural architecture of the subject [41]. This process is further improved through the additional contribution of visual elements [42].

Of course, it can be hard to import all these concepts into archaeology and paleontology, because of the general lack of information associated with fossil species or extinct populations. In general, most studies in human evolution and ergonomics have dealt with the anatomical evolution of the human hand [43,44]. Biomechanics can indeed provide some interesting clues to the structural relationships between hand features and stone tool properties [45]. Nonetheless, there are limitations in these approaches, and there are still many disagreements on the evolutionary changes associated with the evolution of the manipulative abilities in the human lineage [46–48]. In fact, the gross hand anatomy can supply poor and incomplete information on hand functions, the fossil record is largely fragmented, and the evolution of the hand has probably experienced mosaic, parallel and independent phylogenetic diversification [49]. Indeed, because of these limitations, the most we can do is integrate anatomical and behavioural differences in order to provide speculative but reasonable hypotheses on the evolution of the body–tool perceptual and cognitive system [50,51].

With these caveats in mind, we can consider what kind of perceptual or structural features can be associated with the special shape of the stone balls, namely with tools that tend towardsa spherical geometry. On the one hand, we can consider the effects and factors associated with their regular geometry, providing a gross comparison with choppers and handaxes. This comparison is relevant because all these kinds of tool share an important feature: they are handled (and sensed) with the whole hand. It is likely that most Lower Paleolithic tools were handled with the whole hand while, after technological specialisation and increased complexity (especially during and after the Middle Paleolithic), most tools were sensed with the fingers and then with the fingertips, a perceptual change that was probably associated with consistent cognitive differences [52,53]. It is worth noting that the variation in stone ball proportions seems to occur in accordance with the same rule of variation in the (modern) human palm. A preliminary survey on the relationships between stone ball size and palm size shows that if we plot the length and width variation in the adult palm [52] and in a sample of stone balls [30], we can see that the two proportions display the same allometric pattern (Figure 4). That is, stone ball variability matches human hand palm variability (at least when considering modern humans). Results of a recent experimental study further support this distinction; in the framework of this experiment, six individuals between the ages of 25 and 60 broke 10 limb bones of cows and pigs for ten minutes using different replicas of SSBs. Two women and four men participated; two had experience in bone-breaking activities, while four had no experience. In the experiment it was shown that when using SSBs in percussion activities, they are handled by all the participants with the entire palm (personal observation). Conversely, when a hammerstone is held in knapping, it is held with the fingers and not in direct contact with the palm. This pattern suggests the possible existence of a structural correspondence, which underlines that, when investigating human technological evolution, we should consider the hand–tool system as the functional and structural unit, instead of analysing the two parts separately [54].

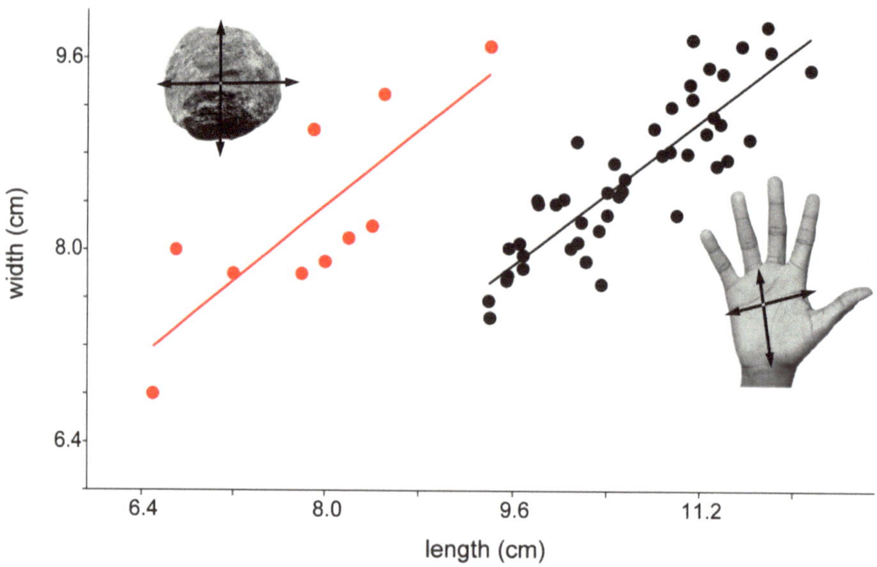

Figure 4. Hand palm (black dots) and stone ball (red dots) length and width. A preliminary analysis of covariance supports differences in the intercept ($p < 0.0001$) but not in the slope ($p = 0.80$). Namely, despite their differences in size, the modern human palm and the stone balls do vary with the same pattern of proportions. In particular, stone balls are shorter than the human palm, but display a similar range of width. Data from [30,52].

When comparing choppers with handaxes in terms of shape, we can see that the main pattern polarising their morphological variability is the degree of elongation, while the second factor involved is the longitudinal position of the maximum thickness [53]. The first factor (elongation) separated the two tool typologies (although with a certain degree of overlap), while the second source of variation (the longitudinal position of the maximum thickness) is shared by both types of tools [55]. If we plot a perfect sphere into this shape space, we can see that it is projected right in the middle of the chopper distribution (Figure 5). Therefore, at least for these two main geometric features, stone balls were probably intended as a special case of choppers, at least according to their gross spatial properties. It is worth noting that according to our experimental study, these two tool types are also held in a similar way—in direct contact with the palm—and the thumb is not in line with and close to the other fingers but instead a few centimetres away from them, so as to increase the grasping aperture. This pattern may have implications regarding the distribution of these two tool types in LP sites, although in some sites they coexist (e.g., Revadim, Evron, see [56,57], while in others (e.g., Qesem Cave, Jaljulia, see [58,59]) they do not.

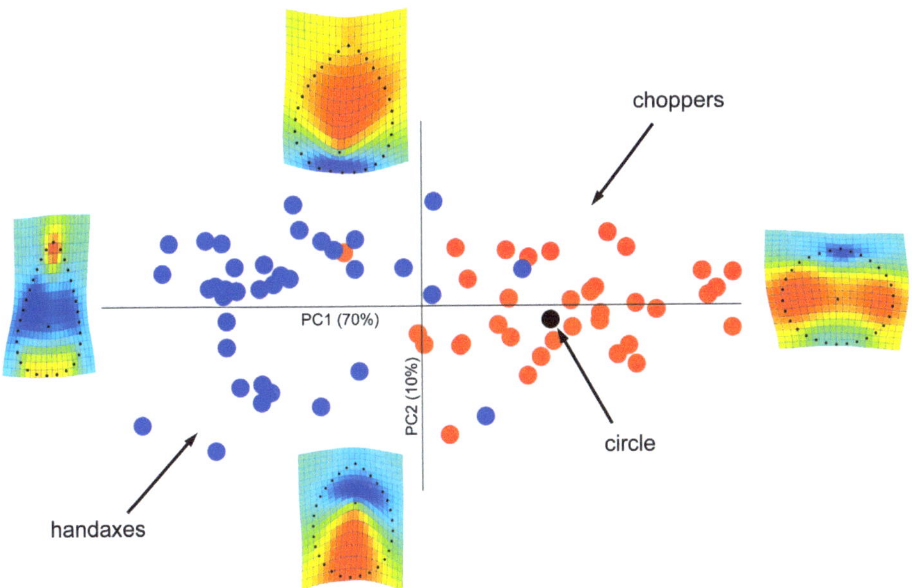

Figure 5. A principal component analysis of the chopper (red dots) and handaxe (blue dots) shape (including the outline and the position of the maximum thickness) showing that the first axis of variation is due to elongation, while the second is due to the longitudinal position of the maximum thickness. A perfect sphere (black dot), in this case, lies in the centre of the choppers' range. Data from [55].

In sum, when dealing with the evolutionary meaning of stone balls, we should also take into consideration their features in terms of affordance (*sensu* Gibson, 1979 [60]) and perceptual properties. This means that, apart from their implications as tools to be made and to be used, we should also consider their special perceptual aspects, associated with body–tool integration at the somatic and visual level. If they are a "special case" of chopper-like tools in terms of geometry, we can probably apply to their physical and perceptual interpretation most of the conclusions considered valid for Oldowan technology. Instead, if their spherical regularity adds to the perceptual feedback, we should consider the possibility of additional embodied aspects. Regularity (including symmetry, ergonomic intentional design, or even symbolism) is, indeed, a concept that has been often proposed to be associated with some kind of cognitive complexity. However, regular designs in nature are often due to spontaneous (unintentional) spatial adjustments that follow basic geometrical rules, and the existence of regular forms should hence not be interpreted, *per se*, as the result of a conscious or abstract process. However, at the same time, the distinctive morphology of this implement, i.e., the spherical shape with the high, obtuse-angled ridges, which were a useful outline for bone-breaking activities, and therefore may have been conceptualised with purposeful intent, was preserved and passed down over hundreds of thousands of years. Indeed, as discussed below, the outstanding geographical and chronological distribution of stone balls merits attention and deserves further inquiries into their possible structural and functional role in the scenario of human technological extension.

2.2. SSBs and Socio-Cultural Processes

As SSBs display recurring and recognizable common features that represent a universal phenomenon (e.g., the selectivity in rock type, the round morphology, and the intersecting negatives and ridges), should we regard them as cultural products (as suggested for the handaxe, see Wynn and Gowlett [61]), or as genetic products (as suggested by Corbey et al. [62]). Corbey et al. propose that LP handaxe production would have involved both genetic transmission and social learning. Some aspects, such as raw material selection, the manufacturing process, and basic design principles would have been under genetic control, while the fine-tuning phase involved social learning—mainly with specific role models rather than as a direct transmission within a group [62].

Wynn and Gowlett [61] argue that sequential tasks such as knapping stone tools practiced by Lower Paleolithic humans relied on a complex neural network that evolved for object manipulation, incorporating portions of the frontal, parietal and temporal lobes (as observed in modern stone knappers, see Hecht et al. [63]). For handaxe shaping, the working memory mechanism also played an important role, as suggested by Stout et al. [64]. According to this approach, the knapping process was an ergonomically guided procedure derived by cognitive features, for which the neural resources of the human object-manipulation network were self-sufficient. While it could have developed again and again over the course of the Stone Age, the Acheulian record appears to reflect a continuing tradition in many ways, as suggested by Wynn, Gowlett and others [65]. Moreover, it has been argued that Lower Paleolithic humans were capable of transmitting the necessary knowledge from generation to generation [58,66–68] (, which was evident even earlier at Oldowan sites [69,70].

How do SSBs fit into the overall Lower Paleolithic picture? The dispersion of SSBs can reflect processes of social learning and cultural transmission or, alternatively, processes of convergent evolution. Among humans, learning and sharing are enhanced [71], leading to the accumulation of information and the development of tools, beliefs, and practices that enable adaptation to new environments and habitats [72]. As part of this process, innovations are gradually assimilated into a population's knowledge base through mechanisms of social learning—a phenomenon known as 'cumulative culture' [73]. Some aspects of it require a highly accurate transmission of knowledge [74,75], and the question arises as to when, why, and how high-fidelity learning mechanisms arose in prehistoric times. As mentioned, some scholars view the Lower Paleolithic period as a turning point in the evolution of social learning and knowledge transmission mechanisms [76]. While some suggest [71] that technological behaviours of the LP constitute low-precision solutions in response to a trigger associated with a specific social or external change, others suggest that early Paleolithic technologies were learned via detailed copy [77,78].

Based on the current data available, we link the development of SSBs with an increased consumption and extraction of fat of large-medium sized herbivores in the Lower Paleolithic period. Indeed, SSBs may also have been used as percussion tools for processing vegetal material, perhaps similarly to handaxes, which were primarily used for butchering (i.e., [79,80]), but were occasionally used for woodworking [81]. However, current functional and experimental data associate these items with marrow extraction activities, as well as some contextual data linking these items with faunal remains [17,82]. We thus argue that these are designated tools shaped in a complex procedure—a technological solution invented during the LP in order to respond to the growing need for fat consumption [83,84]. They were gradually assimilated into the LP population's knowledge base during the Acheulian, enriching their repertoire through mechanisms of social learning. Given the technological complexity and sequential nature of the shaping process of this implement, we raise the possibility that the knowledge about it was preserved and passed down from generation to generation, and formed the basis for the development of later technologies.

3. What Is Next for the Study of Shaped Stone Balls?

After decades of studying SSBs, fundamental enigmas still remain. The next step, we suggest, requires a multidisciplinary approach to better understand the circumstances behind the production and long-term use of SSBs, and their eventual replacement by other technologies and stone implements. A type of research that would integrate fields representing a meeting point will take us a step forward and provide a significant evolutionary perspective [85].

Such research can include 'traditional' archeological analysis, with an emphasis on techno-morpho-functional analysis, in order to deepen our understanding of the shaping process and use of SSBs worldwide, while examining aspects of cultural transmission, technological persistency, and tradition versus change. Experimental archaeology is an important part of the picture. Experimental procedure can contribute to the recognition of specific techniques of SSB production as well as the identification of particular techno-types in the assemblages. These will allow us to identify similarities in shaping strategies and detect their presence even if the SSBs themselves are absent from the assemblage. It will form the basis for a general discussion on cognition and spheroid geometry, and ergonomic and perceptual properties, while referring to haptics and aspects of technological extension. Finally, cultural evolution mathematical models will allow us to estimate the rate of dispersal and routing of SSBs throughout the Old World while discussing social learning mechanisms possibly involved in the process.

Hopefully, future in-depth integrative studies will contribute to a deeper understanding of this enigmatic tool type and the cognitive abilities of the ancient humans who used it throughout the Lower Paleolithic period and beyond. While exploring extinct species necessarily involves many limitations [85], these multidisciplinary studies are expected to provide a detailed reconstruction of behaviours related to the application of early stone tool technologies and their broader cultural and evolutionary implications reflected in one of the oldest stone tools made and used by humans.

Author Contributions: Conceptualisation, E.A., J.B.P. and E.B.; writing—original draft preparation, E.A., J.B.P. and E.B.; writing—review and editing, E.A., J.B.P. and E.B. All authors have read and agreed to the published version of the manuscript.

Funding: E.A. is supported by CARE archaeological foundation, research grant of 2021. J.B.P. is funded by Project HAR2016-76760-C3-2-P financed by AEI and FEDER. E.B. is funded by the Spanish Government (PID2021-122355NB-C33) and by the Italian Institute of Anthropology (ISITA).

Data Availability Statement: The archaeological material is available at the archaeology department of Tel Aviv University, Israel.

Conflicts of Interest: The authors declare no conflict of interest.

References

1. Cabanès, J.; Borel, A.; Preysler, J.B.; Lourdeau, A.; Moncel, M.-H. Palaeolithic polyhedrons, spheroids and bolas over time and space. *PLoS ONE* **2022**, *17*, e0272135. [CrossRef]
2. Diez-Martín, F.; Sánchez, P.; Domínguez-Rodrigo, M.; Mabulla, A.; Barba, R. Were Olduvai Hominins making butchering tools or battering tools? Analysis of a recently excavated lithic assemblage from BK (Bed II, Olduvai Gorge, Tanzania). *J. Anthr. Archaeol.* **2009**, *28*, 274–289. [CrossRef]

3. de la Torre, I.; Benito-Calvo, A.; Arroyo, A.; Zupancich, A.; Proffitt, T. Experimental protocols for the study of battered stone anvils from Olduvai Gorge (Tanzania). *J. Archaeol. Sci.* **2013**, *40*, 313–332. [CrossRef]
4. García-Vadillo, F.J.; Canals-Salomó, A.; Rodríguez-Álvarez, X.P.; Carbonell-Roura, E. The large flake Acheulean with spheroids from Santa Ana Cave (Cáceres, Spain). *J. Archaeol. Sci. Rep.* **2022**, *41*, 103265. [CrossRef]
5. Jones, P.R. Results of experimental work in relation to the stone industries of Olduvai Gorge. In *Olduvai Gorge*; Leakey, M.D., Roe, D.A., Eds.; Cambridge University Press: Cambridge, UK, 1994; Volume 5.
6. Mora, R.; de la Torre, I. Percussion tools in Oduvai Beds I and II (Tanzania): Implication for early human activities. *J. Anthropol. Archaeol.* **2005**, *24*, 179–192. [CrossRef]
7. Sahnouni, M.; Parés, J.M.; Duval, M.; Cáceres, I.; Harichane, Z.; van Der Made, J.; Pérez-González, A.; Abdessadok, S.; Kandi, N.; Derradji, A.; et al. 1.9-million-and 2.4-million-year-old artifacts and stone tool–cutmarked bones from Ain Boucherit, Algeria. *Science* **2018**, *362*, 1297–1301. [CrossRef]
8. Santonja, M.; Panera, J.; Rubio-Jara, S.; Pérez-González, A.; Uribelarrea, D.; Domínguez-Rodrigo, M.; Mabulla, A.Z.; Bunn, H.T.; Baquedano, E. Technological strategies and the economy of raw materials in the TK (Thiongo Korongo) lower occupation, Bed II, Olduvai Gorge, Tanzania. *Quat. Int.* **2014**, *322–323*, 181–208. [CrossRef]
9. Willoughby, P.R. Spheroids and battered stones in the African Early Stone Age. *World Archaeol.* **1985**, *17*, 44–60. [CrossRef]
10. Barkai, R.; Gopher, A. On anachronism: The curious presence of Spheroids and Polyhedrons at Acheulo–YabrudianQesem Cave, Israel. *Quat. Int.* **2016**, *398*, 118–128. [CrossRef]
11. Bar-Yosef, O.; Goren-Inbar, N.; Gilead, I. *The Lithic Assemblages of 'Ubeidiya: A Lower Palaeolithic Site in the Jordan Valley*; Hebrew University of Jerusalem: Jerusalem, Israel, 1993.
12. Sharon, G.; Feibel, C.; Alperson-Afil, N.; Harlavan, Y.; Feraud, G.; Ashkenazi, S.; Rabinovich, R. New evidence for the northern Dead Sea rift Acheulian. *PaleoAnthropology* **2010**, *2010*, 79–99.
13. Shemer, M.; Crouvi, O.; Shaar, R.; Ebert, Y.; Matmon, A.; Horwitz, L.K.; Eisenmann, V.; Enzel, Y.; Barzilai, O.; ASTER Team. Geochronology, paleogeography, and archaeology of the Acheulian locality of 'Evron Landfill in the western Galilee, Israel. *Quat. Res.* **2019**, *91*, 729–750. [CrossRef]
14. Yang, S.X.; Huang, W.W.; Hou, Y.M.; Yuan, B.Y. Is the Dingcun lithic assembly a "chopper-chopping tool industry", or "Late Acheulian"? *Quat. Int.* **2014**, *321*, 3–11. [CrossRef]
15. Barsky, D.; Vergès, J.-M.; Sala, R.; Menéndez, L.; Toro-Moyano, I. Limestone percussion tools from the late Early Pleistocene sites of Barranco León and Fuente Nueva 3 (Orce, Spain). *Philos. Trans. R. Soc. B Biol. Sci.* **2015**, *370*, 20140352. [CrossRef] [PubMed]
16. Titton, S.; Barsky, D.; Bargalló, A.; Serrano-Ramos, A.; Vergès, J.M.; Toro-Moyano, I.; Sala-Ramos, R.; Solano, J.G.; Jimenez Arenas, J.M. Subspheroids in the lithic assemblage of Barranco León (Spain): Recognizing the late Oldowan in Europe. *PLoS ONE* **2020**, *15*, e0228290. [CrossRef]
17. Walker, N. Through the crystal ball: Making sense of spheroids in the Middle Stone Age. *S. Afr. Archaeol. Bull.* **2008**, *63*, 12–17.
18. Clark, J.D. The stone ball: Its associations and use by prehistoric man in Afrika. In *Congrès Panafricain de Préhistoire*; Balout, L., Ed.; Arts et Métiers Graphiques: Paris, France, 1952; pp. 403–417.
19. Kleindienst, M.R. Components of the East African Acheulean assemblage: An analytic approach. In *Actes du 4e Congrès Panafricain de Préhistoire et de l'Étude du QuaternaireTervuren*; Mortelmans, G., Ed.; Musée Royal de l'Afrique Centrale: Tervuren, Belgium, 1962.
20. Leakey, M.D. Excavations in Beds I and II, 1960–1963. In *Olduvai Gorge*; Cambridge University Press: Cambridge, UK, 1971; Volume 3.
21. Sahnouni, M.; Schick, K.; Toth, N. An experimental investigation into the nature of faceted limestone "spheroids" in the Early Palaeolithic. *J. Archaeol. Sci.* **1997**, *24*, 701–713. [CrossRef]
22. De Weyer, L. An Early Stone Age in Western Africa? Spheroids and polyhedrons at Ounjougou, Mali. *J. Lithic Stud.* **2017**, *4*. [CrossRef]
23. Texier, P.J.; Roche, H. Polyèdre, sub-sphéroïde, sphéroïde et bola: Des segments plus oumoins longs d'unemêmechaîneopératoire. *Cahier. Noir.* **1995**, *7*, e40.
24. Clark, J.D. The cultures of the Middle Palaeolithic/Middle Stone Age. *Camb. Hist. Afr.* **1982**, *1*, 248–341. [CrossRef]
25. Isaac, B. Throwing and human evolution. *Afr. Archaeol. Rev.* **1987**, *5*, 3–17. [CrossRef]
26. Leakey, L.S.B. *The Stone Age Cultures of Kenya Colony*; Cambridge University Press: Cambridge, UK, 1931.
27. Yustos, P.S.; Diez-Martín, F.; Díaz, I.M.; Duque, J.; Fraile, C.; Domínguez, M. Production and use of percussive stone tools in the Early Stone Age: Experimental approach to the lithic record of Olduvai Gorge, Tanzania. *J. Archaeol. Sci. Rep.* **2015**, *2*, 367–383. [CrossRef]
28. Schick, K.D.; Toth, N. Early Stone Age technology in Africa: A review and case study into the nature and function of spheroids and subspheroids. In *Integrative Paths to the Past. Paleoanthropological Advances in Honor of F. Clark Howell*; Prentice Hall: Upper Saddle River, NJ, USA, 1994; pp. 429–449.
29. Roussel, M.; Bourguignon, L.; Soressi, M. Las 'bolas'o 'boules de caliza'musterienses:¿Percutores?, elejemplo del 'fasonado'de las raederasbifaciales de Quina de Chez-Pinaud (Jonzac, Francia). In *La investigación experimental aplicada a la Arqueología*; Morgado, A., Baena, J., García, D., Eds.; Departamento de Prehistoria y Arqueología de la Universidad de Granada: Granada, Spain, 2011; pp. 69–76.

30. Assaf, E.; Caricola, I.; Gopher, A.; Rosell, J.; Blasco, R.; Bar, O.; Zilberman, E.; Lemorini, C.; Baena, J.; Barkai, R.; et al. Shaped stone balls were used for bone marrow extraction at Lower Paleolithic Qesem Cave, Israel. *PLoS ONE* **2020**, *15*, e0230972. [CrossRef] [PubMed]
31. Assaf, E.; Preysler, J.B. Raw material selectivity in Lower Paleolithic shaped stone ball production: Experimental research. *Archaeol. Anthr. Sci.* **2022**, *14*, 24. [CrossRef]
32. Baena, I.; Rubio, D.; Dominguez-Rodrigo, M. Testing cognitive skills in Early Pleistocene hominins: An analysis of the concepts of hierarchization and predetermination in the lithic assemblages of Type Section (Peninj, Tanzania). In *Stone Tools and Fossil Bones: Debates in the Archaeology of Human Origins*; Cambridge University Press: Cambridge, UK, 2012; pp. 245–309.
33. Wynn, T. Archaeology and cognitive evolution. *Behav. Brain Sci.* **2002**, *25*, 389–402. [CrossRef]
34. Shea, J.J. Occasional, obligatory, and habitual stone tool use in hominin evolution. *Evol. Anthr. Issues News Rev.* **2017**, *26*, 200–217. [CrossRef]
35. Bruner, E. Evolving Human Brains: Paleoneurology and the Fate of Middle Pleistocene. *J. Archaeol. Method Theory* **2021**, *28*, 76–94. [CrossRef]
36. Borghi, A.M.; Cimatti, F. Embodied cognition and beyond: Acting and sensing the body. *Neuropsychologia* **2010**, *48*, 763–773. [CrossRef]
37. Lederman, S.J.; Klatzky, R.L. Haptic perception: A tutorial. *Atten. Percept. Psychophys.* **2009**, *71*, 1439–1459. [CrossRef]
38. Kappers, A.M.; Bergmann Tiest, W.M. Haptic perception. *Wiley Interdiscip. Rev. Cogn. Sci.* **2013**, *4*, 357–374. [CrossRef]
39. Heed, T.; Buchholz, V.N.; Engel, A.K.; Röder, B. Tactile remapping: From coordinate transformation to integration in sensorimotor processing. *Trends Cogn. Sci.* **2015**, *19*, 251–258. [CrossRef]
40. Ackerley, R.; Kavounoudias, A. The role of tactile afference in shaping motor behaviour and implications for prosthetic innovation. *Neuropsychologia* **2015**, *79*, 192–205. [CrossRef] [PubMed]
41. Turvey, M.T.; Carello, C. Obtaining information by dynamic (effortful) touching. *Philos. Trans. R. Soc. B Biol. Sci.* **2011**, *366*, 3123–3132. [CrossRef]
42. Lacey, S.; Stilla, R.; Sreenivasan, K.; Deshpande, G.; Sathian, K. Spatial imagery in haptic shape perception. *Neuropsychologia* **2014**, *60*, 144–158. [CrossRef]
43. Marzke, M.W. Precision grips, hand morphology, and tools. *Am. J. Phys. Anthropol.* **1997**, *102*, 91–110. [CrossRef]
44. Marzke, M.W.; Marzke, R.F. Evolution of the human hand: Approaches to acquiring, analysing and interpreting the anatomical evidence. *J. Anat.* **2000**, *197*, 121–140. [CrossRef] [PubMed]
45. Rolian, C.; Lieberman, D.E.; Zermeno, J.P. Hand biomechanics during simulated stone tool use. *J. Hum. Evol.* **2011**, *61*, 26–41. [CrossRef] [PubMed]
46. Tocheri, M.W.; Orr, C.M.; Jacofsky, M.C.; Marzke, M.W. The evolutionary history of the hominin hand since the last common ancestor of Pan and Homo. *J. Anat.* **2008**, *212*, 544–562. [CrossRef]
47. Almécija, S.; Smaers, J.B.; Jungers, W.L. The evolution of human and ape hand proportions. *Nat. Commun.* **2015**, *6*, 7717. [CrossRef]
48. Kivell, T.L. Evidence in hand: Recent discoveries and the early evolution of human manual manipulation. *Philos. Trans. R. Soc. B Biol. Sci.* **2015**, *370*, 20150105. [CrossRef]
49. Kivell, T.L.; Baraki, N.; Lockwood, V.; Williams-Hatala, E.M.; Wood, B.A. Form, function and evolution of the human hand. *Am. J. Biol. Anthr.* **2022**, *181*, 6–57. [CrossRef]
50. Bruner, E.; Lozano, M.; Lorenzo, C. Visuospatial integration and human evolution: The fossil evidence. *J. Anthr. Sci.* **2016**, *94*, 81–97. [CrossRef]
51. Bruner, E. *Cognitive Archaeology, Body Cognition, and the Evolution of Visuospatial Perception*; Academic Press: London, UK, 2023.
52. Bruner, E.; Fedato, A.; Silva-Gago, M.; Alonso-Alcalde, R.; Terradillos-Bernal, M.; Fernández-Durantes, M.A.; Martín-Guerra, E. Cognitive archeology, body cognition, and hand–tool interaction. *Prog. Brain Res.* **2018**, *238*, 325–345. [PubMed]
53. Bruner, E.; Fedato, A.; Silva-Gago, M.; Alonso-Alcalde, R.; Terradillos-Bernal, M.; Fernández-Durantes, M.A.; Martín-Guerra, E. Visuospatial integration and hand-tool interaction in cognitive archaeology. In *Processes of Visuospatial Attention and Working Memory. Current Topics in Behavioral Neurosciences*; Hodgson, T., Ed.; Springer: Cham, Switzerland, 2018; Volume 41, pp. 13–36.
54. Silva-Gago, M.; Fedato, A.; Rios-Garaizar, J.; Bruner, E. A preliminary survey on hand grip and hand-tool morphometrics in three different stone tools. *J. Archaeol. Sci. Rep.* **2019**, *23*, 567–573. [CrossRef]
55. Silva-Gago, M.; Fedato, A.; Terradillos-Bernal, M.; Alonso-Alcalde, R.; Martín-Guerra, E.; Bruner, E. Not a matter of shape: The influence of tool characteristics on electrodermal activity in response to haptic exploration of Lower Palaeolithic tools. *Am. J. Hum. Biol.* **2022**, *34*, e23612. [CrossRef]
56. Chazan, M. Butchering with small tools: The implications of the Evron Quarry assemblage for the behaviour of *Homo erectus*. *Antiquity* **2013**, *87*, 350–367. [CrossRef]
57. Marder, O.; Milevski, I.; Rabinovich, R. The Revadim Quarry. In *Hadashot Arkheologiyot: Excavations and Surveys in Israel*; Israel Antiquities Authority: Jerusalem, Israel, 2007; Volume 119.
58. Assaf, E. Throughout the generations: Learning processes and knowledge transmission mechanisms as reflected in lithic assemblages of the terminal Lower Paleolithic Levant. *J. Archaeol. Sci. Rep.* **2021**, *35*, 102772. [CrossRef]

59. Shemer, M.; Greenbaum, N.; Taha, N.; Brailovsky-Rokser, L.; Ebert, Y.; Shaar, R.; Falgueres, C.; Voinchet, P.; Porat, N.; Faershtein, G.; et al. Late Acheulian Jaljulia—Early human occupations in the paleo-landscape of the central coastal plain of Israel. *PLoS ONE* **2022**, *17*, e0267672. [CrossRef]
60. Gibson, J.J. *The Ecological Approach to Visual Perception*; Houghton Mifflin: Boston, MA, USA, 1979.
61. Wynn, T.; Gowlett, J. The handaxe reconsidered. *Evol. Anthropol. Issues News Rev.* **2018**, *27*, 21–29. [CrossRef]
62. Corbey, R.; Jagich, A.; Vaesen, K.; Collard, M. The acheuleanhandaxe: More like a bird's song than a beatles' tune? *Evol. Anthropol. Issues News Rev.* **2016**, *25*, 6–19. [CrossRef]
63. Hecht, E.E.; Gutman, D.A.; Khreisheh, N.; Taylor, S.V.; Kilner, J.; Faisal, A.A.; Bradley, B.A.; Chaminade, T.; Stout, D. Acquisition of Paleolithic toolmaking abilities involves structural remodeling to inferior frontoparietal regions. *Anat. Embryol.* **2015**, *220*, 2315–2331. [CrossRef]
64. Stout, D.; Hecht, E.; Khreisheh, N.; Bradley, B.; Chaminade, T. Cognitive Demands of Lower Paleolithic Toolmaking. *PLoS ONE* **2015**, *10*, e0121804. [CrossRef] [PubMed]
65. Finkel, M.; Barkai, R. The Acheulean Handaxe Technological Persistence: A Case of Preferred Cultural Conservatism? *Proc. Prehist. Soc.* **2018**, *84*, 1–19. [CrossRef]
66. Shipton, C. Imitation and Shared Intentionality in the Acheulean. *Camb. Archaeol. J.* **2010**, *20*, 197–210. [CrossRef]
67. Tehrani, J.J.; Riede, F. Towards an archaeology of pedagogy: Learning, teaching and the generation of material culture traditions. *World Archaeol.* **2008**, *40*, 316–331. [CrossRef]
68. Lycett, S.J.; Gowlett, J.A. On questions surrounding the Acheulean 'tradition'. *World Archaeol.* **2008**, *40*, 295–315. [CrossRef]
69. Delagnes, A.; Roche, H. Late Pliocene hominid knapping skills: The case of Lokalalei 2C, West Turkana, Kenya. *J. Hum. Evol.* **2005**, *48*, 435–472. [CrossRef] [PubMed]
70. Nonaka, T.; Bril, B.; Rein, R. How do stone knappers predict and control the outcome of flaking? Implications for understanding early stone tool technology. *J. Hum. Evol.* **2010**, *59*, 155–167. [CrossRef]
71. Tennie, C.; Braun, D.R.; Premo, L.S.; McPherron, S.P. The Island test for cumulative culture in the Paleolithic. In *The Nature of Culture: Based on an Interdisciplinary Symposium 'the Nature of Culture'*; Springer: Tübingen, Germany, 2016; pp. 121–133.
72. Boyd, R.; Richerson, P.J.; Henrich, J. The cultural niche: Why social learning is essential for human adaptation. *Proc. Natl. Acad. Sci. USA* **2011**, *108*, 10918–10925. [CrossRef]
73. Laland, K.N.; Hoppitt, W. Do animals have culture? *Evol. Anthropol. Issues News Rev.* **2003**, *12*, 150–159. [CrossRef]
74. Acerbi, A.; Mesoudi, A. If we are all cultural Darwinians what's the fuss about? Clarifying recent disagreements in the field of cultural evolution. *Biol. Philos.* **2015**, *30*, 481–503. [CrossRef]
75. Tennie, C.; Call, J.; Tomasello, M. Ratcheting up the ratchet: On the evolution of cumulative culture. *Philos. Trans. R. Soc. B Biol. Sci.* **2009**, *364*, 2405–2415. [CrossRef] [PubMed]
76. Torres Navas, M. *Aprendizaje Durante el Achelense y el Musteriense: Estudio Tecnoexperimental de los Contextos de Captación del Centro Peninsular (Yacimientos de Los Ahijones, Los Berrocales y El Cañaveral)*; Universidad Autónoma de Madrid: Madrid, Spain, 2022; Available online: https://repositorio.uam.es/handle/10486/703630 (accessed on 30 May 2023).
77. Stout, D.; Rogers, M.J.; Jaeggi, A.V.; Semaw, S. Archaeology and the Origins of Human Cumulative Culture: A Case Study from the Earliest Oldowan at Gona, Ethiopia. *Curr. Anthr.* **2019**, *60*, 309–340. [CrossRef]
78. Stout, D.; Hecht, E.E. Evolutionary neuroscience of cumulative culture. *Proc. Natl. Acad. Sci. USA* **2017**, *114*, 7861–7868. [CrossRef] [PubMed]
79. Barkai, R. The elephant in the handaxe: Lower Palaeolithic ontologies and representations. *Camb. Arch. J.* **2021**, *31*, 349–361. [CrossRef]
80. Zupancich, A.; Shemer, M.; Barkai, R. Biface use in the Lower Paleolithic levant: First insights from late acheulean Revadim and jaljulia (Israel). *J. Archaeol. Sci. Rep.* **2021**, *36*, 102877. [CrossRef]
81. Domínguez-Rodrigo, M.; Serrallonga, J.; Juan-Tresserras, J.; Alcala, L.; Luque, L. Woodworking activities by early humans: A plant residue analysis on Acheulian stone tools from Peninj (Tanzania). *J. Hum. Evol.* **2001**, *40*, 289–299. [CrossRef]
82. Rivals, F.; Blasco, R.; Rosell, J.; Efrati, B.; Gopher, A.; Barkai, R. Seasonality, duration of the hominin occupations and hunting grounds at Middle Pleistocene Qesem Cave (Israel). *Archaeol. Anthr. Sci.* **2021**, *13*, 205. [CrossRef]
83. Ben-Dor, M.; Sirtoli, R.; Barkai, R. The evolution of the human trophic level during the Pleistocene. *Am. J. Phys. Anthr.* **2021**, *175*, 27–56. [CrossRef]
84. Plummer, T.W.; Oliver, J.S.; Finestone, E.M.; Ditchfield, P.W.; Bishop, L.C.; Blumenthal, S.A.; Lemorini, C.; Caricola, I.; Bailey, S.E.; Herries, A.I.R.; et al. Expanded geographic distribution and dietary strategies of the earliest Oldowan hominins and *Paranthropus*. *Science* **2023**, *379*, 561–566. [CrossRef]
85. Bruner, E. Prehistory, neuroscience, and evolutionary anthropology: A personal journey. *J. Anthropol. Sci.* **2022**, *100*, 173–192. [CrossRef]

quaternary

MDPI

Article

The Magnitude of a Practice: Collection and Recycling of Patinated 'Old' Flint Items During the Levantine Late Lower Paleolithic

Bar Efrati [1,2,*] and Ran Barkai [1,*]

[1] Department of Archaeology and Ancient Near Eastern Cultures, Tel Aviv University, P.O. Box 39040, Tel Aviv 6697801, Israel
[2] School of Archaeology, University of Oxford, 1 South Parks Road, Oxford OX1 3TG, UK
* Correspondence: bar.efrati@arch.ox.ac.uk (B.E.); barkaran205@gmail.com (R.B.)

Abstract: This study examines the prevalent practice of recycling patinated flint tools ("double patina") of 18 lithic assemblages from three Late Lower Paleolithic sites in Israel. Determined as recycled from 'old' patinated items using visual observation, these tools, bearing both old, patinated surfaces and new modifications, offer insights into lithic strategies, cultural behaviors, and memory preservation. The study shows that the collection and recycling of 'old' patinated items into new tools was ubiquitously practiced, ranging from 41% at Late Acheulian Jaljulia and 11–17% at Acheulo-Yabrudian Qesem Cave. Two main recycling methods were identified, with variations across sites reflecting diverse cultural norms and functional needs (Type A–B). The type-B recycling trajectory was found to be the most prominent, as it prioritizes the preservation of the tool's original appearance, patinated surfaces, and old scars. Following these features, the study additionally suggests that type-B recycling likely stemmed from necessity, cultural preferences, and a choice to connect with the past and preserve it, thus emphasizing the complex interplay of practicality, culture, and memory in the Late Lower Paleolithic period.

Keywords: Levantine Late Lower Paleolithic; lithic recycling; double patina; lithic technology; traditional ecological knowledge; cultural and behavioral implications

Citation: Efrati, B.; Barkai, R. The Magnitude of a Practice: Collection and Recycling of Patinated 'Old' Flint Items During the Levantine Late Lower Paleolithic. *Quaternary* **2024**, *7*, 58. https://doi.org/10.3390/quat7040058

Academic Editors: James B. Innes and José Javier Baena Preysler

Received: 14 September 2024
Revised: 6 November 2024
Accepted: 11 December 2024
Published: 16 December 2024

1. Introduction

The recycling of lithic artifacts throughout the Paleolithic Period has been gaining more attention in recent years and is exemplified by the special volume on lithic recycling in the Paleolithic Period, published in Quaternary International, titled "The Origins of Recycling: A Paleolithic Perspective" [1]. This growing interest is further evidenced by the publication of a large number of subsequent studies [2–12]. The practice of recycling also extends beyond the Paleolithic Period and is found in later archeological contexts worldwide, including Prehistoric, protohistoric, and historical periods [13–22]. The study of recycling in The Paleolithic record is instrumental in enhancing our understanding of prehistoric human behavior and cultural worldviews and practices, ecological understanding, availability of fresh material, technological complexity and flexibility, problem-solving, and decision-making mechanisms, as well as other important facets of human behavior such as attention and conscious thinking [23,24].

The traditional definition of recycling implies the use of an existing 'old' artifact for a purpose that is new and different from the original one [24]. However, another component considered a strong proxy for defining recycling is the ability to detect a time gap between two use events, even if they are similar in appearance or task and use, as they are now different in a temporal context. Thus, maintenance (e.g., resharpening) and recycling should and can be discerned from one another [1,24]. While a maintenance procedure is aimed at extending the use life of an item, in the case of recycling, the stages of modification are discerned by a phase of discarding between them. Thus, recycling does not extend the original use phase of the item in a temporal sense [1,24–28]. The main subject of the current study, the "double patina" phenomenon, i.e., the collection and recycling of existing 'old' items covered in patina into new items, serves as widely accepted evidence of recycling when detected in prehistoric flint assemblages. In most cases, the second phase, i.e., the later modifications on the 'old' item, can be distinguished from the original old flaked patinated surfaces both by color differentiation and by the technological chaîne opératoire/reduction sequence the items underwent [27,29,30].

Flint patination and patination of other stone artifacts have been a focus of interest since the end of the 19th century [31]. Since then, studies on flint and rock patination have varied in subject and terminology [32,33]. Relevant to the current study and as suggested above, patina differences in human-made lithic finds became part of an attempt to document evidence for lithic recycling and, as a result also, lithic procurement strategies and lithic economy. Thus, the current study and its introductions would not discuss patination as a chemical phenomenon, nor will it elaborate on patination chemical variation or its multiple ways of development. The study will focus on the practice of collecting and recycling 'old' items.

"Double patina" items [29], as they are often called in prehistoric research, refer to recycled 'old' human-made items discarded after manufacture (sometimes after being subsequently used) and covered in patina by various environmental agents. After some unknown time, these patinated items were collected by later prehistoric groups and shaped in a manner that exposed fresh surfaces. This modification process fits the term recycling. While the time it takes for patination to form on silica (i.e., flint) is not definitive, patina is assumed to form over thousands of years. In the case of flint, Thiry et al. [34] suggest patina occurs over periods of 20–50,000 years, while thick weathering can be developed in less than 130,000 years, and friable weathering could be formed in several decades [34].

The collection and recycling of 'old' patinated items is a phenomenon that is prevalent at many Early to Upper Paleolithic sites worldwide and has gained more and more recognition in recent years [3,5,6,14,27,30,35–40]. The collection and recycling of old artifacts were also documented in ethnographic studies of contemporary hunter-gatherer groups [13,41–47]. These studies suggest that recycling was fully integrated into the provisioning strategies of these groups, influencing waste disposal strategies and being a part of their cultural worldviews and practices. Ethnographic examples demonstrate how certain Indigenous groups are aware of past human presence in the same environments, evidenced by 'old' human-made objects available in the landscape. These studies demonstrate how such objects are at times collected and brought home for use with or without further modifications and how their antiquity is also recognized [23,42,46,47]. Moreover, the same collected 'old' objects may have been perceived as belonging to entities in the past (may it be past humans, cosmic beings, or "ancestors") [43,46–49]. Alternatively, other cases suggest old items were believed to be given by the landscape, by "nature" [41,45,47].

Recycled patinated 'old' items have recently undergone extensive study at the Late Acheulian open-air site of Revadim and the Acheulo-Yabrudian site of Qesem Cave, Israel [36,37,50]. This study aims to extend the discussion by observing 18 lithic assemblages from both studied sites while also introducing new evidence regarding the practice at a third, more recently excavated site: the Late Acheulian open-air site of Jaljulia, Israel. This study will demonstrate and discuss the magnitude of collecting and recycling 'old' patinated human-made items, presenting the capacity of the practice at each site using preliminary analysis and suggesting interpretations regarding collection methods. The study will also show why this practice constitutes a conscious and complex behavior that goes beyond practical necessity, explicitly emphasizing the noted technological trajectory of collecting and recycling 'old' patinated items by using them as blanks for making new tools by minimally shaping them, resulting in a recycled tool that still preserves the original outlook of the old collected item, with its varying colors and old patinated scars present and dominant [50].

2. The Archeological Sites

2.1. Revadim

Revadim is a multi-layered open-air site located on the southern coastal plain of Israel (Figure 1) [51]. Four main areas (A–D) and two trenches (stratigraphically connecting areas B and C) were excavated during four seasons of excavation, comprising an area of 250 m^2. Areas C and B were the main excavated areas (Figure 2). Paleomagnetism results showed normal polarity, thus, indicating that the sequence is younger than 780 kyr. Uranium–thorium analysis conducted on carbonates that cover flint items suggests that the estimated minimum dates of the site range between 500 and 300 kyr [51]. Lithic and faunal analyses suggest that all layers at the site date to the Late Acheulian [2,7,51–58].

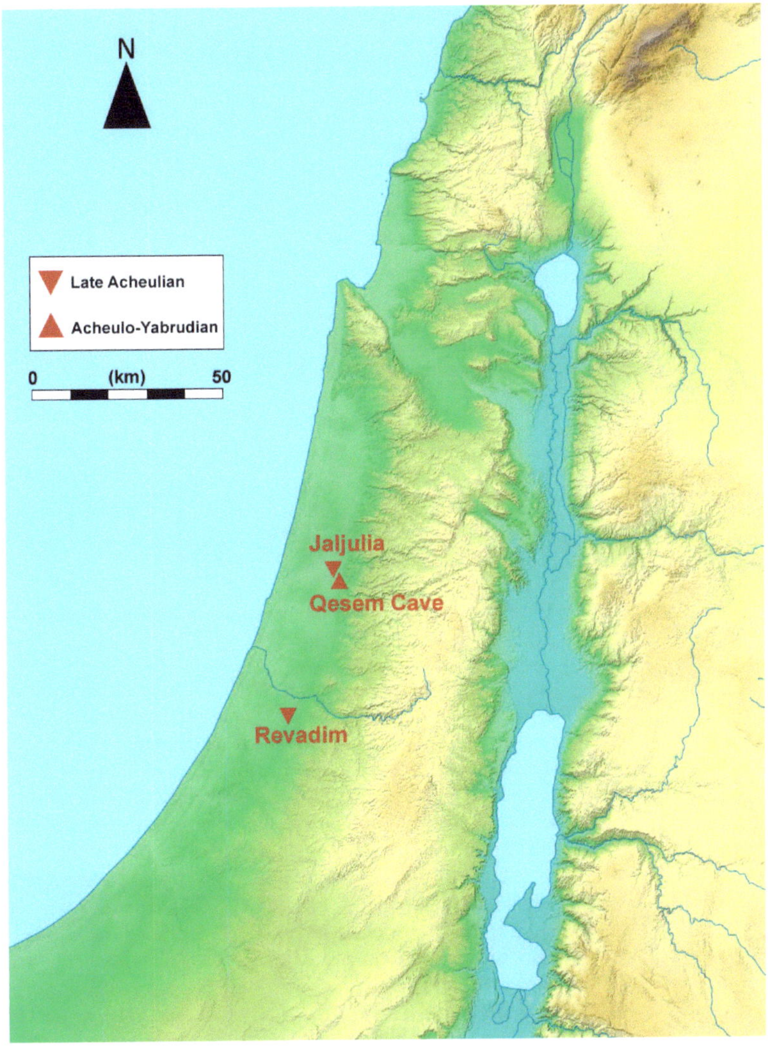

Figure 1. Geographical location of Revadim, Jaljulia and Qesem Cave, Israel.

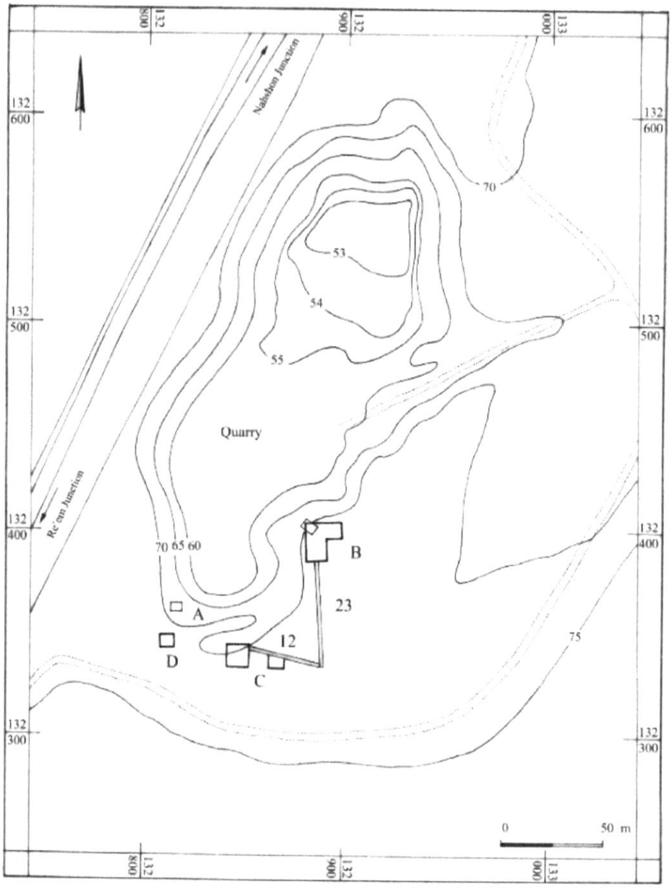

Figure 2. The site of Revadim, featuring the excavated areas designated as A through D. As published in [25].

The lithic assemblages are rich. Various core technologies were detected and include three main sequences: the production of moderately large and medium flakes; the production of pre-determined items from prepared cores, including proto-Levallois cores and discoidal cores; and the production of small flakes by recycling "parent" flakes (cores-on-flakes) [2,7,9,10,55,59,60]. Lithic recycling, in general, is a recurrent practice at the site, and aside from the production of small flakes from cores-on-flakes [2,9,10,59], recycling is evident in all archeological contexts in two other technological and technical forms: recycling of handaxes into cores [7], and the collection and recycling of old patinated flaked items ("double patina"; Figure 3) [2,7,37]. The tool category comprises mostly retouched flake tools ("retouched flakes") and includes items produced by various methods that were further retouched in different ways [7]. Alongside retouched flakes, the tool category consists of other tool types including handaxes, chopping tools and scrapers [55–57,60–63]. Animal bones were also recycled into bone tools, suggested to be the earliest bone tools in the Levant [54].

Extensive use-wear and residue analyses were conducted on the Revadim lithic material, with a particular focus on the lithic material of Area C Layer 3, which provided exceptional results on the functionality of some categories of items and tools due to an excellent state of preservation [12,64]. Among others, functional analyses were conducted on the lithic recycling trajectories available at the site [9,10,64,65], including a study on recycled 'old' patinated tools [37], where the life cycles and functional traits of 28 recycled tools made on 'old' patinated tools were traced. Use-wear results were detected on the new and old edges of eight tools. Six of these items were found to have been in use for different purposes during their first and second life cycles. The results provided additional evidence, this time from a functional perspective, that the recycled 'old' patinated items underwent two life cycles, further supporting the technological assessment that these items are old, shaped tools that were abandoned and then collected and recycled to function as tools again. The studied recycled patinated tools were functionally used for several purposes during their first life cycles as the 'old' patinated active edges bear evidence of cutting, scraping, mixed activities, and chopping. Interestingly, most items were used for scraping during their second life cycle. Moreover, most items were used to work with soft to medium materials, and residue analysis further revealed that their main function was to process animal carcasses [37].

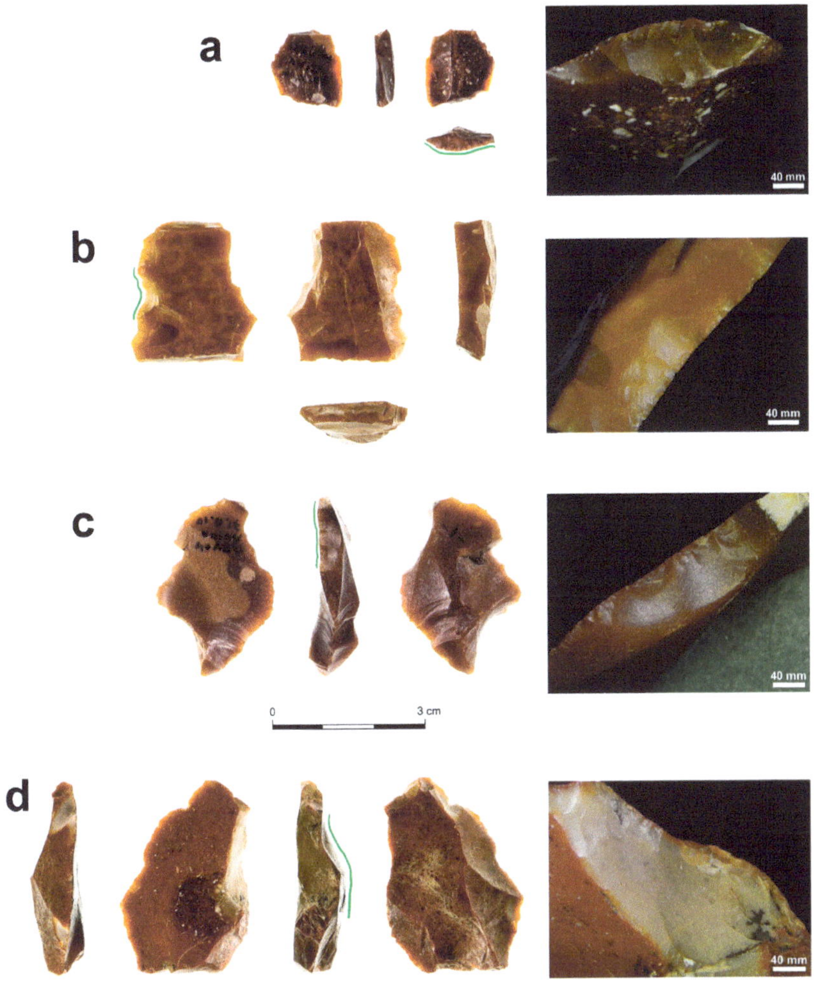

Figure 3. (**left**) Recycled patinated tools from Revadim (items **a**–**d**). Marked in green are the locations from which a close-up caption (**right**) was taken to allow a better observation of patination differences.

2.2. Jaljulia

Jaljulia is a more recently discovered Late Acheulian site located on the central part of the coastal plain of Israel, on the ancient route of Wadi Qana (Figure 1) [66]. Approximately 80 m^2 of Late Acheulian deposits (Areas A–E, G) were excavated by the Israel Antiquities Authority in collaboration with the Institute of Archeology at Tel Aviv University (Figure 4). Techno-typological analysis conducted so far on the lithic assemblages presents classic attributes associating the material with the Late Acheulian industries. Furthermore, paleomagnetic analysis and TT-OSL and pIR-IR290 analyses on quartz and feldspar grains, as well as ESR analysis on quartz grains, provided an absolute chronological frame for the primary human activity on-site of ca. 500–300/200,000 kyr [66].

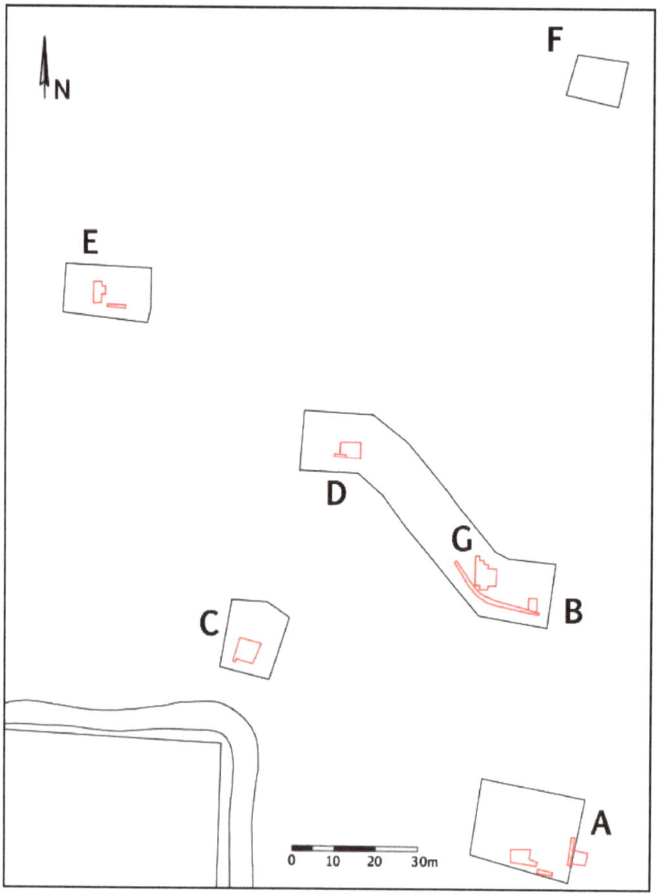

Figure 4. The site of Revadim, featuring the excavated areas designated as A through G.

The lithic assemblages are rich and dominated mainly by flake production, flake tools, and handaxes [66,67]. Core technology varies and includes three primary reduction sequences, similar to those observed at Revadim: the production of moderately large and medium flakes, the production of pre-determined items from prepared cores (including proto-Levallois and discoidal), and the production of small flakes by recycling "parent" flakes [7,66,68–70]. As is the case at Revadim, handaxes were recycled to cores [2] and old patinated flaked items were collected and recycled at Jaljulia (Figure 5) [7,69,70].

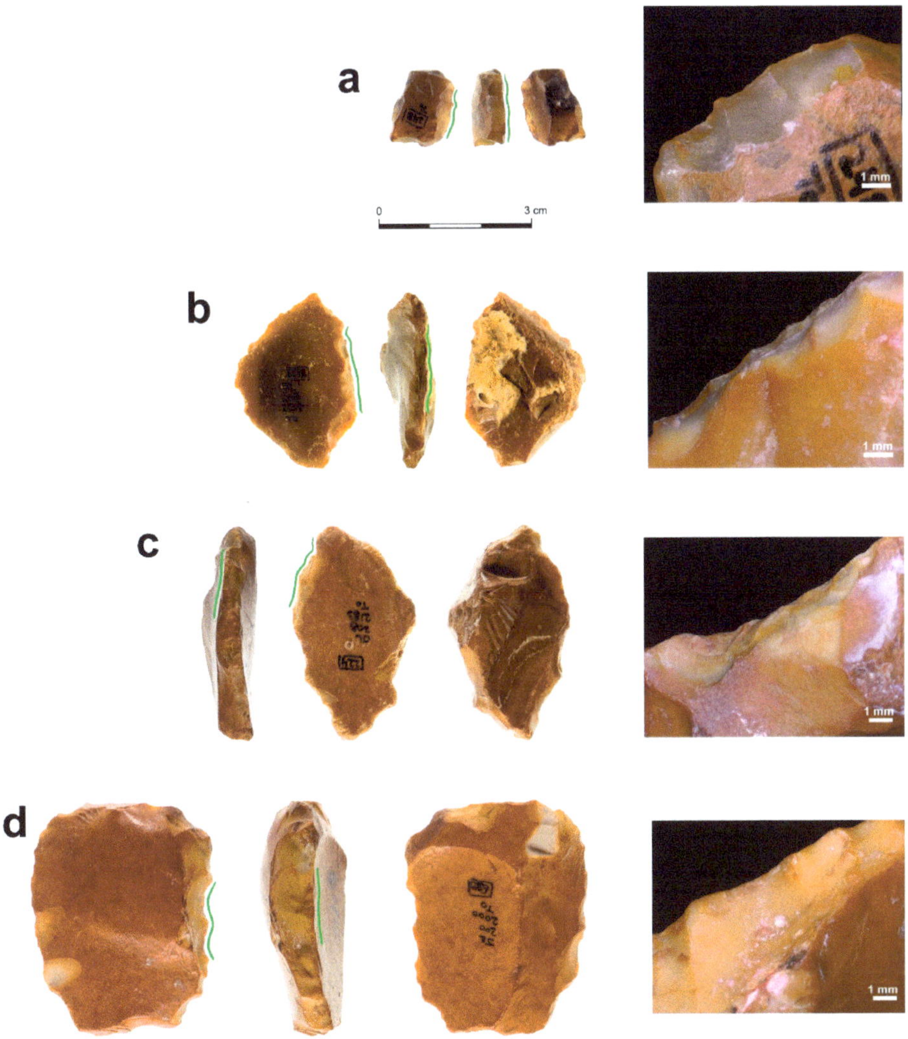

Figure 5. (**left**) Recycled patinated tools from Jaljulia (items **a**–**d**). Marked in green are the locations from which a close-up caption (**right**) was taken to allow a better observation of patination differences.

2.3. Qesem Cave

Qesem Cave is situated about 12 km east of Tel Aviv, Israel, on the western foothills of the Samaria Hills (Figures 1 and 6) [71–74]. So far, excavations have exposed a stratigraphic sequence of over 11 m anthropogenic deposits, including a large central hearth [75]. Bedrock has yet to be reached. U/Th analysis of speleothems from the cave, thermoluminescence (TL), and electron spin resonance (ESR) analyses date the site to ca. 420–200,000 kyr [76–80]. Both absolute dating and the site's lithic assemblages assign Qesem to the Acheulo-Yabrudian Cultural Complex (hereafter, AYCC) of the Late Lower Paleolithic Levant. The lithic assemblages are dominated by the Amudian blade industry [74,81–87]. The Yabrudian industry, dominated by Quina and demi-Quina scrapers, is also evident in three stratigraphically and spatially distinct areas of the site, presented in six contexts [24,82,83,85,88–93]. Handaxes are rare, and only a few were retrieved within the Amudian and Yabrudian contexts. These are argued to have been collected from older sites outside the cave and then brought in to be used again [94,95].

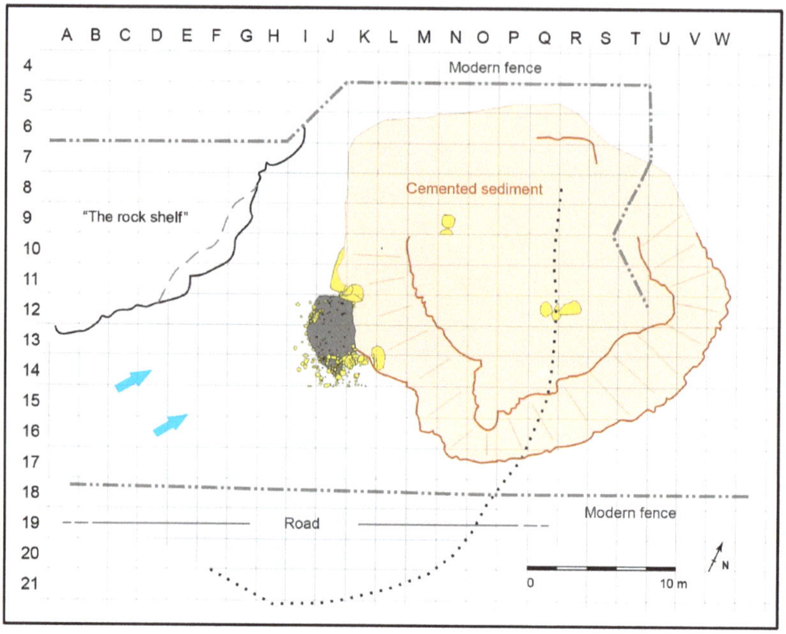

Figure 6. The site of Qesem Cave features the upper sequence of the cave, the hearth area (dark grey feature), and the location of the rock shelf where the lower sequence of the cave begins. As published in [83].

Aside from collection and re-use of old handaxes, lithic recycling is a recurrent practice at the cave and is present in all of the archeological contexts in multiple technological and technical forms: handaxes recycled to cores, scrapers recycled to cores [24] collection and re-use of old shaped stone balls (spheroids/polyhedrons) [96–99], the production of small blades and flakes with sharp edges from "parent" flakes and blades (cores-on-flakes), which were extensively studied by both technological and functional means [8,11,12,24,71,85,100–102], and the collection and recycling of old patinated flaked items ("double patina"; Figure 7) [36,50,89].

Qesem was a mosaic of environmental and ecological regions [103,104], abundant in numerous fresh lithic sources of various types. A systematic study on flint procurement and exploitation strategies at Qesem Cave indicated the use of a large variety of flint types (>90), some of which were selected for specific types of tools or technological trajectories [105–107]. Moreover, earlier studies provided evidence for lithic procurement by both surface collection and quarrying from specific, designated, primary sub-surface sources [89,108,109]; thus, the collection of handaxes, stone balls, and patinated 'old' lithic items were part of the acquisition of lithic materials from the surface [36]. Animal bones were also recycled and served as a material for toolmaking, as more than 20 fragments of soft flint retouchers were found [110–112].

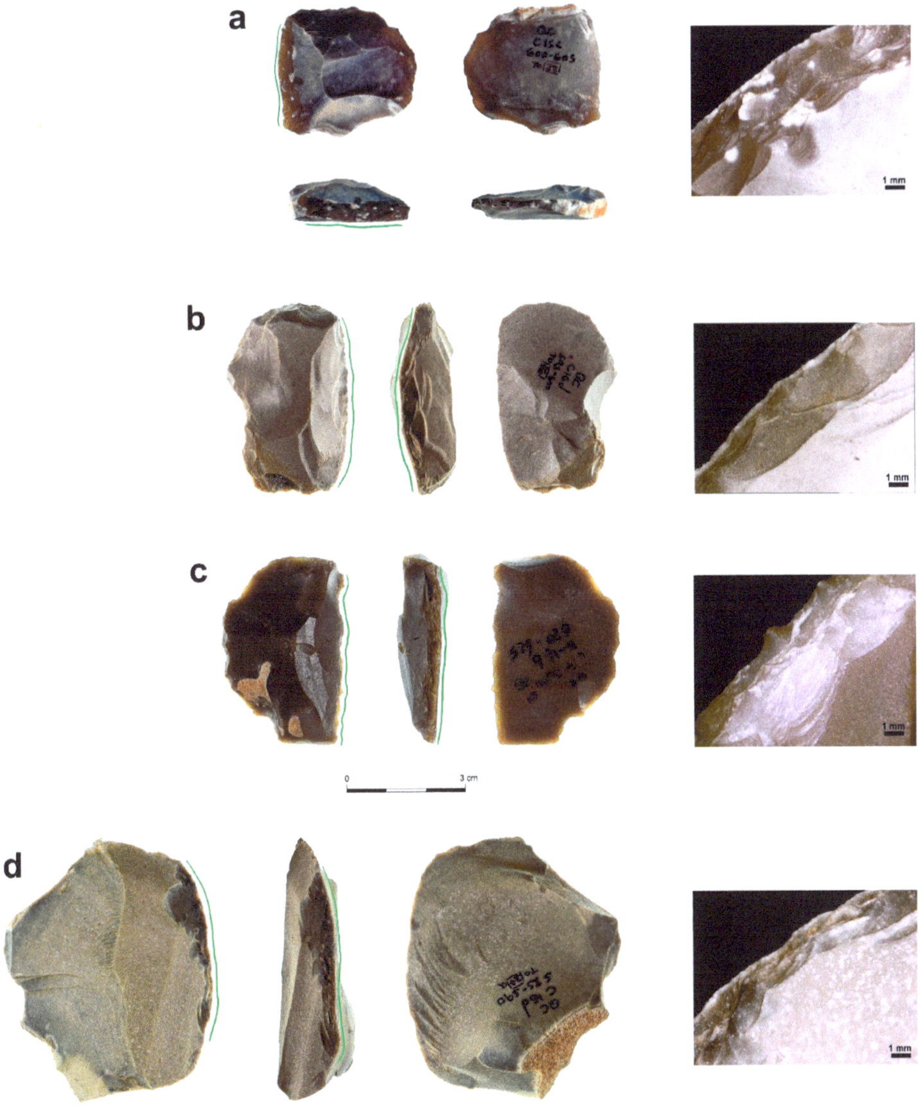

Figure 7. (**left**) Recycled patinated Quina and Demi-Quina Scrapers from Qesem Cave (Items **a**–**d**). Marked in green are the locations from which a close-up caption (**right**) was taken to allow a better observation of patination differences.

3. Recycling 'Old' Patinated Human-Made Flint Items

The presence of recycled 'old' patinated tools was brought to our attention during fieldwork and material analysis. Initially, we were not looking for recycling, and it was not on our research agenda. However, it seems that recycling has found us. These items appear together with items made from fresh, unpatinated flint in all the lithic assemblages of all three studied sites, some of which have been previously mentioned or studied [2,7,24,36,37,50,64,70,74,89,113]. The patina presented on the 'old' tools in all three sites varies in type, color, and texture and differs in color and texture from the natural color of the flint (Figures 3, 5 and 7). So far, the preliminary analysis of recycled 'old' patinated tools from each site revealed two main types of items:

- *Type-A* recycled patinated tools consist of recycled tools detached from 'old' patinated flaked blanks/cores, thus, exhibiting the oldest modified patinated surfaces only on their dorsal face or striking platforms (Figure 8). Their ventral face was detached from the 'old' parent item, revealing the fresh color of the flint, or ventral surfaces that exhibit a less patinated condition (see, for example, Figure 8b) and, thus, counted as a new modification.
- *Type-B* recycled patinated tools consist of completely patinated flaked items that served as blanks to create the new tool. These recycled items preserve and exhibit most of the patinated surfaces of the 'old' items. In most cases, Type B recycled tools preserve almost all of the collected 'old' items, keeping them nearly intact, as the new modifications are few but specific aimed at creating/reshaping a working edge. They, thus, preserve most of the morphological and visual proprieties of the collected 'old' patinated item (Figure 9).

We suggest that old patinated items were collected and recycled for both functional and non-functional reasons. As for functional reasons, we suggest that chosen items were collected and recycled as they were suitable in shape, form, and size for blank production, or alternatively, suitable in shape and form, as is, for the creation of the new desired tool. Previous studies on Qesem Cave and Revadim suggested that 'old' patinated flaked items were selected for their knapping potential, whether they were recognized as being of suitable shape, size, and volume to serve as material mediums for further knapping (type-A) or as being of suitable shape, size, volume, and outline to serve as blanks for the making of new tools (type-B) [36,37]. In the case of recycled patinated type-A tools, a preliminary study indicates that particular morphologies suitable for blank reduction played a role in the selection of the old patinated items. For example, the original patinated scars and ridges would have facilitated the further production of blanks, while previous patinated ridges may have also helped guide new detachments. These patterns were mainly observed in lithic categories such as core trimming elements, cores, and what we defined as "recycling products" (i.e., small flakes and blades detached from cores-on-flakes) [36].

Selection of 'old' flaked items based on knapping potential and desired end-item is even more notable in the case of recycled patinated type-B tools, where all of the tools were recycled in a manner that preserves most of their patinated surfaces and scar patterns. The new modifications were mostly limited to shaping a new working edge. Such specific but minimal reshaping strongly suggests that their collectors wished to preserve the 'old' collected items as-is, as much as possible, while still providing them with new functional potential. Unfortunately, the lack of standardization in ad hoc tools (whether produced by recycling or from fresh flint) makes it impossible to determine, via analysis or technological comparison, what functional properties the collectors of the 'old' patinated items might have sought. Nonetheless, it is reasonable to suggest that the 'old' patinated blank types of varying size and shape were chosen for recycling because their morphology was suitable for creating the desired tool [36,37]. Furthermore, the correlation between form and function noted on the sample of type-B tools from layer C-3, Revadim supports our view that old patinated items were intentionally collected and modified to create new edges, which, in the case of the studied sample, were suitable for scraping activities [37]. In the case of more standardized technological trajectories practiced at each site, we assume that 'old' blanks for making type-B tools (such as recycled Quina and demi-Quina scrapers made on patinated blanks from Qesem Cave [50]) were chosen for their specific shape and morphology. However, further analysis is required to confirm or refute this hypothesis.

Nevertheless, functional reasons do not include a shortage of lithic materials. It has often been suggested that flint recycling is a result of scarcity in lithic materials, a way to maximize lithic resource profitability, including the collection and recycling of 'old' patinated items [13,14,30,39,114,115]. However, lithic recycling has also been documented in areas abundant with lithic materials. Lithic abundance is also evident in the vicinity of all discussed sites. Not only was the supply of fresh flint at all three sites constant and abundant, but the inhabitants also invested intentional effort in the ongoing acquisition of fresh material and, in the case of Qesem Cave, even from distant sources (15–25 km from the site) [105,107]. At Jaljulia, large quantities of flint nodules were transported by the stream, and at the site, inhabitants were physically sitting on gravel deposits extremely rich in fresh flint nodules [69]. In addition, tools made of fresh flint and typologically identified as similar to recycled patinated equivalent items were found in all the studied assemblages, hand-in-hand with recycled patinated items in the same contexts and even dominated the assemblages at each site [36,37,50].

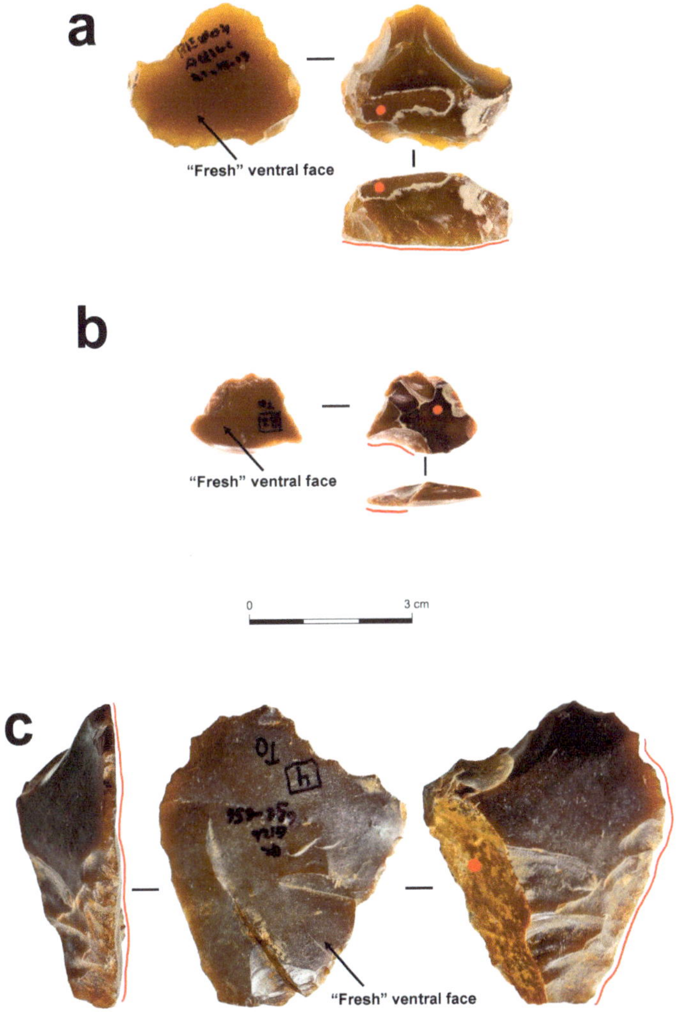

Figure 8. Type-A recycled patinated old items from Revadim (**a**), Jaljulia (**b**), and Qesem Cave (**c**). The patinated 'old' surfaces on the items' dorsal faces are noted (**red dot**), while the ventral surfaces clearly exhibit fresh flint or a different (**black arrow**), later type of patina matching the tools' retouch (**red line**).

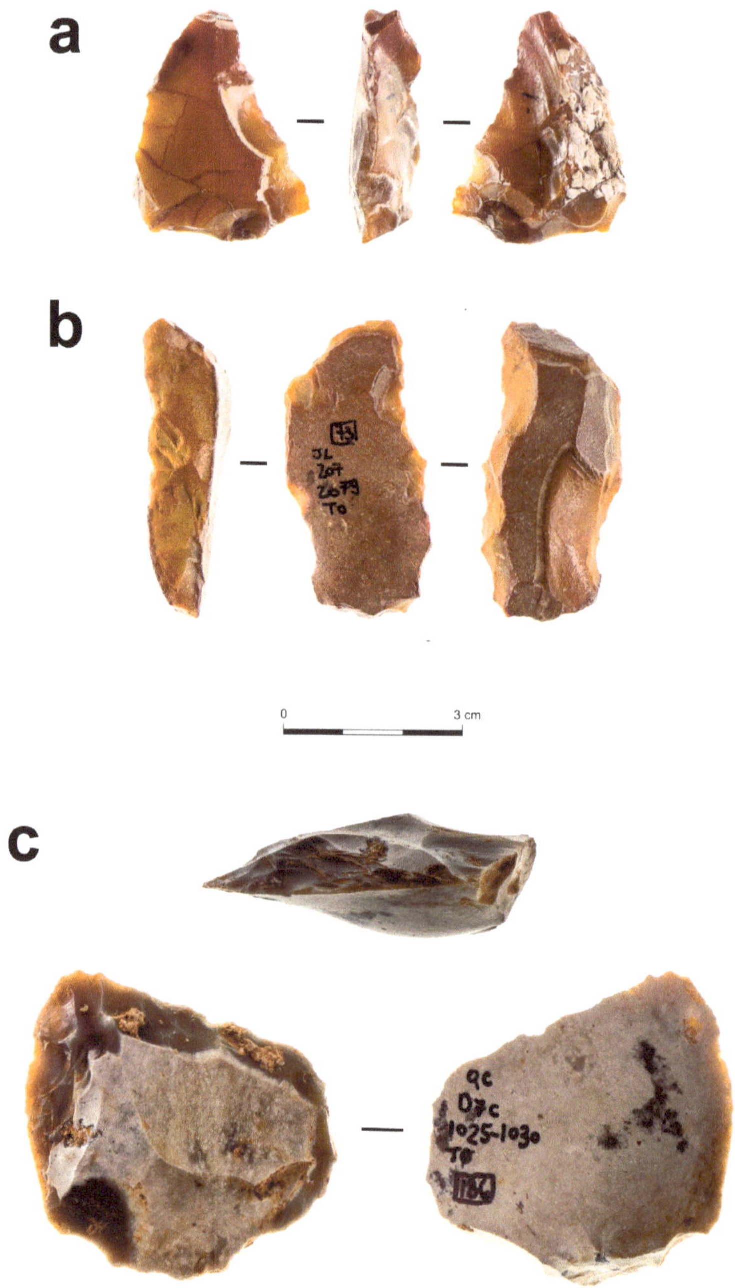

Figure 9. Type-B recycled patinated old items from Revadim (**a**), Jaljulia (**b**), and Qesem Cave (**c**).

We, therefore, contend that lithic recycling practices should also be viewed in cultural and social terms [24,27,105–108,116–118]. The same can be said regarding the collection of 'old' patinated items as workable materials at Revadim, Jaljulia, and Qesem Cave. We believe this practice should be viewed in the framework of traditional (in our case, ancient) ecological knowledge and use of resources and as a set of behaviors that combined necessities and cultural choices. Traditional ecological knowledge is referred to in the literature as "a cumulative body of knowledge, practice, and belief, evolving by adaptive processes and handed down through generations by cultural transmission about the relationship of living beings (including humans) with one another and with their environment. It is a subset of Indigenous knowledge, which is local knowledge held by Indigenous people or local knowledge unique to a given culture or society" [119]. Thus, we suggest that these choices were based on cultural settings, worldviews, perceptions, and economic and functional preferences, which eventually enhanced and dictated relationships and interactions between humans and their environment, including flint and existing flint tools as material mediums. Our interpretations have been published in the past [23,37,50] and will be further elaborated in the discussion part in accordance with the manuscript's results.

4. Materials and Methods

A total of 18 assemblages from the three studied sites were selected for this study based on the availability of detailed lithic analyses and their good state of preservation. Each assemblage represents a distinct archeological context (for a complete list and description of assemblages, see Supplementary Information). All selected assemblages were techno-typologically analyzed using methodology and terminology from previous studies [2,55,74,120,121] and following established definitions [122,123].

Recycled 'old' patinated tools were identified using only the naked eye, separated, and counted. No additional microscopic tools were used. Tools were classified and counted as recycled from 'old' patinated items if they carried flaked surfaces covered in patina, as well as later scars that cut through the patinated modified surfaces. These later scars either revealed the fresh color of the flint or were covered with a different later patina. When 'old' flaked surfaces could not be confidently distinguished from old cleavage surfaces or natural surfaces (e.g., cortex/neo-cortex), the items were not included in the PPF category. Hence, the numbers presented should be regarded as minimum estimations of the phenomenon in each site. This methodology follows previous studies [35,36] and the definition of Goodwin [29] for 'double patina': items "that have been modified again, thus leaving newer scars in unpatinated, or less patinated, condition" [29].

In the case of Qesem Cave, a light translucent patina was formed on some of the "fresh" items knapped and left on-site, perhaps due to specific post-depositional conditions in the cave [102]. This type of patina can only be observed microscopically; hence, it was not included in our analysis. In addition, the data gathered from the Amudian and Yabrudian assemblages at Qesem cave will be presented separately and compared due to characteristic differences between the industries.

5. Results

The analyzed tool sample reveals that patinated 'old' items were extensively collected and recycled into various types of tools at all three sites (Table 1): at Revadim, where four distinct assemblages were analyzed (amounting to a total of 3436 analyzed tools), 754 items were identified as recycled patinated items, constituting 22% of the analyzed tools. The extent of the practice at Jaljulia was even more evident. Out of two distinct assemblages (n = 4036 tools), 1661 tools were identified as recycled patinated 'old' items, constituting 41% of the analyzed tools at the site. Finally, Qesem cave presented smaller quantities, where the tools of 12 distinct assemblages assigned to the Amudian (n = 3660) or Yabrudian (n = 2001) industries were analyzed. A total of 741 recycled patinated tools were identified at Qesem Cave, constituting 11% and 17% of total tools assigned to each industry, Amudian and Yabrudian, respectively.

The quantities of type-A versus type-B recycled patinated tools differ noticeably between the Late Acheulian sites and AYCC Qesem Cave (Table 1; Figure 10). At Revadim and Jaljulia, most recycled patinated items were made on 'old' blanks (type-B), constituting very high percentages (82% for Revadim and 90% for Jaljulia) of the recycled patinated tool category. At Revadim, this tool category comprises 18% of all the tools, and 3% of total débitage, and at Jaljulia, 37% of all the tools and 11% of the total débitage. At Qesem Cave, however, most recycled patinated tools were detached from 'old' human-made blanks (type-A), constituting 49% (Yabrudian) and 83% (Amudian) of the recycled patinated tool category and a small percentage of the total debitage (0.2% and 2% for Amudian and Yabrudian, respectively).

Table 1. A general breakdown of recycled patinated tools (including a breakdown of types A–B) available in Revadim, Jaljulia, and Qesem Cave (Amudian and Yabrudian). The cells displaying data about types A–B are shown in greyscale to visually represent the capacity of each type across the different sites in comparison to one another.

Site	Revadim	Jaljulia	Qesem Amudian	Qesem Yabrudian
Débitage	24,342	13,094	29,811	10,028
Total number of tools	3436	4036	3660	2001
Recycled patinated tools	754	1661	407	334
Recycled patinated tools out of the tool category	22%	41%	11%	17%
Tools made from fresh flint	2682	2375	3253	1667
Fresh tools out of the tool category	78%	59%	89%	83%
Recycled tools knapped from old patinated items (type-A)	139	163	337	165
Type-A out of total débitage	1%	1%	1%	2%
Type-A out of the tool category	4%	4%	9%	8%
Type-A out of the recycled patinated tools	18%	10%	83%	49%
Recycled tools on patinated blanks (type-B)	615	1498	70	169
Type-B out of total débitage	3%	11%	0.2%	2%
Type-B out of the tool category	18%	37%	2%	8%
Type-B out of the recycled patinated tools	82%	90%	17%	51%

Jaljulia has the most type-B recycled patinated tools (n = 1498) compared with the analyzed samples from the other sites. Revadim (n = 615) and the Yabrudian assemblages (n = 169) at Qesem follow, while the Amudian assemblages of Qesem have the smallest number (n = 70; Table 1). A comparison of the percentage of these recycled tools out of the entire tool category in each assemblage from each site further emphasizes this finding (Table 2): Jaljulia and Revadim have the highest percentage out of the entire tool category, ranging from 27 to 47% for the Jaljulia assemblages and 14 to 35% for Revadim. The Yabrudian assemblages of Qesem follow; however, they are still far behind, as type-B recycled patinated tools in the Yabrudian assemblages comprise only 6–10% of the tool category, with the Amudian assemblages showing the lowest percentages (1–6%). Most type-B recycled patinated tools in Jaljulia and Revadim are ad hoc tools: retouched flakes, notches, varia tools, and retouched fragments (Table 3). Although the percentages for these categories vary between Revadim and Jaljulia, both sites show much higher percentages than both the Amudian and the Yabrudian assemblages at Qesem.

Most type-B recycled patinated tools at Qesem Cave are scrapers, and most are found in the Yabrudian assemblages (Table 3; Figure 11). However, while the Yabrudian assemblages contain the highest percentage of type-B recycled patinated scrapers, the percentages of these scrapers at Revadim and Jaljulia (about 1% out of the entire tool category and 3% out of the recycled patinated tools) are similar to the Amudian assemblages of Qesem. However, if we look at the percentage of these scrapers out of the entire scraper category at each site, we see that it is higher in the Revadim and Jaljulia assemblages than in the Amudian assemblages at Qesem (44% and 43% at Revadim and Jaljulia, respectively, as opposed to only 12% at Amudian Qesem; Table 3).

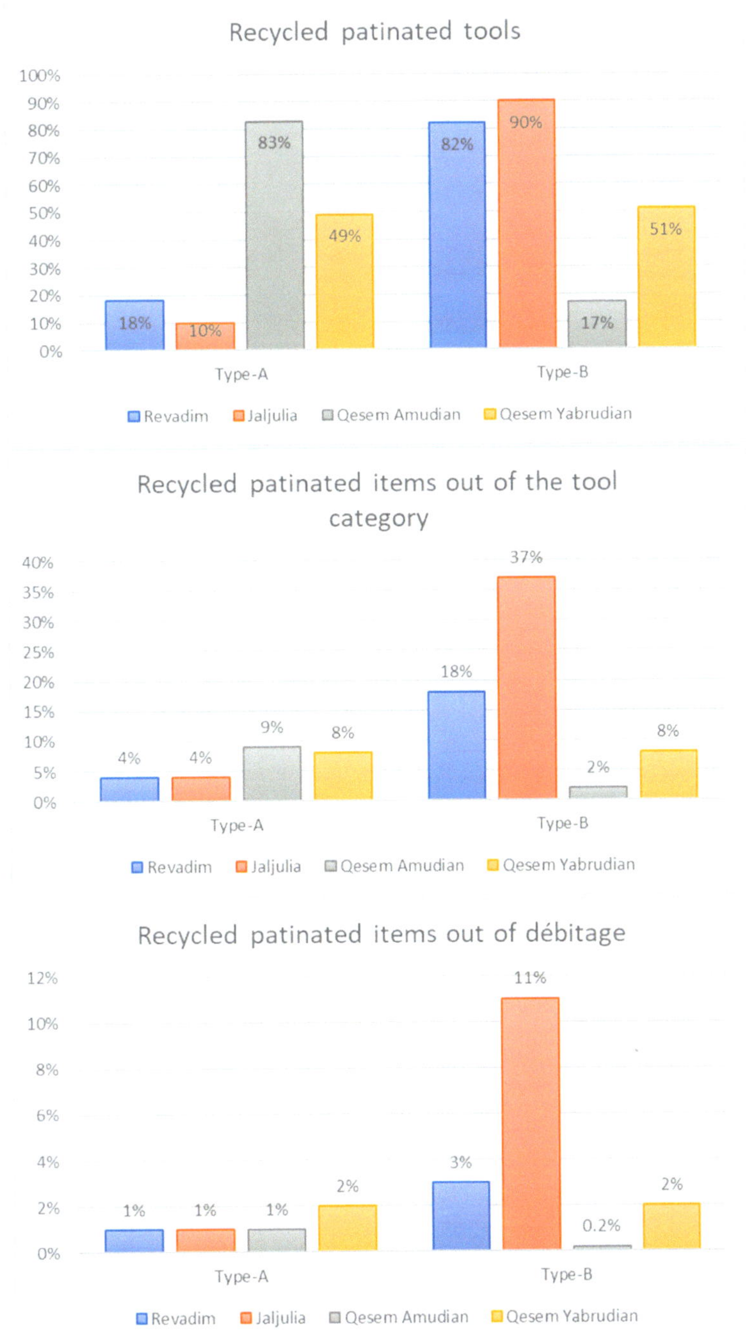

Figure 10. A visual representation of type-A versus type-B recycled 'old' patinated tools at each site. (**top**) A general breakdown of type-A and type-B recycled tools at each site. (**middle**) The capacity of type-A and type-B within the tool category at each site. (**bottom**) The capacity of type-A and type-B within the débitage of each site.

Table 2. Breakdown of recycled patinated tools made on patinated blanks (type-B) from Revadim, Jaljulia, and Qesem Cave. The cells displaying data about type B are shown in greyscale to visually represent the capacity of each type across the different assemblages and sites.

Site	Context	Débitage	Tools	Recycled Patinated Tools	Recycled Tools on Patinated Blanks (Type-B)	Type-B out of Total Débitage	Type-B out of the Tool Category	Type-B out of Patinated Tools
Revadim	Area B, Locus 20	3571	558	204	196	5%	35%	96%
	Area B, Locus 21	984	108	34	30	3%	28%	88%
	Layer C-3	18,048	2546	461	345	2%	14%	75%
	Layer C-5	1739	224	55	44	3%	20%	80%
Jaljulia	Area B	6998	2018	623	543	8%	27%	87%
	Area DII	6096	2018	1038	955	16%	47%	92%
Qesem Cave	Unit I (Amudian)	2889	363	44	15	0.5%	4%	34%
	K/10 (Amudian)	1278	178	20	1	0.1%	1%	5%
	G/19–20 (Amudian)	1582	380	46	2	0.1%	1%	4%
	Hearth (Amudian)	3307	283	27	3	0.1%	1%	11%
	South of Hearth (Amudian)	2877	158	24	10	0.3%	6%	42%
	Southern Area (Amudian)	6145	597	93	12	0.2%	2%	13%
	Shelf (Amudian)	7396	1199	68	10	0.1%	1%	15%
	Deep Shelf (Amudian)	4337	502	85	17	0.4%	3%	20%
	Unit I (Yabrudian)	455	109	23	8	2%	7%	35%
	Southwestern Area (Yabrudian)	1394	294	32	19	1%	6%	59%
	Shelf (Yabrudian)	4305	887	156	89	2%	10%	57%
	Deep Shelf (Yabrudian)	3874	711	123	53	1%	7%	43%

Table 3. Breakdown of six selected tool categories made on patinated blanks (type-B) from Revadim, Jaljulia, and Qesem Cave. Some of the cells displaying data about type B are shown in greyscale to visually represent the capacity of items across the different tool categories at each sites.

Site	Revadim	Jaljulia	Qesem Cave Amudian	Qesem Cave Yabrudian
Tools	3436	4036	3360	2001
Recycled patinated tools	754	1661	407	334
Recycled tools on patinated blanks (type-B)	615	1498	70	169
Retouched flakes	1385	1660	1163	500
Recycled patinated retouched flakes	265	738	147	69
Recycled retouched flakes on patinated blanks (type-B retouched flakes)	206	666	15	24
Type-B retouched flakes out of the tool category	6%	17%	0.4%	1%
Type-B retouched flakes out of the recycled patinated tools	27%	40%	4%	7%
Type-B retouched flakes out of the retouched flakes category	15%	40%	1%	5%
Type-B retouched flakes out of the recycled patinated retouched flakes	78%	90%	10%	35%
Notches	554	335	70	52
Recycled patinated notches	157	163	1	10
Recycled notches on patinated blanks (type-B notches)	137	144	0	4
Type-B notches out of the tool category	4%	4%	0%	0.2%
Type-B notches out of the recycled patinated tools	18%	9%	0%	1%
Type-B notches out of the notches category	25%	43%	0%	8%
Type-B notches out of the recycled patinated notches	87%	88%	0%	40%
Retouched fragments	750	848	767	378
Recycled patinated retouched fragments	155	242	79	31
Recycled retouched fragments on patinated blanks (type-B retouched fragments)	118	207	12	14
Type-B retouched fragments out of the tool category	3%	5%	0.3%	1%
Type-B retouched fragments out of the recycled patinated tools	16%	12%	3%	4%
Type-B retouched fragments out of the retouched fragments Category	16%	24%	2%	4%
Type-B retouched fragments out of the recycled patinated retouched fragments.	76%	86%	15%	45%
Varia tools	212	503	57	43

Table 3. *Cont.*

Site	Revadim	Jaljulia	Qesem Cave Amudian	Qesem Cave Yabrudian
Recycled patinated varia tools	54	292	10	5
Recycled varia tools on patinated blanks (type-B varia)	46	276	2	4
Type-B varia out of the tool category	1%	7%	0.1%	0.2%
Type-B varia out of the recycled patinated tools	6%	17%	0.5%	1%
Type-B varia out of the varia tools category	22%	55%	4%	9%
Type-B varia out of the recycled patinated varia tools	85%	95%	20%	80%
Scrapers	59	108	170	461
Recycled patinated scrapers	29	52	40	153
Recycled scrapers on patinated blanks (type-B scrapers)	26	46	21	115
Type-B scrapers out of the tool category	1%	1%	0.6%	6%
Type-B scrapers out of the recycled patinated tools	3%	3%	5%	34%
Type-B scrapers out of the scrapers category	44%	43%	12%	25%
Tpe-B scrapers out of the recycled patinated scrapers	90%	88%	53%	75%

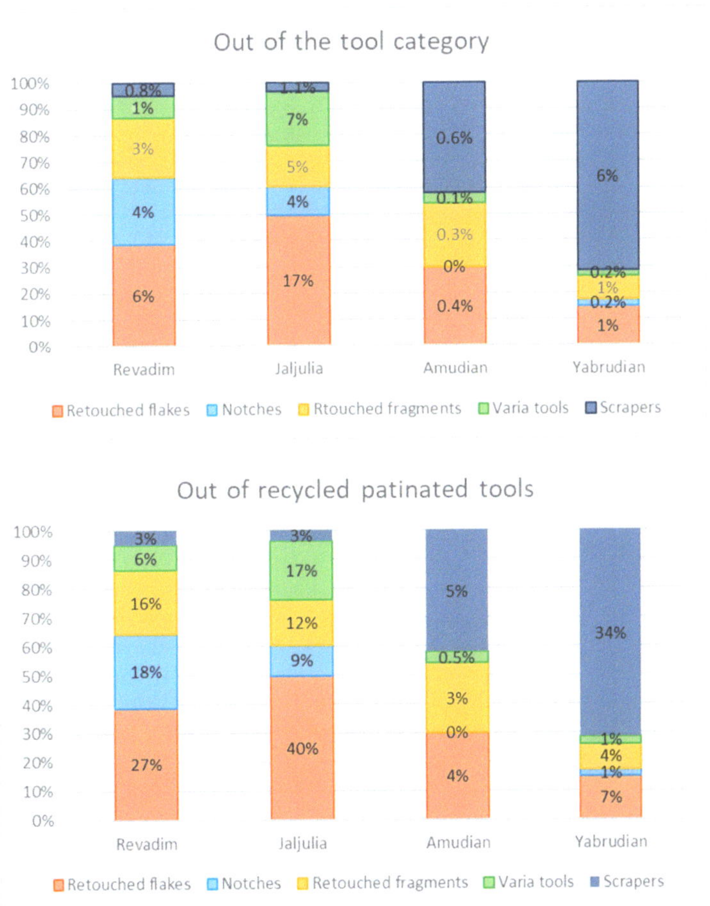

Figure 11. A visual representation of type-B recycled patinated tools from selected tool categories: retouched flakes, notches, retouched fragments, varia tools, and scrapers. (**top**) General breakdown of type-B recycled patinated tools within each tool category at each site, (**bottom**) General breakdown of type-B recycled patinated tools out of the total of recycled patinated tools at each site.

Overall, we can see from these results that, although the majority of tools at each site were made from fresh unpatinated flint (Table 1), recycled flaked patinated items comprise a significant portion of the tools in all the studied assemblages from all three sites.

6. Discussion and Conclusions

The study shows that old, patinated items were regularly collected and recycled into new tools at the following three late Lower Paleolithic sites in Israel: Revadim, Jaljulia, and Qesem Cave. These recycled flint items, which exhibit two life cycles with a clear time gap between them [2,36,37,50,89], were used alongside fresh flint and other items made by various technological means of recycling, such as the production of small flakes/blades from "parent" flakes, and the recycling of handaxes and scrapers into cores [2,7–11,24,59,66,85,100–102]. The current, more extensive sample is a significant addition to previously presented evidence from Revadim and Qesem Cave [36,37,50]. The results add important new evidence for the prevalence of the practice in different archeological contexts of the sites and throughout time.

As previously suggested [37], the items could have been collected during recurrent or extended stays at each site and locality, brought to the site from elsewhere during excursions, or selected from the arsenal of old flaked flint items available on-site from prior older settlements. We suggest that a combination of different collection strategies of old patinated items were practiced at each site. For example, in Qesem Cave, 'old' flaked patinated items are assumed to have been collected from the surroundings outside of the cave and brought in to be recycled and used again, most probably from older sites in different environments and localities. The collection of old blanks from within the cave does not seem to have been an option in the case the items discussed here, as the cave's patina is a light translucent patina that could not be identifiable using the naked eye [36,102] and could not be matched with the very distinct and present variations in patinas identified on the 'old' patinated blanks that were chosen to be recycled into tools. The considerable variation in color of the patinated surfaces found at Qesem may also support this assumption (Figure 7a) [36,50].

However, in the case of Revadim and Jaljulia, which are both open-air sites composed of multiple localities that were inhabited recurrently throughout the Late Acheulian, it is more likely that patinated flaked items were also collected from older localities within the sites themselves, alongside collection of old items from the vicinity of the site, and then brought to the designated contexts studied here [37]. The slight variation in color of the patinated surfaces (mainly yellow to orange in hue) suggests that the items were collected from similar environments (Figures 3 and 5). As the same types and colors of patina cover most of the lithic assemblages of both sites that are not considered recycled, we suggest that the option of collecting 'old' patinated blanks from older localities within the area of the sites is the most probable in the case of both Revadim and Jaljulia. Since lithic material was not in short supply, and fresh items that show no signs of patina or post-patina modifications are abundantly present in the same contexts and make the majority of the assemblage, we argue that these items were selected intentionally, especially given that conceiving and creating a new tool from an existing patinated item, whose shape and size are already fixed, is not necessarily an easier task and would have posed challenges to the knapper that would not be encountered when creating a tool from fresh material [37,50].

The variation in types/categories of tools in the samples from each site strengthens our previous suggestion that 'old' patinated items were considered workable materials. At both Revadim and Jaljulia, old patinated items served as blanks for new tools more often than they served as material from which new blanks were created, whereas Qesem Cave represents a different pattern, where the type-A recycled tools are far more common. The significant variations in recycling 'old' tools and in the frequency of recycled tools detached from old patinated items (type-A) versus that of recycled tools made on old patinated blanks (type-B) at Qesem in comparison to the Late Acheulian sites can perhaps be explained as a matter of preference, as the recycling practice and both trajectories were available in both studied cultural complexes. The variations may also have stemmed from differences in cultural settings of prehistoric Late Acheulian people versus Acheulo-Yabrudian, which also dictated a change in worldviews and practices. However, a more holistic comparison of additional Acheulian and Acheulo-Yabrudian sites is required to more confidently support this hypothesis, which would also benefit from a comparison of open-air versus cave sites to examine whether the nature of the site was a component in the cultural preferences of the inhabitants. Nevertheless, the presence of both types of recycling old patinated items at all three sites and the quantitative differences in the modes of recycling at each site indicate that such preferences did exist, and we believe that cultural worldviews and practices played a role in dictating the preferred recycling trajectory—inter and intra-site alike.

The type-B trajectory is the most curious mode of recycling old patinated blanks. Made on completely patinated flaked blanks, the new tools underwent precise and minor modifications aimed at creating a new working edge while still preserving the old items' original morphology and appearance (Figure 9). The question of why the 'old' items were so minimally reworked cannot be definitively answered. We give our interpretation that does not contradict the item's functional aspects in the collection process of potential workable 'old' blanks or in their manner of recycling to achieve a new tool. We suggest that minimal reworking of the 'old' blanks might have also stemmed from the desire to preserve memories and experiences in a multidimensional sensorial manner that allows the new users to participate and engage with the past and present through seeing, touching, relating, and using [23,37,50,124]. The preservation of memory and mindful experience were both visual and physical: the old item's original appearance, old scars, and colorful patinated surfaces were preserved as much as possible, as was their new functionality [23].

Whether appearing in high or low quantities, this group of recycled tools presents interesting inter and intra-site quantitative patterns. At Revadim and Jaljulia, most type-B tools are ad hoc (fragments and retouched notches), as further evidenced by the large quantity of retouched tool fragments found at both sites. The intra-site differences are more clearly manifested in the complete pieces: Jaljulia is oriented towards more notches recycled in this manner, while Revadim is oriented towards more retouched flakes. Qesem, where almost all the recycled patinated type-B tools are scrapers from the Yabrudian assemblages, presents a completely different pattern. Their frequency in comparison to the Amudian assemblages is too great to be attributable merely to the characteristic dominance of scrapers in Yabrudian contexts. They might, thus, represent a preference of past people rather than a known bias. By choosing to recycle most patinated 'old' blanks at Qesem into scrapers in a manner that preserves the old item's morphology as well as the color and texture of the patina, the Qesem knappers might have been paying tribute to the vital role of this tool, in their daily practice as well as in their perception and worldviews. In addition, it seems that the inhabitants of Qesem Cave made some of the recycled patinated scrapers from old scrapers. Figure 12 shows two scrapers from Qesem Cave that show clear evidence of an older, patinated Quina retouch, which the new Quina retouch cuts. A future technological study coupled with functional analysis may be able to suggest if the recycled scrapers were recycled to pay tribute not only to the memory of the 'old' tool and its history but also regarding its function during its multiple life histories [37].

We previously suggested that collecting and recycling 'old' patinated items into new tools in such a specific manner was an intentional behavior not motivated solely by functional reasons or shortage, but also for their perceived potency and the experiences they may have evoked during the encounter with them [23,37,50]. In this paper, we presented new evidence on the practice of collecting and recycling 'old' patinated items into tools from Jaljulia. These new data strengthen the understanding of the magnitude and capacity of the practice during the Lower Paleolithic. As mentioned, we suggest that this mode of lithic recycling should also be viewed in cultural and social terms, in the framework of ancient ecological knowledge and use of resources and as a set of behaviors that combined necessities and cultural choices [24,27,105–108,116–119], based on cultural settings, worldviews, perceptions, and economic and functional preferences, which eventually dictated relationships and interactions between humans and their environment, including flint and existing flint tools as material mediums [23,37,50]:

First, we suggest that collecting and recycling existing old items, specifically in the type-B method, is a different endeavor than conceiving and creating a new tool from an existing patinated item of predetermined size and shape. We do not contend that one recycling trajectory is more straightforward to create than the other, nor that collecting and recycling older patinated items, in general, is an easier or harder task than making tools from fresh material. Rather, creating a tool from fresh versus existing material is different. Each affords the knapper different options for shaping the material as desired, correcting mistakes, and overcoming any material faults. Thus, we suggest that there was a shift in perception, a conscious decision-making process behind the recycling process of type-B tools, which, in our opinion, suggests the practice was not solely driven by mere practicality and efficiency. Producing a technologically identical end product that is functional yet visually preserves the original, with its multiple types of memories and experiences, requires a mindful shift from the standard technical procedures and decisions a knapper will follow to create a tool from fresh unaltered flint. The material from which the knapper begins the creation process is already a finished product, with limited options for styling and reshaping. The case of type-B Quina and demi-Quina scrapers made from older scrapers from Qesem Cave (Figure 12), as well as examples from Revadim [37], can further strengthen the conscious choice of old blanks wished to be recycled for functional reasons as well as reasons that stem from cultural worldviews.

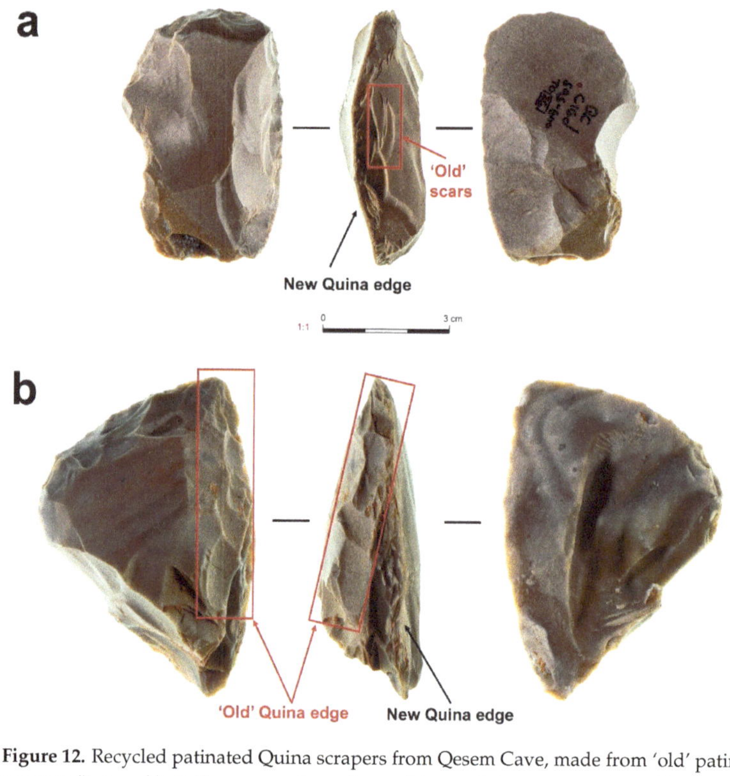

Figure 12. Recycled patinated Quina scrapers from Qesem Cave, made from 'old' patinated Quina scrapers (items **a**,**b**). Both scrapers show clear evidence of the presence of an older, patinated Quina retouch (marked in red), which the new Quina retouch cuts (marked with a black arrow).

Thus, we see this required shift from the standard procedure as a conscious and mindful cognitive process along the lines of Wynn et al. [125] and Malafouris [126,127]. We suggest that the process of recycling existing old items was a conscious process that was not internal but rather very much embodied, embedded, enactive, and extended (4E cognition), having very much to do with the engagement with the old object's features and properties to afford a successful end product from an already finished product. While 4E cognition is relevant in the case of flint knapping in general, starting from the Lower Paleolithic Period, we suggest that the case of recycling existing old objects into new tools using minimal procedure (type-B tools) makes somewhat of an extreme example of how the cognitive process is equally dependent on the material, how perception and mind are shaped during human–material co-engagement, and perhaps even how we can think about mindful consciousness and perception in early prehistoric periods, as, again, in the case of this study the material is already a finished product that may limit if the traditional thinking process and technical procedures of flint reduction and knapping were used.

Second, we further consider the technical process of the type-B recycling trajectory to be conceptually and technically analogous to the Readymade technique, known from the world of modern plastic art [50], where existing objects are given new meaning and function. In collecting 'old' items and recycling them using readymade-like concepts and techniques into new tools, prehistoric people during the Late Lower Paleolithic exhibited a complex and conscious decision-making process that went well beyond mere practicality and efficiency.

Third, we further suggest that the practice served as a preservation technique. Readymade-like in its approach, it reflects a complex interplay of necessity, cultural choices, and memory preservation. The aim of preservation in the new tools was potentially the deep past, whether of old private/communal memories, to pay homage to older groups and entities (human or natural), or to mark familiar and important locations. We propose that the practice served to preserve old memories through attention to and acknowledgment/awareness of old makers and users or even the mnemonic memories the item may have evoked in the new collector. Imbued with potency due to the deep impression they made on the collector at the moment of encounter, these items would also have evoked consequent experiences that the collector might have wished to acknowledge [23], while attributing new meaning and function to the collected patinated items through their insertion into a new functional and cultural sphere. By preserving the properties of the 'old' patinated items in the new tool, past people may also (alongside the functional purpose of the tool) have been trying to preserve and highlight the item's itineraries and accumulated significance [128–130]. This purposeful pattern might have stemmed from the inhabitants' desire to preserve the memory of older groups and entities (cosmological or natural) [6,17,22,41,43–47,50], familiar/important locations (even their own settled locality, as is the case of Revadim and Jaljulia, where 'old' flaked items were likely collected from the sites themselves) [17,37,43,47,50,130–133].

Fourth, the 'old' blanks preserved in type-B recycled tools might have been imbued with a perceived potency and chosen for their evoked experiences, connecting the past and present through a multidimensional sensorial manner. Therefore, we also suggest that 'old' patinated items were intentionally selected and integrated into the practical tool sets of the site's inhabitants due to items' perceived potency and the experience [22] gained during their previous life cycle [23,37,50]. Additionally, these items may have captured the attention of the collector and user due to their personal memories, practices, and preferences. In other words, "old" items may have been chosen, collected, and recycled to be preserved and reintegrated into functional and social realms for personal sentimental, mnemonic, esthetic, or cultural significance that the collector and/or user may hold [18,50,134–137].

We, thus, see the collection and recycling of double-patinated items as a functional process as well as a preservation technique that is readymade-like in that it "protects" the old features of the collectible item while also creating its new functional and visual attributes [23,37,50]. The intensity of the practice and the visual appearance of the items (i.e., preserving old with new) speak for themselves. Moreover, variations in recycling practices across the three sites may suggest that cultural settings, worldviews, perceptions, and economic and functional preferences can influence these practices.

Supplementary Materials: The following supporting information can be downloaded at: https://www.mdpi.com/article/10.3390/quat7040058/s1, SI text: The studied assemblages; Figure S1: Site plan of Revadim showing the location of the excavated areas and the studied contexts (numbered as in text); Figure S2: Top plan of Revadim Area B, Layer B2 showing the location of loci 20-21 (numbered as in text); Figure S3: Stratigraphic plan of Area C, showing the stratigraphic location of Layers C3 and C5 in the sequence; Figure S4: Site plan of Jaljulia, showing the location of the excavated areas and the studied contexts (numbered as in text); Figure S5: Site plan of Qesem Cave, showing the location of the studied areas (numbered as in text); Table S1: A list of the studied assemblages and additional information.

Author Contributions: B.E. conceived and designed the research, processed the lithic material to identify recycled double-patinated items and conceived, designed, and wrote the manuscript text. R.B. contributed to the design of the research and the design and writing of the manuscript text. All authors have read and agreed to the published version of the manuscript.

Funding: The material analysis of Revadim and Jaljulia was supported by the joint ISF-UGC Joint Scientific Research Grant (Israel-India program) entitled "The Global Culture: Lower Paleolithic Acheulean Adaptations at the Two Ends of Asia" 2712/16 and the Israel Science Foundation grant 321/19 entitled "Late Acheulean Jaljulia—A lithic perspective". The material analysis of Qesem was supported by the DFG research grant (UT 41/4-1) entitled "Cultural and Biological Transformations in the Late Middle Pleistocene (420–200 ka Ago) at Qesem Cave. Israel: In Search for a Post-Homo Erectus Lineage in the Levantine Corridor".

Data Availability Statement: The data supporting this study are included in the article and the Supplementary Material.

Acknowledgments: The excavation of Revadim was conducted by Ofer Marder and Ianir Milevsky on behalf of the Israel Antiquities Authority. The excavation of Jaljulia was conducted by Ma'ayan Shemer on behalf of the Israel Antiquities Authority. The excavation of Qesem Cave was conducted by Ran Barkai and Avi Gopher on behalf of the Tel Aviv University Institute of Archaeology. We wish to thank our editor, Sharon Kessler, for providing valuable comments to improve this manuscript. We would also like to thank Flavia Venditti (UT), Sasha Flit (TAU), and Pavel Shrago for the photographs used in this article.

Conflicts of Interest: The authors declare no conflicts of interest.

References

1. Barkai, R.; Lemorini, C.; Vaquero, M. The Origins of Recycling: A Paleolithic Perspective. *Quat. Int.* **2015**, *361*, 1–3. [CrossRef]
2. Agam, A.; Barkai, R. Small Flake Acheulian: Further Insights into Lithic Recycling at Late Acheulian Revadim, Israel. *Tel Aviv* **2018**, *45*, 170–192. [CrossRef]
3. Brumm, A.; Pope, M.; Leroyer, M.; Emery, K. Hominin Evolution and Stone Tool Scavenging and Reuse in the Lower Paleolithic. In *Squeezing Minds from Stones: Cognitive Archaeology and the Evolution of the Human Mind*; Overmann, K.A., Coolidge, F.L., Eds.; Oxford University Press: Oxford, UK, 2019; pp. 149–178.
4. Delpiano, D.; Heasley, K.; Peresani, M. Assessing Neanderthal Land Use and Lithic Raw Material Management in Discoid Technology. *J. Anthropol. Sci.* **2018**, *96*, 89–110. [CrossRef]
5. Mayor, A.; Sossa-Ríos, S.; Vaquero, M.; Hernández, C.M. Recycling and Implications for Neanderthal Productive Strategies in the Stratigraphic Unit Viii of the El Salt Rockshelter (Alcoi, Eastern Iberia). *Lithic Technol.* **2024**, 1–18. [CrossRef]
6. Ota, S.B.; Srivastava, N.; Pandey, S. An Early Case of Lithic Recycling in India: Evidence from the Acheulian Site at Damdongri, Madhya Pradesh. *Curr. Sci.* **2020**, *118*, 123–132. [CrossRef]
7. Rosenberg-Yefet, T.; Shemer, M.; Barkai, R. Acheulian Shortcuts: Cumulative Culture and the Use of Handaxes as Cores for the Production of Predetermined Blanks. *J. Archaeol. Sci. Rep.* **2021**, *36*, 102822. [CrossRef]
8. Venditti, F. *Understanding Lithic Recycling at the Late Lower Palaeolithic Qesem Cave, Israel: A Functional and Chemical Investigation of Small Flakes*; Archaeopress: Oxford, UK, 2019.
9. Venditti, F.; Agam, A.; Barkai, R. Techno-Functional Analysis of Small Recycled Flakes from Late Acheulian Revadim (Israel) Demonstrates a Link between Morphology and Function. *J. Archaeol. Sci. Rep.* **2019**, *28*, 102039. [CrossRef]

10. Venditti, F.; Cristiani, E.; Nunziante-Cesaro, S.; Agam, A.; Lemorini, C.; Barkai, R. Animal Residues Found on Tiny Lower Paleolithic Tools Reveal Their Use in Butchery. *Sci. Rep.* **2019**, *9*, 13031. [CrossRef]
11. Venditti, F.; Nunziante-cesaro, S.; Parush, Y.; Gopher, A.; Barkai, R. Recycling for a Purpose in the Late Lower Paleolithic Levant: Use-Wear and Residue Analyses of Small Sharp Flint Items Indicate a Planned and Integrated Subsistence Behavior at Qesem Cave (Israel). *J. Hum. Evol.* **2019**, *131*, 109–128. [CrossRef] [PubMed]
12. Venditti, F. Flint Recycling in the Lower Paleolithic Levant: A Microscopic Investigation of Small Recycled Flakes at Qesem Cave (Israel). *Mitteilungen Der Ges. Für Urgesch.* **2020**, *29*, 15–52. [CrossRef]
13. Amick, D.S. Investigating the Behavioral Causes and Archaeological Effects of Lithic Artifact Recycling. In *Tools Versus Cores: Alternative Approaches to Stone Tool Analysis*; McPherron, S.P., Ed.; Cambridge Scholars Publishing: Newcastle, UK, 2007; pp. 223–252.
14. Amick, D.S. The Recycling of Material Culture during the Paleolithic and Today. *Quat. Int.* **2015**, *361*, 4–20. [CrossRef]
15. Amick, D.S. *Chipped-Stone Crescents from the Terminal Pleistocene–Early Holocene of Far Western North America and the Transverse Projectile Point Hypothesis*; Springer: New York, NY, USA, 2024. [CrossRef]
16. Galup, S.M. *Postclassic Maya Lithic Tool Maintenance, Recycling, and Consumption Patterns at Laguna de On Island*; Institute for Mesoamerican Studies, University of Albany: Albany, NY, USA, 2007.
17. Hocsman, S.; Alderete, M. Similar No Es Necesariamente Algo Idéntico: Reclamación de Un Diseño de Punta de Proyectil En El Tránsito a La Producción de Alimentos En Antofagasta de La Sierra (Privincia de Catamarca, Argentina). *Arqueología* **2022**, *28*, 9935. [CrossRef]
18. Jennings, B. Repair, Recycle or Re-Use? Creating Mnemonic Devices Through the Modification of Object Biographies During the Late Bronze Age in Switzerland. *Camb. Archaeol. J.* **2014**, *24*, 163–176. [CrossRef]
19. Parush, Y.; Yerkes, R.W.; Efrati, B.; Barkai, R.; Gopher, A. Recycling in the Neolithic, Evidence of Flint Recycling at Ein-Zippori, Israel. *J. Lithic Stud.* **2018**, *5*, 1–32. [CrossRef]
20. Taxel, I. Same Medium, Different Messages: Stamped Storage Jar Handles in Judea from the Iron Age to the Early Islamic Period as Evidencing Regional Memory and Acquaintance with the Past. *Time Mind* **2020**, *13*, 35–57. [CrossRef]
21. Weitzel, C.; Mazzia, N.; Hocsman, S.; Pazzi, F. Recycled Fishtail Points in the Argentinian Pampa: Everyday Tools on Distinctive Paleoamerican Objects? *PaleoAmerica* **2022**, *8*, 130–144. [CrossRef]
22. Whyte, T.R. Gifts of the Ancestors: Secondary Lithic Recycling in Appalachian Summit Prehistory. *Am. Antiq.* **2014**, *79*, 679–696. [CrossRef]
23. Efrati, B. Material Sense: Perceptual Experience in Stone and Mineral Selection for Tool-Making. *Archaeologies* **2024**, *20*, 24–63. [CrossRef]
24. Parush, Y.; Assaf, E.; Slon, V.; Gopher, A.; Barkai, R. Looking for Sharp Edges: Modes of Flint Recycling at Middle Pleistocene Qesem Cave, Israel. *Quat. Int.* **2015**, *361*, 61–87. [CrossRef]
25. Camilli, E.L.; Ebert, J.I. Artifcat Reuse and Recycling in Continuous Surface Distributions and Implications for Interpreting Land Use Patterns. In *Space, Time and Archaeological Landscapes*; Rossignol, J., Wandsnider, L.A., Eds.; Plenum Press: New York, NY, USA, 1992; pp. 113–136.
26. Schiffer, M.B. *Behavioral Archaeology: Principles and Practice*; Equinox Publishing Ltd.: London, UK; Oakville, ON, Canada, 1976.
27. Vaquero, M. New Perspectives on Recycling of Lithic Resources Using Refitting and Spatial Data. *Quartär* **2011**, *58*, 113–130. [CrossRef]
28. Vaquero, M.; Alonso, S.; García-Catalán, S.; García-Hernández, A.; Gómez de Soler, B.; Rettig, D.; Soto, M. Temporal Nature and Recycling of Upper Paleolithic Artifacts: The Burned Tools from the Molí Del Salt Site (Vimbodí i Poblet, Northeastern Spain). *J. Archaeol. Sci.* **2012**, *39*, 2785–2796. [CrossRef]
29. Goodwin, A.J.H. Chemical Alteration (Patination) of Stone. *S. Afr. Archaeol. Bull.* **1960**, *15*, 67–76. [CrossRef]
30. Romagnoli, F. A Second Life: Recycling Production Waste during the Middle Palaeolithic in Layer L at Grotta Del Cavallo (Lecce, Southeast Italy). *Quat. Int.* **2015**, *361*, 200–211. [CrossRef]
31. Judd, J.W. On the Unmaking of Flints. *Proc. Geol. Assoc.* **1887**, *10*, 217–226. [CrossRef]
32. Nadel, D.; Gordon, D. Patination of Flint Artefacts: Evidence from Bikta, A Submerged Prehistoric Occurrence in the Sea of Galilee, Israel. *Mitekufat Haeven J. Isr. Prehist. Soc.* **1993**, *25*, 145–162.
33. Purdy, B.A.; Clark, D.E. Weathering of Inorganic Materials: Dating and Other Applications. *Adv. Archaeol. Method Theory* **1987**, *11*, 211–254. [CrossRef]
34. Thiry, M.; Fernandes, P.; Milnes, A.; Raynal, J.P. Driving Forces for the Weathering and Alteration of Silica in the Regolith: Implications for Studies of Prehistoric Flint Tools. *Earth-Sci. Rev.* **2014**, *136*, 141–154. [CrossRef]
35. Belfer-Cohen, A.; Bar-Yosef, O. Paleolithic Recycling: The Example of Aurignacian Artifacts from Kebara and Hayonim Caves. *Quat. Int.* **2015**, *361*, 256–259. [CrossRef]
36. Efrati, B.; Parush, Y.; Ackerfeld, D.; Gopher, A.; Barkai, R. Seeing Colors: Collecting and Modifying Patinated Flaked Items at Middle Pleistocene Qesem Cave, Israel. *Mitekufat Haeven J. Isr. Prehist. Soc.* **2019**, *49*, 32–62.
37. Efrati, B.; Barkai, R.; Nunziante Cesaro, S.; Venditti, F. Function, Life Histories, and Biographies of Lower Paleolithic Patinated Flint Tools from Late Acheulian Revadim, Israel. *Sci. Rep.* **2022**, *12*, 2885. [CrossRef] [PubMed]
38. Jagher, R. Nadaouiyeh Aïn Askar, an Example of Upper Acheulean Variability in The Levant. *Quat. Int.* **2016**, *411*, 44–58. [CrossRef]

39. Peresani, M.; Boldrin, M.; Pasetti, P. Assessing the Exploitation of Double Patinated Artifacts from the Late Mousterian: Implications for Lithic Economy and Human Mobility in Northern Italy. *Quat. Int.* **2015**, *361*, 238–250. [CrossRef]
40. Shimelmitz, R. The Recycling of Flint Throughout the Lower and Middle Paleolithic Sequence of Tabun Cave, Israel. *Quat. Int.* **2015**, *361*, 34–45. [CrossRef]
41. Brumm, A. Lightning Teeth and Ponari Sweat: Folk Theories and Magical Uses of Prehistoric Stone Axes (and Adzes) in Island Southeast Asia and the Origin of Thunderstone Beliefs. *Aust. Archaeol.* **2018**, *84*, 37–55. [CrossRef]
42. Dittert, A.E. *Culture Change in the Cebolleta Mesa Region, Central Western New Mexico*; University of Arizona: Tucson, AZ, USA, 1959.
43. Gould, R.A. Chipping Stones in the Outback. *Nat. Hist. Mag.* **1968**, *77*, 42–49.
44. Gould, R.A. *Living Archaeology*; Cambridge University Press: New York, NY, USA, 1980.
45. Janowski, M.; Barton, H. Reading Human Activity in the Landscape: Stone and Thunderstones in the Kelabit Highlands. *Indones. Malay World* **2012**, *40*, 354–371. [CrossRef]
46. Sedig, J.W. An Analysis of Non-Utilitarian Stone Point Function in the US Southwest. *J. Anthropol. Archaeol.* **2014**, *34*, 120–132. [CrossRef]
47. Wandsnider, L. Long-Term Land Use, Formation Processes, and the Structure of the Archaeological Landscape: A Case Study from Southwestern Wyoming. Ph.D. Dissertation, University of New Mexico, Albuquerque, NM, USA, 1989.
48. Russell, F. The Pima Indians. In *Twenty-Sixth Annual Report of the Bureau of American Ethnology to the Secratary of the Smithsonian Institution 1915-1916*; Holmes, W.H., Ed.; United States Government Printing Office: Washington, DC, USA, 1908; pp. 3–390.
49. Sedig, J.W. *Getting to the Point: An Examination of Projectile Point Use in the Northern American Southwest, A.D. 900-1300*; University of Colorado: Boulder, CO, USA, 2010.
50. Efrati, B. Memory Scrapers: Readymade Concepts and Techniques as Reflected in Collecting and Recycling Patinated Lower Palaeolithic Items at Qesem Cave, Israel. *Camb. Archaeol. J.* **2021**, *31*, 337–347. [CrossRef]
51. Marder, O.; Malinsky-Buller, A.; Shahack-Gross, R.; Ackermann, O.; Ayalon, A.; Bar-Matthews, M.; Goldsmith, Y.; Inbar, M.; Rabinovich, R.; Hovers, E. Archaeological Horizons and Fluvial Processes at the Lower Paleolithic Open-Air Site of Revadim (Israel). *J. Hum. Evol.* **2011**, *60*, 508–522. [CrossRef] [PubMed]
52. Gvirtzman, G.; Wieder, M.; Marder, O.; Khalaily, H.; Rabinovich, R.; Ron, H. Geological and Pedological Aspects of an Early-Paleolithic Site: Revadim, Central Coastal Plain, Israel. *Geoarchaeology-An. Int. J.* **1999**, *14*, 101–126. [CrossRef]
53. Marder, O.; Gvirtzman, G.; Ron, H.; Khalaily, H.; Wieder, M.; Bankirer, R.; Rabinovich, R.; Porat, N.; Saragusti, I. The Lower Paleolithic Site of Revadim Quarry, Preliminary Finds. *J. Isr. Prehist. Soc.* **1999**, *28*, 21–53. [CrossRef]
54. Rabinovich, R.; Ackermann, O.; Aladjem, E.; Barkai, R.; Biton, R.; Milevski, I.; Solodenko, N.; Marder, O. Elephants at the Middle Pleistocene Acheulian Open-Air Site of Revadim Quarry, Israel. *Quat. Int.* **2012**, *276–277*, 183–197. [CrossRef]
55. Rosenberg-Yefet, T.; Barkai, R. Early in the Late Acheulian: Techno-Typological Analysis of the Earliest Lithic Assemblage from Lower Paleolithic Revadim, Israel. *Mitekufat Haeven J. Isr. Prehist. Soc.* **2019**, *49*, 6–31.
56. Solodenko, N. On Tools and Elephants: An Analysis of the Lithic Assemblage from Area B of the Late Acheulian Site Revadim Quarry. Ph.D. Dissertation, Tel Aviv University, Tel Aviv, Israel, 2010.
57. Solodenko, N.; Zupancich, A.; Cesaro, S.N.; Marder, O.; Lemorini, C.; Barkai, R. Fat Residue and Use-Wear Found on Acheulian Biface and Scraper Associated with Butchered Elephant Remains at the Site of Revadim, Israel. *PLoS ONE* **2015**, *10*, e0118572. [CrossRef]
58. Zupancich, A.; Solodenko, N.; Rosenberg-Yefet, T.; Barkai, R. On the Function of Late Acheulean Stone Tools: New Data from Three Specific Archaeological Contexts at the Lower Palaeolithic Site of Revadim, Israel. *Lithic Technol.* **2018**, *43*, 255–268. [CrossRef]
59. Agam, A.; Marder, O.; Barkai, R. Small Flake Production and Lithic Recycling at Late Acheulian Revadim, Israel. *Quat. Int.* **2015**, *361*, 46–60. [CrossRef]
60. Malinsky-Buller, A.; Grosman, L.; Marder, O. A Case of Techno-Typological Lithic Variability & Continuity in the Late Lower Palaeolithic. *Before Farming Archaeol. Anthropol. Hunt.-Gatherers* **2011**, *2011*, 1–32. [CrossRef]
61. Cohen, A. The Handaxes from Revadim Late Acheulian Quarry: Similarities and Differences in Time and Space. Ph.D. Dissertation, Tel Aviv University, Tel Aviv, Israel, 2018.
62. Malinsky-Buller, A. The Muddle in the Middle Pleistocene: The Lower-Middle Paleolithic Transition from the Levantine Perspective. *J. World Prehistory* **2016**, *29*, 1–78. [CrossRef]
63. Marder, O.; Milevski, I.; Matskevich, Z. The Handaxes of Revadim Quarry: Typo-Technological Considerations and Aspects of Intra-Site Variability. In *Axe Age: Acheulian Toolmaking from Quarry to Discard*; Goren-Inbar, N., Sharon, G., Eds.; Equinox: London, UK, 2006; pp. 223–242.
64. Venditti, F.; Barkai, R.; Cesaro, S.N.; Agam, A. The Role of Side-Scrapers and Cortical Flakes in Late Acheulian Toolkits: Results of a Techno-Functional Analysis from Revadim (Israel). *Lithic Technol.* **2022**, *47*, 243–265. [CrossRef]
65. Marinelli, F.; Lemorini, C.; Barkai, R. Lower Palaeolithic Small Flakes and Megafauna: The Contribution of Experimental Approach and Use-Wear Analysis to Reveal the Link. In *Human-Elephant Interactions: From Past to Present*; Konidaris, G.E., Barkai, R., Tourloukis, V., Harvati, K., Eds.; Tübingen University Press: Tübingen, Germany, 2021; pp. 169–192.
66. Shemer, M.; Greenbaum, N.; Taha, N.; Brailovsky-Rokser, L.; Ebert, Y.; Shaar, R.; Falguères, C.; Voinchet, P.; Porat, N.; Faershtein, G.; et al. Late Acheulian Jaljulia-Eraly Human Occupations in the Paleo-Landscape of the Central Coastal Plain of Israel. *PLoS ONE* **2022**, *17*, e0267672. [CrossRef]

67. Muller, A.; Barkai, R.; Shemer, M.; Grosman, L. 3D Morphology of Handaxes from Late Acheulean Jaljulia: A Flexible Reduction Strategy in the Lower Paleolithic Levant. *Archaeol. Anthropol. Sci.* **2022**, *14*, 206. [CrossRef]
68. Zupancich, A.; Shemer, M.; Barkai, R. Biface Use in the Lower Paleolithic Levant: First Insights from Late Acheulean Revadim and Jaljulia (Israel). *J. Archaeol. Sci. Rep.* **2021**, *36*, 102877. [CrossRef]
69. Agam, A.; Rosenberg-Yefet, T.; Wilson, L.; Shemer, M.; Barkai, R. Flint Type Analysis at Late Acheulian Jaljulia (Israel), and Implications for the Origins of Prepared Core Technologies. *Front. Earth Sci.* **2022**, *10*, 858032. [CrossRef]
70. Rosenberg-Yefet, T.; Shemer, M.; Barkai, R. Lower Paleolithic Winds of Change: Prepared Core Technologies and the Onset of the Levallois Method in the Levantine Late Acheulian. *Front. Earth Sci.* **2022**, *10*, 847358. [CrossRef]
71. Barkai, R.; Gopher, A. Cultural and Biological Transformations in the Middle Pleistocene Levant: A View from Qesem Cave, Israel. In *Dynamics of Learning in Neanderthals and Modern Humans*; Akazawa, T., Nishiaki, Y., Aoki, K., Eds.; Springer: Tokyo, Japan, 2013; pp. 115–137.
72. Barkai, R.; Rosell, J.; Blasco, R.; Gopher, A. Fire for a Reason: Barbecue at Middle Pleistocene Qesem Cave, Israel. *Curr. Anthropol.* **2017**, *58*, S314–S328. [CrossRef]
73. Barkai, R.; Blasco, R.; Rosell, J.; Gopher, A. A Land of Flint and Fallow Deer: Human Persistence at Middle Pleistocene Qesem Cave. In *Crossing the Human Threshold: Dynamic Transformation and Persistent Places During the Middle Pleistocene*; Pope, M., McNab, J., Gamble, C., Eds.; Routledge: London, UK, 2018; pp. 60–82.
74. Gopher, A.; Barkai, R.; Shimelmitz, R.; Khalaily, M.; Lemorini, C.; Hershkovitz, I.; Stiner, M. Qesem Cave: An Amudian Site in Central Israel. *Mitekufat Haeven J. Isr. Prehist. Soc.* **2005**, *35*, 69–92.
75. Shahack-Gross, R.; Berna, F.; Karkanas, P.; Lemorini, C.; Gopher, A.; Barkai, R. Evidence for the Repeated Use of a Central Hearth at Middle Pleistocene (300ky Ago) Qesem Cave, Israel. *J. Archaeol. Sci.* **2014**, *44*, 12–21. [CrossRef]
76. Barkai, R.; Gopher, A.; Lauritzen, S.E.; Frumkin, A. Uranium Series Dates from Qesem Cave, Israel, and the End of the Lower Palaeolithic. *Nature* **2003**, *423*, 977–979. [CrossRef] [PubMed]
77. Falguères, C.; Richard, M.; Tombret, O.; Shao, Q.; Bahain, J.J.; Gopher, A.; Barkai, R. New ESR/U-Series Dates in Yabrudian and Amudian Layers at Qesem Cave, Israel. *Quat. Int.* **2016**, *398*, 6–12. [CrossRef]
78. Falguères, C.; Barkai, R.; Tombret, O.; Gopher, A. New ESR/U-Series Dates of the Lowest Acheulo-Yabrudian Levels of Qesem Cave. *Quat. Geochronol.* **2022**, *69*, 101266. [CrossRef]
79. Gopher, A.; Ayalon, A.; Bar-Matthews, M.; Barkai, R.; Frumkin, A.; Karkanas, P.; Shahack-Gross, R. The Chronology of the Late Lower Paleolithic in the Levant Based on U-Th Ages of Speleothems from Qesem Cave, Israel. *Quat. Geochronol.* **2010**, *5*, 644–656. [CrossRef]
80. Mercier, N.; Valladas, H.; Falguères, C.; Shao, Q.; Gopher, A.; Barkai, R.; Bahain, J.J.; Vialettes, L.; Joron, J.L.; Reyss, J.L. New Datings of Amudian Layers at Qesem Cave (Israel): Results of TL Applied to Burnt Flints and ESR/U-Series to Teeth. *J. Archaeol. Sci.* **2013**, *40*, 3011–3020. [CrossRef]
81. Barkai, R.; Gopher, A. Innovative Human Behavior between Acheulian and Mousterian: A View from Qesem Cave, Israel. In *The Lower and Middle Palaeolithic in the Middle East and Neighbouring Regions*; Le Tensorer, J.-M., Jagher, R., Otte, M., Eds.; Etudes et Recherches Archéologiques de l'Université de Liège (ERAUL): Liége, Belgium, 2011; pp. 49–57.
82. Barkai, R.; Lemorini, C.; Shimelmitz, R.; Lev, Z.; Stiner, M.C.; Gopher, A. A Blade for All Seasons? Making and Using Amudian Blades at Qesem Cave, Israel. *Hum. Evol.* **2009**, *24*, 57–75.
83. Gopher, A.; Parush, Y.; Assaf, E.; Barkai, R. Spatial Aspects as Seen from a Density Analysis of Lithics at Middle Pleistocene Qesem Cave: Preliminary Results and Observations. *Quat. Int.* **2016**, *398*, 103–117. [CrossRef]
84. Lemorini, C.; Stiner, M.C.; Gopher, A.; Shimelmitz, R.; Barkai, R. Use-Wear Analysis of an Amudian Laminar Assemblage from the Acheuleo-Yabrudian of Qesem Cave, Israel. *J. Archaeol. Sci.* **2006**, *33*, 921–934. [CrossRef]
85. Parush, Y.; Gopher, A.; Barkai, R. Amudian versus Yabrudian under the Rock Shelf: A Study of Two Lithic Assemblages from Qesem Cave, Israel. *Quat. Int.* **2016**, *398*, 13–36. [CrossRef]
86. Shimelmitz, R.; Barkai, R.; Gopher, A. Systematic Blade Production at Late Lower Paleolithic (400–200 Kyr) Qesem Cave, Israel. *J. Hum. Evol.* **2011**, *61*, 458–479. [CrossRef] [PubMed]
87. Shimelmitz, R.; Barkai, R.; Gopher, A. Regional Variability in Late Lower Paleolithic Anudian Blade Technology: Analyzing New Data from Qesem, Tabun and Yabrud I. *Quat. Int.* **2016**, *398*, 37–60. [CrossRef]
88. Agam, A.; Zupancich, A. Interpreting the Quina and Demi-Quina Scrapers from Acheulo-Yabrudian Qesem Cave, Israel: Results of Raw Materials and Functional Analyses. *J. Hum. Evol.* **2020**, *144*, 102798. [CrossRef]
89. Lemorini, C.; Bourguignon, L.; Zupancich, A.; Gopher, A.; Barkai, R. A Scraper's Life History: Morpho-Techno-Functional and Use-Wear Analysis of Quina and Demi-Quina Scrapers from Qesem Cave, Israel. *Quat. Int.* **2016**, *398*, 86–93. [CrossRef]
90. Lev, Z. Techno-Typological Analysis of a Yabrudian Assemblage from Qesem Cave, Israel. Master's Thesis, Tel Aviv University, Tel Aviv, Israel, 2010.
91. Rivals, F.; Blasco, R.; Rosell, J.; Efrati, B.; Gopher, A.; Barkai, R. Seasonality, Duration of the Hominin Occupations and Hunting Grounds at Middle Pleistocene Qesem Cave (Israel). *Archaeol. Anthropol. Sci.* **2021**, *13*, 205. [CrossRef]
92. Zupancich, A. Understanding the Use of Quina Scrapers at Middle Pleistocene Qesem Cave (Israel), and the Implication for the Study of the Quina Phenomenon in the Levant and Beyond. Ph.D. Dissertation, Tel Aviv University, Tel Aviv, Israel, 2019.
93. Zupancich, A.; Lemorini, C.; Gopher, A.; Barkai, R. On Quina and Demi-Quina Scraper Handling: Preliminary Results from the Late Lower Paleolithic Site of Qesem Cave, Israel. *Quat. Int.* **2016**, *398*, 94–102. [CrossRef]

94. Agam, A.; Wilson, L.; Gopher, A.; Barkai, R. Flint Type Analysis of Bifaces from Acheulo-Yabrudian Qesem Cave (Israel) Suggests an Old Acheulian Origin. *J. Paleolit. Archaeol.* **2019**, *3*, 719–754. [CrossRef]
95. Barkai, R.; Gopher, A.; Solodenko, N.; Lemorini, C. An Amudian Oddity: A Giant Biface from Late Lower Palaeolithic Qesem Cave. *Tel Aviv* **2013**, *40*, 176–186. [CrossRef]
96. Assaf, E.; Baene Preysler, J. Raw Material Selectivity in Lower Paleolithic Shaped Stone Ball Production: Experimental Research. *Archaeol. Anthropol. Sci.* **2022**, *14*, 24. [CrossRef]
97. Assaf, E.; Caricola, I.; Gopher, A.; Rosell, J.; Blasco, R.; Bar, O.; Zilberman, E.; Lemorini, C.; Baena, J.; Barkai, R.; et al. Shaped Stone Balls Were Used for Bone Marrow Extraction at Lower Paleolithic Qesem Cave, Israel. *PLoS ONE* **2020**, *15*, e0230972. [CrossRef]
98. Assaf, E.; Baena Preysler, J.; Bruner, E. Lower Paleolithic Shaped Stone Balls—What Is Next? Some Cultural-Cognitive Questions. *Quaternary* **2023**, *6*, 51. [CrossRef]
99. Barkai, R.; Gopher, A. On Anachronism: The Curious Presence of Spheroids and Polyhedrons at Acheulo-Yabrudian Qesem Cave, Israel. *Quat. Int.* **2016**, *398*, 118–128. [CrossRef]
100. Assaf, E.; Parush, Y.; Gopher, A.; Barkai, R. Intra-Site Variability in Lithic Recycling at Qesem Cave, Israel. *Quat. Int.* **2015**, *361*, 88–102. [CrossRef]
101. Barkai, R.; Lemorini, C.; Gopher, A. Palaeolithic Cutlery 400,000-200,000 Years Ago: Tiny Meat-Cutting Tools from Qesem Cave, Israel. *Antiquity* **2010**, *84*, 118–128.
102. Lemorini, C.; Venditti, F.; Assaf, E.; Parush, Y.; Barkai, R.; Gopher, A. The Function of Recycled Lithic Items at Late Lower Paleolithic Qesem Cave, Israel: An Overview of the Use-Wear Data. *Quat. Int.* **2015**, *361*, 103–112. [CrossRef]
103. Maul, L.C.; Smith, K.T.; Barkai, R.; Barash, A.; Karkanas, P.; Shahack-Gross, R.; Gopher, A. Microfaunal Remains at Middle Pleistocene Qesem Cave, Israel: Preliminary Results on Small Vertebrates, Environment and Biostratigraphy. *J. Hum. Evol.* **2011**, *60*, 464–480. [CrossRef] [PubMed]
104. Sánchez-Marco, A.; Blasco, R.; Rosell, J.; Gopher, A.; Barkai, R. Birds as Indicators of High Biodiversity Zones Around the Middle Pleistocene Qesem Cave, Israel. *Quat. Int.* **2016**, *421*, 23–31. [CrossRef]
105. Agam, A. Late Lower Paleolithic Lithic Procurement and Exploitation Strategies: A View from Acheulo-Yabrudian Qesem Cave (Israel). *J. Archaeol. Sci. Rep.* **2020**, *33*, 102447. [CrossRef]
106. Agam, A. *Flint Procurement and Exploitation Strategies in the Late Lower Paleolithic Levant: A View from Acheulo-Yabrudian Qesem Cave (Israel)*; Archaeopress Publishing Ltd: Oxford, UK, 2021.
107. Wilson, L.; Agam, A.; Barkai, R.; Gopher, A. Raw Material Choices in Amudian versus Yabrudian Lithic Assemblages at Qesem Cave: A Preliminary Evaluation. *Quat. Int.* **2016**, *398*, 61–69. [CrossRef]
108. Verri, G.; Barkai, R.; Bordeanu, C.; Gopher, A.; Hass, M.; Kaufman, A.; Kubik, P.; Montanari, E.; Paul, M.; Ronen, A.; et al. Flint Mining in Prehistory Recorded by In Situ-Produced Cosmogenic 10Be. *Proc. Natl. Acad. Sci. USA* **2004**, *101*, 7880–7884. [CrossRef] [PubMed]
109. Verri, G.; Barkai, R.; Gopher, A.; Hass, M.; Kubik, P.W.; Paul, M.; Ronen, A.; Weiner, S.; Boaretto, E. Flint Procurement Strategies in the Late Lower Palaeolithic Recorded by In Situ Produced Cosmogenic 10Be in Tabun and Qesem Caves (Israel). *J. Archaeol. Sci.* **2005**, *32*, 207–213. [CrossRef]
110. Blasco, R.; Rosell, J.; Cuartero, F.; Fernández Peris, J.; Gopher, A.; Barkai, R. Using Bones to Shape Stones: MIS 9 Bone Retouchers at Both Edges of the Mediterranean Sea. *PLoS ONE* **2013**, *8*, e76780. [CrossRef] [PubMed]
111. Rosell, J.; Blasco, R.; Peris, J.F.; Carbonell, E.; Barkai, R.; Gopher, A. Recycling Bones in the Middle Pleistocene: Some Reflections from Gran Dolina TD10-1 (Spain), Bolomor Cave (Spain) and Qesem Cave (Israel). *Quat. Int.* **2015**, *361*, 297–312. [CrossRef]
112. Zupancich, A.; Nunziante-Cesaro, S.; Blasco, R.; Rosell, J.; Cristiani, E.; Venditti, F.; Lemorini, C.; Barkai, R.; Gopher, A. Early Evidence of Stone Tool Use in Bone Working Activities at Qesem Cave, Israel. *Sci. Rep.* **2016**, *6*, 37686. [CrossRef] [PubMed]
113. Venditti, F.; Agam, A.; Tirillò, J.; Nunziante-Cesaro, S.; Barkai, R. An Integrated Study Discloses Chopping Tools Use from Late Acheulean Revadim (Israel). *PLoS ONE* **2021**, *16*, e0245595. [CrossRef] [PubMed]
114. Hiscock, P. Reduction, Recycling, and Raw Material Procurement in Western Arnhem Land, Australia. In *Lithic Materials and Paleolithic Societies*; Adams, B., Blades, B.S., Eds.; Blackwell Publishing Ltd.: Chichester, UK, 2009; pp. 78–93.
115. Hiscock, P. Recycling in the Haua Fteah Sequence of North Africa. *Quat. Int.* **2015**, *361*, 251–255. [CrossRef]
116. Baena Preysler, J.; Ortiz Nieto-Márquez, I.; Torres Navas, C.; Bárez Cueto, S. Recycling in Abundance: Re-Use and Recycling Processes in the Lower and Middle Paleolithic Contexts of the Central Iberian Peninsula. *Quat. Int.* **2015**, *361*, 142–154. [CrossRef]
117. Dibble, H.; Mcpherron, S. The Missing Mousterian. *Curr. Anthropol.* **2006**, *47*, 777–803. [CrossRef]
118. Thiébaut, C.; Claud, E.; Mourre, V.; Chacón, M.G.; Asselin, G.; Brenet, M.; Paravel, B. The Recycling and Reuse of Cores and Bifaces during the Middle Paleolithic in Western Europe: Functional and Cultural Inrerpretations. *Palethnology* **2010**, 3–41. [CrossRef]
119. Berkes, F. *Biodiversity, Religion Traditions. Encyclopedia of Biodiversity*, 3rd ed.; Academy Press: Cambridge, MA, USA, 2024; Volume 3, pp. 18–29. [CrossRef]
120. Assaf, E.; Barkai, R.; Gopher, A. Another Side of the Amudian Industry at Qesem Cave, Israel: The Southern Area Lithic Assemblage. *Lithic Technol.* **2017**, *42*, 161–178. [CrossRef]
121. Barkai, R.; Gopher, A.; Shimelmitz, R. Middle Pleistocene Blade Production in the Levant: An Amudian Assemblage from Qesem Cave, Israel. *Eurasian Prehistory* **2005**, *3*, 39–74.

122. Debénath, A.; Dibble, H.L. *Handbook of Paleolithic Typology: Lower and Middle Paleolithic of Europe, Vol. 1: Lower and Middle Paleolithic of Europe*; The University Museum, University of Pennsylvania: Philadelphia, PA, USA, 1994.
123. Inizan, M.-L.; Reduron-Ballinger, M.; Roche, H.; Tixier, J. *Technology and Terminology of Knapped Stone*; Cercle de Recherches et d'Etudes Préhistoriques: Nanterre, Franch, 1999.
124. Harries, J. A Stone That Feels Right in the Hand: Tactile Memory, the Abduction of Agency and Presence of the Past. *J. Mater. Cult.* **2017**, *22*, 110–130. [CrossRef]
125. Wynn, T.; Overmann, K.A.; Malafouris, L. 4E Cognition in the Lower Palaeolithic. *Adapt. Behav.* **2021**, *29*, 99–106. [CrossRef]
126. Malafouris, L. Knapping Intentions and the Marks of the Mental. In *The Cognitive Life of Things: Recasting the Boundaries of the Mind*; Malafouris, L., Renfrew, C., Eds.; McDonald Institute Monographs: Cambridge, UK, 2010; pp. 13–22.
127. Malafouris, L. How Does Thinking Relate to Tool Making? *Adapt. Behav.* **2021**, *29*, 107–121. [CrossRef]
128. Hahn, H.P.; Weiss, H. Introduction: Biographies, Travels and Itineraries of Things. In *Mobility, Meaning & Transformations of Things: Shifting Contexts of Material Culture through Time and Space*; Hahn, H.P., Weiss, H., Eds.; David Brown Book Company: Oakville, ON, Canada, 2013; pp. 1–14.
129. Kopytoff, I. The Cultural Biography of Things. Commoditization as Process. In *The Social Life of Things: Commodities in Cultural Perspective*; Appadurai, A., Ed.; Cambridge University Press: Cambridge, UK, 1986; pp. 64–91.
130. Boivin, N. From Veneration to Exploitation: Human Engagement with the Mineral World. In *Soils, Stones and Symbols: Cultural Perceptions of the Mineral World*; Boivin, N., Owoc, M.A., Eds.; Routledge: Abingdon, UK, 2004; pp. 1–30.
131. Brumm, A. "The Falling Sky": Symbolic and Cosmological Associations of the Mt William Greenstone Axe Quarry, Central Victoria, Australia. *Camb. Archaeol. J.* **2010**, *20*, 179–196. [CrossRef]
132. Reimer, R. The Social Importance of Volcanic Peaks for the Indigenous Peoples of British Columbia. *J. Northwest Anthropol.* **2018**, *52*, 4–35.
133. Taçon, P.S.C. The Power of Stone: Symbolic Aspects of Stone Use and Tool Development in Western Arnhem Land, Australia. *Antiquity* **1991**, *65*, 192–207. [CrossRef]
134. Arthur, K.W. *The Lives of Stone Tools: Crafting the Status, Skill, and Identity of Flintknappers*; The University of Arizona Press: Tucson, AZ, USA, 2018.
135. Crowell, A.L. Sea Mammals in Art, Ceremony, and Belief: Knowledge Shared by Yupik and Iñupiaq Elders. In *Gifts from the Ancestors: Ancient Ivories of Bering Strait*; Fitzhugh, W.W., Hollowell, J., Crowell, A.L., Eds.; Princeton University Art Museum: Princeton, NJ, USA, 2009; pp. 206–225.
136. Currie, G.; Zhu, X. Aesthetic Sense and Social Cognition: A Story from the Early Stone Age. *Synthese* **2019**, *198*, 6553–6572. [CrossRef]
137. Ricoeur, P. *Memory, History, Forgetting*; The University of Chicago Press: Chicago, IL, USA, 2004.

quaternary

MDPI

Article

Stage or Sub-Stage: The Contribution of Small Mammals to the Characterization of Middle Pleistocene Local Climate Variation

Loïc Lebreton [1,2,*] and Juan Manuel López-García [1,2]

[1] Institut Català de Paleoecologia Humana i Evolució Social (IPHES-CERCA), Zona Educacional 4, Campus Sescelades URV (Edifici W3), 43007 Tarragona, Spain; jmlopez@iphes.cat

[2] Departament d'Història i Història de l'Art, Universitat Rovira i Virgili, Avinguda de Catalunya 35, 43002 Tarragona, Spain

* Correspondence: llebreton@iphes.cat

Abstract: In western Europe, the Middle Pleistocene is marked by Acheulean settlement and their diversification after the MIS 12. The Arago cave recovery of numerous human settlements correlate to MIS 14, 13 and MIS 12 making it an important site for the understanding of the Lower Palaeolithic in Southwestern Europe. It is also an important site for the understanding of palaeoenvironments and palaeobiodiversity as it has yielded rich faunal associations. The faunal associations allow us to observe three climatic stages within this study: two cold ones and a mild one. Small vertebrates, with their abundance and their diversity, are particularly useful for observing these periods, which historically have been correlated to glacial or interglacial stages. If the first cold phase, dated 438 ± 31 ka, is correlated to the Marine Isotopic Stage 12 (MIS 12), the correlation of the following phases to isotopic stages can be discussed. They may correspond to climatic variations of the MIS 12. Indeed, the latest studies about palaeoclimatic reconstitution which allow us to define the evolution of the palaeo-temperature show that these differences are relatively small. Therefore, instead of a correlation to MIS 12, 13 and 14, the medium complex of the Arago cave could belong solely to MIS 12. The correlation of these environmental changes to other global data, notably the isotopic curve, is challenging because there are various local factors influencing faunal association. We propose here both hypotheses and discuss the various factors which influence the distribution and the representation of the small vertebrate species present on the site.

Keywords: Pleistocene; palaeoenvironment; Acheulean; small mammals

Citation: Lebreton, L.; López-García, J.M. Stage or Sub-Stage: The Contribution of Small Mammals to the Characterization of Middle Pleistocene Local Climate Variation. *Quaternary* **2023**, *6*, 54. https://doi.org/10.3390/quat6040054

Academic Editor: Ran Barkai

Received: 11 July 2023
Revised: 25 August 2023
Accepted: 15 September 2023
Published: 10 October 2023

1. Introduction

Climate dynamics during the Quaternary are well documented by extra-site proxies such as deep-sea cores or ice cores, providing a clear view of climatic oscillations during this period [1,2]. These climate changes have had an important impact on human populations in Europe, modifying their environment, their ecosystems and also their resource distributions [3]. Whereas continental-scale climate change is well understood, it is the local environment that directly affects human groups by altering their ecosystem and resource availability. However, the correlation between the local environment and global changes can be delicate due to the particularities of the local context, this context is needed to analyse environmental responses during climatic oscillations.

Small mammals serve as reliable indicators due to their abundance and diversity in archaeological sites, their limited migration, and their ability to establish stable populations in small areas, acting as climatic refuges allowing a local scale approach to studying environmental variations [4,5].

Based on this background, the present study uses several methods to characterize different climatic variations and tries to incorporate these into a more global context. To this end, we use different methods to carry out tests at the Arago cave. This cave provides a good foundation for palaeoenvironmental reconstructions of the Middle Pleistocene in the western Mediterranean area. Alternations of cold and temperate periods are traceable, and the numerous layers are correlated to MIS 14, 13 and 12 [6]. This site has a good small vertebrates record that allows us to follow these climatic changes. The location of the site is also important because it lies within the climatic influence of the Pyrenees and the Mediterranean Sea, which may have a significant impact on local environments.

The Arago cave is a key site for understanding the Acheulean occupations in southern Europe because it has provided recurrent occupation layers with Acheulean industry from MIS 14 to MIS 12 [7–9]. This period is important for an appreciation of the diversity of the early Acheulean, as the end of the MIS 12 glaciation marks a threshold after which the archaeological records show an increasing number of occupations, evidence of new subsistence behaviours, and numerous technical innovations [10]. Furthermore, early Acheulean sites witness occasional occupations [11–15]. The recurrence of occupations in the Arago cave allows us to observe technical developments during phases of climatic variation in a similar geographical context. Understanding the environment that enabled human settlement to continue during phases of climatic transition is key to studying the diversity and evolution of human settlements in southwestern Europe.

2. Site Presentation

The Arago cave is a vast karstic cavity located in the valley of Tautavel, in southern France (Figure 1a). The site was occupied several times by groups of hominins during different climatic phases. These phases have been determined through variations in the abundance of faunal species and pollen. The main sedimentary sequence (the Middle Complex) is subdivided into three ensembles, comprising stratigraphic units (SU) ranging from D to R, the last excavated layer (Figure 1b). Ensemble I (SU K to Q) is correlated to MIS 14, with SU P dated to 532 ± 106 ka [16]. In this ensemble, *Rangifer tarandus* is predominant, followed by *Ovis ammon antiqua* and *Cervus elaphus* [17]. Ensemble II (SU H, I and J) is correlated to a temperate phase (MIS 13) mainly on the basis of changes in the representation of the faunal community. Indeed, *Cervus elaphus* is the predominant species, along with *Dama* cf. *clactoniana*. It is noteworthy that *Rangifer tarandus* and *Praeovibos priscus* are still present [17–19]. Ensemble III (SU D to G) is correlated to MIS 12, with SU F dated to 392 ± 43 ka and SU G to 438 ± 31 ka [16]. The predominant species for this ensemble are *Ovis ammon antiqua* and *Equus ferus mosbachensis* [17]. The filling of the Middle Complex of the Arago cave yielded many Acheulean occupations that are among the oldest in Europe, as well as many human remains (*Homo erectus tautavelensis*) [20].

Small vertebrates are abundant in the site. These have been recovered by systematic sieving through 1.2 mm and 0.6 mm mesh. The remains have shown great diversity and highlighted different climatic phases based on variations in the abundance of different species [21–31].

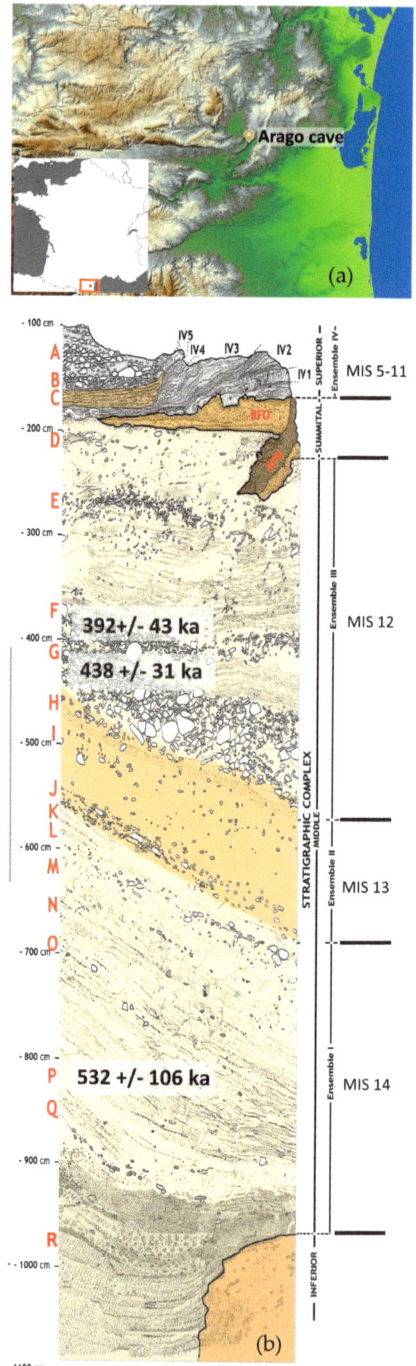

Figure 1. (**a**) Location of the Arago cave. (**b**) Synthetic stratigraphic column of the Arago cave [21], dating and correlation with the MIS come from [16].

3. Materials and Methods

The material used for this study has been previously published in several papers [21,24,25,27–29].

Multivariate analysis makes it possible to use the abundance of each taxon (MNI) by archaeological level to highlight possible associations between rodent species and different archaeological levels. This method enables us to observe associations between levels without attributing any ecological value to the species, dispensing with the principle of actualism. Multivariate analyses are often used to characterize associations of micromammals according to their ecological and climatic affinities [21,24,25,27,32–34]. Correspondence analyses (CA) were carried out using R 3.1.2 software (with the Rcmdr package), with rodent data from Table 1. In order to have a better vision of the grouping between levels, we apply a cluster analysis (UPGMA, Euclid distance) based on the data from Axes 1–3 of the CA with consideration of the stratigraphic position of the layers. This analysis is carried out using Past 4.06 software.

Table 1. Minimum number of individuals (MNI) of the small-mammal species for the different stratigraphic units of the Arago cave. Data come from [21,27–29].

Taxa	D	E	F	G1	G2	G3	G4	H	I	J	K	L	M	N	O	P	R
Crocidura sp.								7	1	1					5	14	
Ericaneus davidi		1	2	1	1	1	1	1		1							
Sorex cf. *araneus*																2	
Sorex sp.																	1
Talpa europaea	1	1	2	1	1	1	3	6	1	3	1						1
Arvicola mosbachensis	2	2	1	2		1		2	2	2	3	5	2	3	8	6	
Allocricetus bursae	23	39	59	2	4	9	15	63	9	30	20	22	22	20	30	39	27
Apodemus sp.	29	16	21	5	6	8	26	62	7	23	7	12	2	6	6	13	14
Castor fiber	2	1		1	1	2	3	5	1	7	3	1	1				
Chionomys nivalis	3		2					1					4		1	2	
Clethrionomys glareolus		1	1			1											
Dicrostonyx sp.	3	3	4	1	1						2						
Eliomys quercinus	3	5	11	2	2	5	9	19	3	7	2	1	2	1		3	3
Glis glis									2								
Hystrix sp.	3			1			1	1									
M. (Iberomys) brecciensis	39	32	38	4	10	22	62	190	8	55	7	3	8	3	8	58	3
Marmota sp.	1	6	1	1	1		1	1	1	1	1	1	1		2	2	
Microtus agrestis/arvalis	25	28	71	3	4	5	6	37	7	30	15	10	17	9	17	35	10
Lasiopodomys gregalis	93	141	213	5	12	6	6	69	13	84	99	106	186	189	420	253	
Alexandromys oeconomus		1	1					4	1	7	6	9	18	14	41	21	
Microtus vaufreyi	73	133	220	13	19	19	27	38	6	15	7	10	10	10	23	47	3
Pliomys episcopalis																	2
Pliomys coronensis	32	83	192	10	18	21	29	81	14	59	43	39	63	84	177	127	15
Sciurus sp.	1	1	2			1	1	1									
Spermophilus sp.	11	31	35	4	1		1	2	1	7	5	2	3	4	6	9	
Ochotona pusilla	1	1	4					7		7	2	1	3	2	2	1	
Oryctolagus cuniculus	5	16	48	7	12	13	40	45	9	37	9	16				7	24
Lepus timidus	1	2	3	1	2		1	1		2	1	6					

To obtain palaeoclimatic parameters, we used the bioclimatic model (BM) and the quantified ecology method (QEM).

The bioclimatic model, first developed by Hernández-Fernández [35] and further developed in subsequent papers [36–39], is based on the assumption that there is a correlation between climate and mammalian communities. This method (using R software, version 3.1.2) allows for an initial approach to climatic parameters, in particular temperature (mean annual temperature, mean temperature of the hottest month, and mean temperature of the coldest month) and precipitation (mean annual precipitation).

The values were adapted for some species. One case was *Marmota* sp. and as there was no single value for the genus *Marmota*, we used the data for *Marmota baibacina*, which is the most ubiquitous species. For *Microtus vaufreyi*, the ecological values of *Microtus atapuerquensis* were attributed to it. For the genus *Pliomys*, the values assigned to subtropical environments were not considered. The genus *Hystrix* was not considered.

The quantified ecology method developed by Jeannet [40] assigns quantitative climatic values (average annual temperature, minimum annual temperature, etc.) to each species according to its geographical area. One of the limitations of this method is that it only considers current species. For some species, ecological data from related species or their descendants can be used. Ecological data from *Microtus cabrerae* are used for *M. (Iberomys) mediterraneus*, and data from *Cricetulus migratorius* for *Allocricetus bursae*. However, this method is based on a large database, and a comparison with the values obtained by the other methods seems interesting.

To compare our local data with global variations we rely on the 'Oscillayer' data. This dataset builds upon interpolated anomalies (Δ layers) between bioclim layers of the present and the Last Glacial Maximum (LGM) that are scaled relative to the Plio-Pleistocene global mean temperature curve, derived from benthic stable oxygen isotope ratios [41]. These data allow us to display the evolution of the climate variables in a global way at 10 ka intervals. In this case, we are interested in the variations in the mean annual temperature. These data were processed using QGIS software (version 3.28.0).

4. Results

4.1. Correspondence Analysis

The correspondence analysis (CA) (Figure 2) sets on axis 1 (50.3% inertia) the species with affinities to open environments in cold and dry climates (*Alexandromys oeconomus, Lasiopodomys gregalis, Chionomys nivalis*) against the species associated with closed environments in temperate climates (*Microtus (Iberomys) brecciensis, Pliomys episcopalis, Glis glis, Apodemus* sp.). The species contributing most to this axis are *L. gregalis* (33.5%) and *A. oeconomus* (5.9%) in the negative part, and *M. (I.) brecciensis* (33.5%) and *Apodemus* sp. (11%) in the positive part. However, we can see that it is more of a general trend and some taxa like *Dicrostonyx* sp., which has an affinity for cold environments, are seen in the positive part of the axis. This is related to the fact that other variables have an influence on the grouping of species, as suggested by the 50% inertia of axis 1. Axis 2 (23.2% inertia) is more difficult to interpret because there does not seem to be a link between the different species that are divided by it.

On this factorial level, the clustering of the major stratigraphic units can be distinguished, in much the same way as in the previous study [27]. Nevertheless, the subdivisions of SU G reveal a convergence of SU G3 and G4 with SU H, I and J of Ensemble II. The same is true for SU R, which joins the levels of Ensemble II attributed to a temperate episode. SU G1 and G2 are grouped with the other stratigraphic units of Ensemble III. The dendrogram produced via the cluster analysis gives us a better view of the groupings of levels. (Figure 3). We can observe a grouping that corresponds to the stratigraphic ensemble except for the clustering of SU G3 and G4 with the other layers of Ensemble II. This leads us to assume that the climatic transition between Ensemble II and Ensemble III occurs between SU G2 and G3 and not between SU G4 and H.

This dendrogram also illustrates the isolation of US R from the rest of the other layers of the Arago cave. This isolation could be a consequence of a sampling bias or of a very different taxonomic composition, notably related to the presence of *P. episcopalis*.

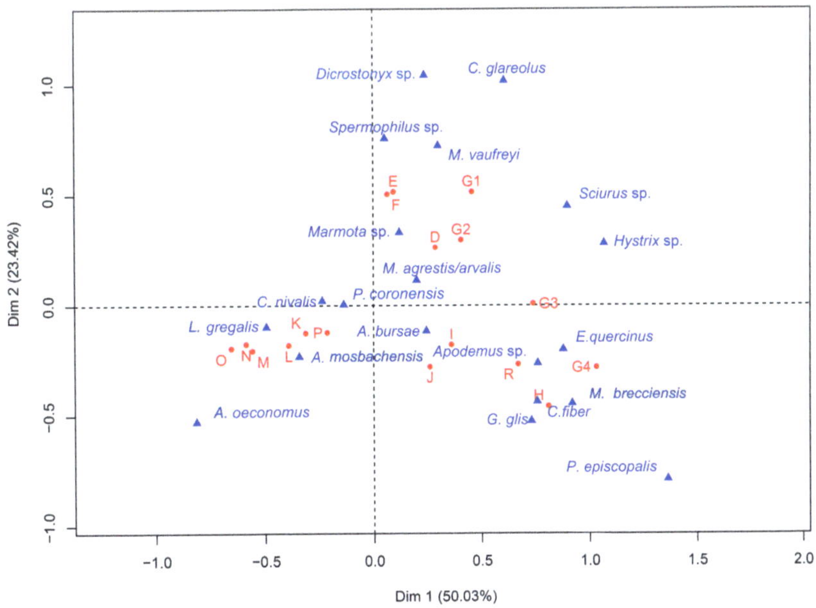

Figure 2. Correspondence analysis (CA) between rodent species and layers of the Arago cave. Axis 1 represents 50.3% inertia and axis 2 represents 23.2% inertia.

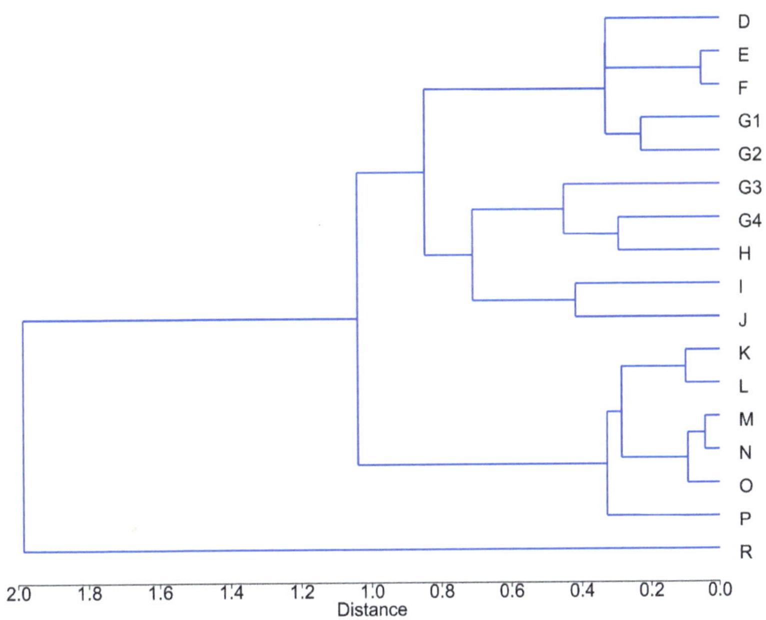

Figure 3. Dendrogram plot obtained from the cluster analysis based on the first three CA axes.

4.2. Climatic Parameters

The bioclimatic model reveals that the mean annual temperature ranges from 6 °C for the coldest stratigraphic unit (O) to 9.5 °C for SU R (Table 2). Apart from SU R, all the mean annual temperatures are around 7 ± 1 °C. The high mean temperature for SU R is related to the absence of species associated with cold environments, such as *L. gregalis*, *A. oeconomus*, and *Dicrostonyx torquatus*. Since *L. gregalis* is ultra-dominant in the overlying stratigraphic units, the absence of this species in SU R is probably not the result of a recording bias. Furthermore, taphonomic studies have shown that the accumulation of small vertebrates in this level is produced by the Eurasian eagle-owl, which is a species with a broad prey spectrum, whose accumulations are a good reflection of the palaeocommunities [42].

Table 2. Climatic parameters obtained by the bioclimatic model (BM, standard error in parentheses) and the quantified ecology method (QEM) for the different stratigraphic units of the Arago cave. MAT = mean annual temperature, T°max = mean temperature of the warmest month, T°min = mean temperature of the coldest month, expressed in °C.

US	T°max		MAT		T°min	
	BM	QEM	BM	QEM	BM	QEM
D	20.6 (±2.6)	21.0	6.9 (±3.4)	7.2	−5.3 (±7.4)	−6.8
E	20.1 (±2.5)	20.8	6.3 (±3.3)	6.3	−6.1 (±7.4)	−8.5
F	19.8 (±2.5)	20.9	6.1 (±3.3)	6.7	−6.2 (±7.4)	−7.8
G1	20.7 (±2.6)	20.7	7.0 (±3.4)	7.3	−5.1 (±7.4)	−6.4
G2	20.5 (±2.5)	20.8	7.1 (±3.3)	7.4	−5.0 (±7.4)	−.6.2
G3	21.4 (±2.7)	21.5	7.2 (±3.6)	8.4	−5.7 (±8)	−4.9
G4	22.1 (±2.7)	21.4	8.1 (±3.6)	8.3	−4.5 (±8)	−4.9
H	21.4 (±2.7)	21.1	7.3 (±3.6)	7.3	−5.4 (±7.9)	−6.7
I	21.8 (±3.6)	21.4	8.1 (±3.5)	8.1	−4.4 (±7.8)	−5.4
J	21.6 (±2.8)	21.2	7.5 (±3.6)	7.3	−5.4 (±7.9)	−6.8
K	20.3 (±2.6)	20.7	6.6 (±3.3)	6.4	−5.7 (±7.4)	−8.2
L	21.6 (±2.8)	21.2	7.5 (±3.6)	7.2	−5.4 (±7.9)	−7.1
M	21.3 (±2.7)	21.8	7.3 (±3.5)	7.8	−5.4 (±7.8)	−6.5
N	21.4 (±2.7)	22	7.8 (±3.5)	7.7	−4.5 (±7.8)	−6.9
O	20.8 (±2.6)	21.8	6.0 (±3.4)	7.8	−7.8 (±7.6)	−6.7
P	21.2 (±2.7)	21.45	7.0 (±3.5)	7.7	−5.5 (±7.7)	−6.3
R	22.6 (±2.9)	21.9	9.5 (±3.7)	9.5	−2.3 (±7.7)	−3.0

The mean temperature values of the warmest and coldest months follow the variations in mean annual temperature. There are no phases with a more temperate or continental climate. The difference between the mean of the warmest and coldest month is around 26 °C. The only stratigraphic unit that stands out is the SU R, which has a higher mean temperature of the coldest month than the others (−2.3 °C), even if this value remains fairly low. For each layer, the summer/gap is still above the 25 °C mark which is the limit of the continental climate [40].

The quantified ecology method shows that the mean annual temperature values range from 6.3 °C (for SU E) to 9.5 °C for SU R. SU G3, G4, and I (8.4 °C, 8.3 °C and 8.1 °C) are among the stratigraphic units with the highest temperatures. SU E, F, and K (6.3 °C, 6.7 °C, and 6.4 °C) have the lowest temperatures (Table 2).

The temperatures of the warmest month show little variation among the stratigraphic units. They range from 20.7 °C for SU G1 to 21.9 °C for SU N. For the temperatures of the coldest month, the variations are more marked. The lowest values are observed for SU E, F, and K (−8.5 °C, −7.8 °C, and −8.2 °C). The highest values are observed for SU G3, G4, and R (−4.9 °C, −4.9 °C, and −3 °C). The summer/winter temperature difference is around 27 °C for the different SU except for SU R with a summer/winter gap of 24.9 °C.

4.3. Synthesis

An initial observation of the fauna list highlights the presence of species with different ecological affinities in each SU. This reflects the potentially great diversity of habitats around the Arago cave. Three groups of stratigraphic units and an isolated layer can be defined using climate reconstruction methods.

The first is composed of SU D to G2. This group is well individualised in the CA projection and the dendrogram (Figures 2 and 3). The lemming (*Dicrostonyx* sp.), which is a characteristic species of the tundra, is present. The stratigraphic units that make up this group have some of the lowest values obtained using the BM and the QEM, particularly for SU E and F. They correspond to a cold phase, the average temperature being between 6.1 °C and 7.2 °C according to the BM and the QEM.

The second group is made up of SU G3 to J. This group stands out from the other stratigraphic units in Ensembles I and III, and is found in the positive part of the CA axis. The stratigraphic units in this group have some of the highest values determined via palaeotemperature reconstruction methods. They seem to correspond to a milder phase than SU D to G2. The calculated mean annual temperature is between 7.2 °C and 8.4 °C according to the BM and QEM. This climate is still colder and more continental than the present climate.

The third group is made up of SU K to P, corresponding to the stratigraphic units of Ensemble I. This group is quite distinct from the other stratigraphic units, and is found in the negative part of axis 1 of the CA. This group presents median palaeotemperature values compared to the other units of the Arago Cave. The average palaeotemperatures calculated for them range from 6 °C to 7.8 °C according to the BM and the QEM. The climate for these stratigraphic units is a little colder than that of SU G3 to J and is similar to SU D to G2

On the CA, the SU R is projected in the positive part of axis 1 and is found with SU G3 to J, which are those whose climate reconstruction is least cold. The dendrogram allows us to clearly separate the SU R from the other layer. This stratigraphic unit represents the warmest and most temperate climatic episode observed at the Arago Cave.

5. Discussion

The reconstructions of the climatic parameters show a similar value for the MAT (mean annual temperature). If they are comparable for the T°max (mean temperature of the warmest month), they are more variable for the T°min (mean temperature of the coldest month). We can also observe that the standard error with the BM is significant for the MAT and the T°max and higher for the T°min. These high values require us to step back from the data. We prefer to base our comparisons only on the MAT, as the standard error of the T°min data is too high. In addition, the values of T°min obtained via the two methods are low for a site located on the Mediterranean coast. These outliers may be intrinsic to the method, which is based on current ecological data for species that may have evolved during the Pleistocene. Furthermore, climatic parameters do not have the same impact on rodent species, as in the case of T°min, which will have less impact on burrowing and hibernating species [40]. However, despite these limits these methods remain the preferred tools for quantifying climatic variation between different layers.

Statistical analyses based on MNI highlight three groups of SU and isolate the SU R from the others. With the reconstruction of the climatic parameters, we can observe slight differences between these groups. These differences are less pronounced with climate reconstruction methods, as only the presence or absence of species is considered. In fact, if we look at each stratigraphic grouping, we see little change in the communities, as the species are broadly similar throughout the stratigraphy, except for the SU R. This is also true for the large mammals, where species associated with cold environments (*Rangifer tarandus*, *Praeovibos priscus*) and species associated with temperate environments (*Dama* cf. *clactoniana*, *Cervus elaphus*) are present in the three ensembles [17].

The presence of non-analogous communities, characterized by the cohabitation of species with different ecological affinities that are not currently associated, has been identified at the Arago cave [21,43]. Such communities may vary in their origins, which are not mutually exclusive. They may correspond to a palaeoecological reality and are in this case associated with a palaeoenvironment that is not currently found [44]. They may also correspond to a contact zone between two climatic influences. The Arago cave is located between two zones with a strong climatic influence. The Pyrenees, which correspond to a continental zone located at altitude, are likely to have served as a refuge for "cold" species during temperate phases. The Mediterranean coastline, on the other hand, corresponds to an area with a milder climate and may have served as a refuge for "temperate" species during cold periods [34,45–47]. They may also generate a time-averaging effect, which is thought to have increased palaeobiodiversity significantly [48,49]. This phenomenon is increased by the proximity of the communities, which allows for rapid recolonization phases.

The local environment has an important impact on the palaeobiodiversity observed at the Arago cave and therefore on the human groups that occupied it. The position of the site offers a definite advantage in terms of access to the resources stretching between the Mediterranean coast and the Pyrenees, highlighted by the wide variety of species exploited [50]. The diversity of ecosystems that can be exploited over a short distance is an argument for a preferential location for settlement. Human groups can exploit different ecological niches and these will react differently to climatic variations. This diversity endowed human settlements with greater resistance and resilience to Pleistocene climate change. This helps to justify the large number of occupations present at the Arago cave.

If the local environment has an impact on diversity, it is also necessary to understand how global changes have influenced palaeocommunities. However, temperature variations can be seen to be very limited among the layers of the Middle Complex (US D to P), and the differences between Ensemble II, interpreted until now as being correlated to MIS 13, and Ensembles I and III are small, of the order of 1 °C for the mean annual temperature. One possible explanation is the relatively low amplitude of global temperature variations along the oxygen isotope series for this period [51]. This could justify less warming during MIS 13 than observed during other interglacials [52]. However, it is complicated to link global data (on benthic record) to our on-site data.

The use of the Oscillayers method allows us to consider the regional context for reconstructing the mean annual temperature. Compared to our data, those one modelled by the Oscillayers method [41] show strong variation between MIS 14, 13, and 12 (Figure 4). Caution must be exercised in interpreting the temperature values at a given point using this type of global model, but it can give us an idea of the scale of climatic variations. It can be seen that the average temperature at the location of the Arago cave varies from 8.7 °C to 13.2 °C (Table 3), a difference of 4.5 °C. These variations are much more marked than those observed in our layers. The variations observed with our data at the Arago cave were more in line with those observed in MIS12.

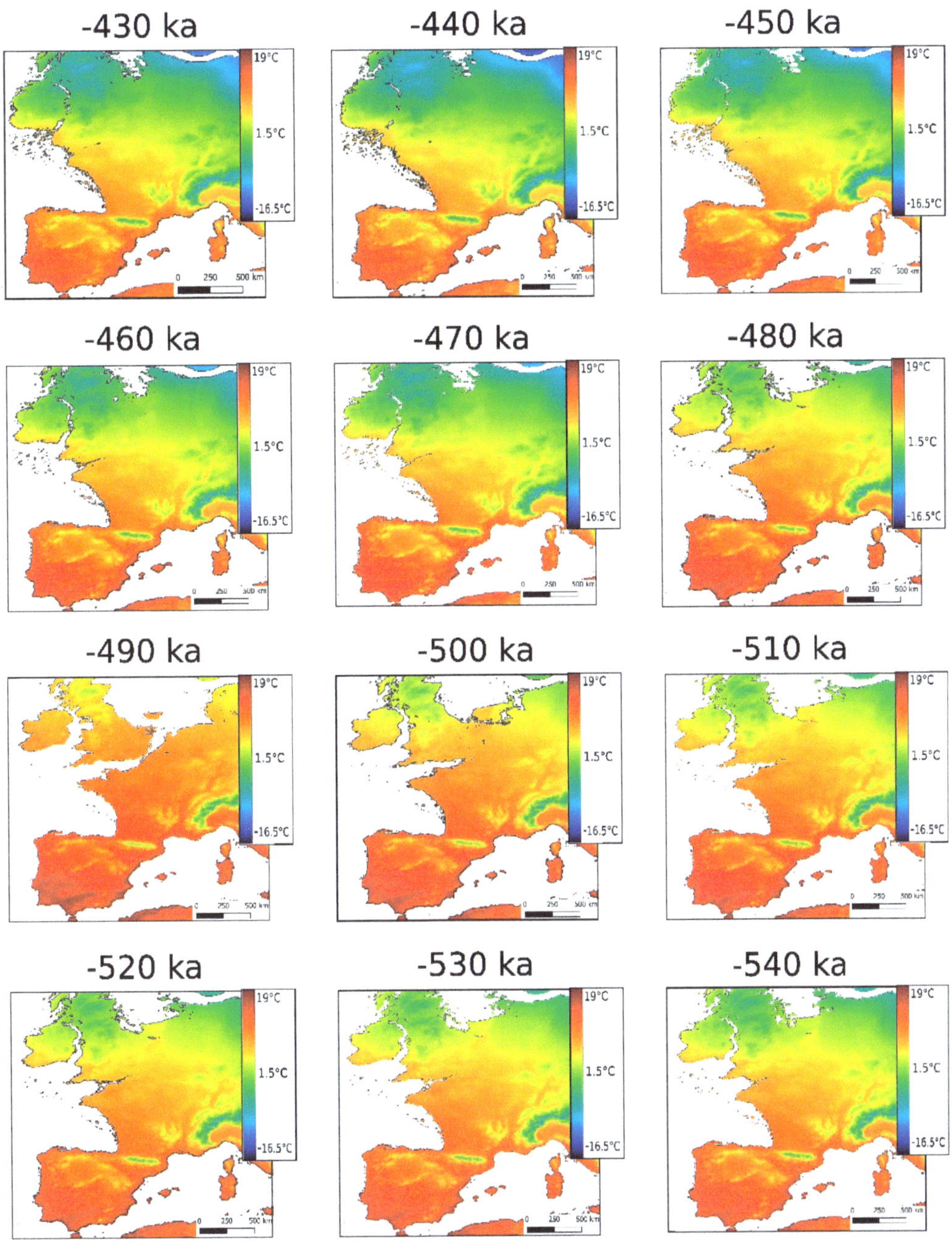

Figure 4. Reconstructed palaeocoastlines and mean annual temperature estimates between −433 ka and −540 ka using the 10 ka stepped outputs from Gamisch [41]. Colour scale based on minimum (−16.5 °C, dark blue) and maximum values (19 °C, dark red).

Table 3. Mean annual temperature (MAT in °C) observed at the location of the Arago cave on each Oscillayer modelled using the Oscillayers method [41].

period	−430 ka	−440 ka	−450 ka	−460 ka	−470 ka	−480 ka
MAT	9.2	8.7	9.1	10	9.8	11.1
period	−490 ka	−500 ka	−510 ka	−520 ka	−530 ka	−540 ka
MAT	13.2	12.1	11.6	11.2	10.7	10.8

In order to better correspond to the global variations, it is necessary to rethink the correlations between the ensembles in the Arago cave. As we have fairly precise dates for SU G, it makes sense to correlate Ensemble III with MIS 12. Previous studies correlate Ensembles II and I with MIS 13 and 14, respectively [16,17]. This applies in particular to the dates from level P, which could correspond to MIS 14. However, the dates obtained for level P are less accurate, as the age given is 532 ± 106 ka [17]. These environmental variations have been interpreted as a change in MIS, whereas there is little change in species and the variations are mainly variations in MNI. However, variations in species representation alone are not necessarily synonymous with climatic variations, particularly for small vertebrates [40]. If the low amplitude of the oxygen isotope series was used as an argument for correlating Ensemble II to the MIS 13, there is no clear argument against correlating Ensemble II with a phase of MIS 12. On the contrary, SU R represents a real discontinuity in the small-vertebrate communities and can be considered a marked environmental change that corresponds better with a change in stage. Variations between Ensembles I, II, and III may then be considered intra-stage variations. This is not in conflict with the dates, which have a very wide confidence interval. Thus, Ensembles I, II, and III of the Arago cave can be correlated to MIS 12, and SU R would mark a stage change and fall within MIS 13.

This calls into question the interpretation of the timing of the filling of the Arago cave. This gives a new perspective to our perceptions of the pace of technical evolution in the Middle Pleistocene.

6. Conclusions

The small-mammal palaeocommunities of the Arago cave indicate the presence of a cold climate throughout the stratigraphic sequence, but with a clear warming in the deepest layer (SU R). The composition of the small-mammal communities is strongly influenced by the local context of the site, with an influence from the Mediterranean coast and the Pyrenees. This double influence has allowed the coexistence of different ecological niches within a restricted space, which may have favored the establishment and maintenance of human groups, endowing the ecosystem with greater resilience and resistance to climatic variations. Environmental reconstruction methods display fairly low climatic variations within the Middle Complex (Ensembles I, II, and III). These variations may correspond better to those observed within MIS 12 than to interstage variations as previously interpreted. Thus, the Middle Complex of the Arago Cave could be correlated to MIS 12 in accordance with the dating, and SU R would mark the possible transition to MIS 13. This has implications for our understanding of Acheulean development in this region, because SU R would then be the only unit that would be correlated with MIS 13. The chronology of the filling of the Arago cave deserves further study in order to gain a better understanding of the evolution of Acheulean behaviour in southwestern Europe.

Author Contributions: Conceptualization, L.L. and J.M.L.-G.; methodology, L.L. and J.M.L.-G.; formal analysis, L.L.; investigation, L.L. and J.M.L.-G.; resources, L.L. and J.M.L.-G.; data curation, L.L. and J.M.L.-G.; writing—original draft preparation, L.L.; writing—review and editing, L.L. and J.M.L.-G.; visualization, L.L. and J.M.L.-G.; supervision, J.M.L.-G.; funding acquisition, L.L. and J.M.L.-G. All authors have read and agreed to the published version of the manuscript.

Funding: The Institut Català de Paleoecologia Humana i Evolució Social (IPHES-CERCA) has received financial support from the Spanish Ministry of Science and Innovation through the "María de Maeztu" program for Units of Excellence (CEX2019-000945-M). This project has received funding from the European Union's Horizon 2020 research and innovation programme under the Marie Skłodowska-Curie grant agreement No 101034349 and the State Research Agency of the Spanish Ministry of Science and Innovation through the Program Maria de Maeztu Unit of Excellence (CEX2019-000945-M). This project was supported by the English correction of Rupert Glasgow. This manuscript is part of a Knowledge Generation project PID2021-122533NB-I00 of the Spanish Ministry of Science and Innovation (MICIN) and project SGR2021-01238 of the Generalitat de Catalunya-AGAUR.

Data Availability Statement: Data sharing not applicable.

Acknowledgments: We also want to thank the two anonymous reviewers for their suggestions and comments that improved the final version of the manuscript.

Conflicts of Interest: The authors declare no conflict of interest.

References

1. Berger, W.H.; Jansen, E. Mid-pleistocene climate shift—The Nansen Connection. In *The Polar Oceans and Their Role in Shaping the Global Environment*; Johannessen, O.M., Muench, R.D., Overland, J.E., Eds.; American Geophysical Union: Washington, DC, USA, 2013; Volume 85, pp. 295–311.
2. Head, M.J.; Gibbard, P.L. Early-Middle Pleistocene transitions: An Overview and Recommendation for the Defining Boundary. In *Early-Middle Pleistocene Transitions: The Land-Ocean Evidence*; Head, M.J., Gibbard, P.L., Eds.; Geological Society of London: London, UK, 2005; pp. 1–18.
3. Discamps, E.; Royer, A. Reconstructing palaeoenvironmental conditions faced by Mousterian hunters during MIS 5 to 3 in southwestern France: A multi-scale approach using data from large and small mammal communities. *Quat. Int.* **2017**, *433*, 64–87.
4. Denys, C. Des référentiels en taphonomie des petits vertébrés: Bilan et perspectives. In *Taphonomie des Petits Vertébrés: Référentiels et Transferts Aux Fossiles*; Laroulandie, V., Mallye, J.-B., Denys, C., Eds.; British Archaeological Reports; International Series 2269; Archaeopress: Oxford, UK, 2011; pp. 7–22.
5. Puzachenko, A.Y.; Markova, A.K. The Scandinavian ice sheet against the Atlantic ocean: How the scandinavian ice sheet affected European small mammal assemblage during the Greenland stadial GS-2.1. *Quat. Sci. Rev.* **2023**, *305*, 108013.
6. Lumley, H.; Fournier, A.; Park, Y.C.; Yokoyama, Y.; Demouy, A. Stratigraphie du remplissage Pléistocène moyen de la Caune de l'Arago à Tautavel. Etude de huit carottages effectués de 1981 à 1983. *L'Anthropologie* **1984**, *88*, 5–18.
7. Barsky, D.; Lumley, H. Early European Mode 2 and the stone industry from the Caune de l'Arago's archeostratigraphical levels "P". *Quat. Int.* **2010**, *223*, 71–86.
8. Barsky, D.; Moigne, A.M.; Pois, V. The shift from typical Western European Late Acheulian to microproduction in unit 'D'of the late Middle Pleistocene deposits of the Caune de l'Arago (Pyrénées-Orientales, France). *J. Hum. Evol.* **2019**, *135*, 102650. [PubMed]
9. Viallet, C.; Grégoire, S.; Perrenoud, C. Break to Rebuild—The First European Evidence of a Fragmented Chaine Opératoire for Handaxe Production (OIS 14, Caune de l'Arago, France). *J. Paleolit. Archaeol.* **2022**, *5*, 1.
10. Blain, H.A.; Fagoaga, A.; Ruiz-Sánchez, F.J.; Garcia-Medrano, P.; Ollé, A.; Jiménez-Arenas, J.M. Coping with arid environments: A critical threshold for human expansion in Europe at the Marine Isotope Stage 12/11 transition? The case of the Iberian Peninsula. *J. Hum. Evol.* **2021**, *153*, 102950. [PubMed]
11. Connet, N.; Soriano, S.; Bertran, P.; Lhomme, V.; Debenham, N. A 400,000 years old milestone of the Acheulian technocomplex in Central-Western France at Londigny (Charente). *J. Archaeol. Sci. Rep.* **2020**, *30*, 102225.
12. Moncel, M.H.; Despriée, J.; Voinchet, P.; Tissoux, H.; Moreno, D.; Bahain, J.J.; Courcimault, G.; Falguères, C. Early evidence of Acheulean settlement in northwestern Europe-La Noira Site, a 700 000 year-old occupation in the center of France. *PLoS ONE* **2013**, *8*, e75529.
13. Moncel, M.H.; Ashton, N. From 800 to 500 ka in Western Europe. The oldest evidence of Acheuleans in their technological, chronological, and geographical framework. In *The Emergence of the Acheulean in East Africa*, 1st ed.; Mussi, M., Gallotti, R., Eds.; Springer: New York, NY, USA, 2018; pp. 215–235.
14. Nicoud, E.; Aureli, D.; Pagli, M.; Villa, V.; Chaussé, C.; Agostini, S.; Zupancich, A. Preliminary data from Valle Giumentina Pleistocene site (Abruzzo, Central Italy): A new approach to a Clactonian and Acheulian sequence. *Quat. Int.* **2016**, *409*, 182–194.
15. Ollé, A.; Lombao, D.; Asryan, L.; García-Medrano, P.; Arroyo, A.; Fernández-Marchena, J.L.; Vallverdú, J. The earliest European Acheulean: New insights into the large shaped tools from the late Early Pleistocene site of Barranc de la Boella (Tarragona, Spain). *Front. Earth Sci.* **2023**, *11*, 1188663.
16. Falguères, C.; Shao, Q.; Han, F.; Bahain, J.J.; Richard, M.; Perrenoud, C.; Moigne, A.M.; Lumley, H. New ESR and U-series dating at Caune de l'Arago, France: A key-site for European Middle Pleistocene. *Quat. Geochronol.* **2015**, *30*, 547–553.
17. Moigne, A.M.; Palombo, M.R.; Belda, V.; Heriech-Briki, D.; Kacimi, S.; Lacombat, F.; Lumley, M.A.; Moutoussamy, J.; Rivals, F.; Quilès, J.; et al. Les faunes de grands mammifères de la Caune de l'Arago (Tautavel) dans le cadre biochronologique des faunes du Pléistocène moyen italien. *L'Anthropologie* **2006**, *110*, 788–831.

18. Magniez, P.; Moigne, A.M.; Testu, A.; Lumley, H. Biochronologie des mammifères quaternaires. Apport des cervidae du site pléistocène moyen de la Caune de l'Arago (Tautavel, Pyrénées-Orientales, France). *Quaternaire* **2013**, *24*, 477–502.
19. Lumley, H. *Caune de l'Arago, Tautavel-en-Roussillon, Pyrénées-Orientales, France: Individualisation des Unités Archéostratigraphiques*; CNRS Editions: Paris, France, 2015; p. 642.
20. Lumley, M.A. L'homme de Tautavel. Un Homo erectus européen évolué. *Homo erectus tautavelensis. L'Anthropologie* **2015**, *119*, 303–348.
21. Lebreton, L.; Desclaux, E.; Hanquet, C.; Moigne, A.M.; Perrenoud, C. Environmental context of the Caune de l'Arago Acheulean occupations (Tautavel, France), new insights from microvertebrates in Q-R levels. *Quat. Int.* **2016**, *411*, 182–192.
22. Chaline, J. L'âge des Hominiens de la Caune de l'Arago à Tautavel (Pyrénées-Orientales), d'après l'étude des Rongeurs. *Comptes Rendus Hebd. Des Séances De L'académie Des Sci.* **1971**, *272*, 1743–1746.
23. Brunet-Lecomte, P. Statut des campagnols souterrains (Rodentia, Arvicolidae) du gisement du Pléistocène moyen de l'Arago à Tautavel (Pyrénées-Orientales, France). *Bull. Mens. De La Société Linnéenne De Lyon* **1990**, *59*, 100–104.
24. Desclaux, E. Les petits vertébrés de la Caune de l'Arago (Tautavel, Pyrénées-orientales): Paléontologie, paléoécologie et taphonomie. Ph.D. Thesis, University of Perpignan, Perpignan, France, 1992.
25. Paunescu, A.C. Les Rongeurs du Pléistocène Inférieur et Moyen de trois grottes du sud-est de la France (Vallonnet, Caune de l'Arago, Baume Bonne). Implications systématiques, biostratigraphiques et paléoenvironnementales. Ph.D. Thesis, University of Perpignan, Perpignan, France, 2001.
26. Brunet-Lecomte, P.; Paunescu, A. Morphologie comparée de la première molaire inférieure du campagnol *Microtus* (*Terricola*) *vaufreyi tautavelensis* (Rodentia, Arvicolidae) du gisement pléistocène moyen de l'Arago (Pyrénées, France) et inférences paléoclimatiques. *Quaternaire* **2004**, *15*, 263–268. [CrossRef]
27. Hanquet, C. Evolution des paléoenvironnements et des paléoclimats au Pléistocène moyen, en Europe méridionale, d'après les faunes de micromammifères. Ph.D. Thesis, University of Perpignan, Perpignan, France, 2011.
28. Lebreton, L.; Desclaux, E.; Hanquet, C.; Cuenca-Bescós, G.; Moigne, A.M.; Perrenoud, C.; Grégoire, S. Variations paléoenvironnementales au sein de l'Unité Archéostratigraphique G (UA G) de la Caune de l'Arago (Tautavel, France): Apport des paléocommunautés de rongeurs. *Quaternaire* **2017**, *28*, 313–321. [CrossRef]
29. Lebreton, L.; Moigne, A.M.; Filoux, A.; Perrenoud, C. A specific small game exploitation for Lower Paleolithic: The beaver (*Castor fiber*) exploitation at the Caune de l'Arago (Pyrénées-Orientales, France). *J. Archaeol. Sci. Rep.* **2017**, *11*, 53–58.
30. Manzano, A. Les Amphibiens et les Reptiles des Sites du Pléistocène Moyen et Supérieur du Pourtour Méditérranéen (Caune de l'Arago, Grotte du Lazaret, Baume Moula-Guercy). Etude D'herpétofaunes et Reconstitutions Paléoclimatiques et Paléoenvironnementales. Ph.D. Thesis, University of Perpignan, Perpignan, France, 2015.
31. Lebreton, L. Approche taphonomique multi-taxons des accumulations de petits vertébrés, implication pour les reconstitutions paléoenvironnementales au Pléistocène. Ph.D. Thesis, University of Perpignan, Perpignan, France, 2018.
32. Marquet, J.C. Paléoenvironnement et Chronologie des Sites du Domaine Atlantique Français d'âge Pléistocène Moyen et Supérieur D'après L'étude des Rongeurs. Ph.D. Thesis, University of Dijon, Dijon, France, 1989.
33. Chaline, J.; Brunet-Lecomte, P.; Campy, M. The last glacial/interglacial record of rodent remains from the Gigny karst sequence in the French Jura used for palaeoclimatic and palaeoecological reconstructions. *Palaeogeogr. Palaeoclimatol. Palaeoecol.* **1995**, *117*, 229–252.
34. Foury, Y.; Desclaux, E.; Daujeard, C.; Defleur, A.; Moncel, M.H.; Raynal, J.P. Evolution des faunes de rongeurs en moyenne vallée du Rhône (rive droite, Ardèche, France) au cours du pléistocène moyen final et du pléistocène supérieur ancien, du mis 6 au mis 4. *Quaternaire* **2016**, *27*, 55–79.
35. Hernández Fernández, M. Bioclimatic discriminant capacity of terrestrial mammal faunas. *Glob. Ecol. Biogeogr.* **2001**, *10*, 189–204.
36. Hernández Fernández, M.; Peláez-Campomanes, P. The bioclimatic model: A method of palaeoclimatic qualitative inference based on mammal associations. *Glob. Ecol. Biogeogr.* **2003**, *12*, 507–517.
37. Hernández Fernández, M.; Peláez-Campomanes, P. Quantitative palaeoclimatic inference based on terrestrial mammal faunas. *Glob. Ecol. Biogeogr.* **2005**, *14*, 39–56.
38. Fernández, M.H.; Sierra, M.Á.; Peláez-Campomanes, P. Bioclimatic analysis of rodent palaeofaunas reveals severe climatic changes in Southwestern Europe during the Plio-Pleistocene. *Palaeogeogr. Palaeoclimatol. Palaeoecol.* **2007**, *251*, 500–526.
39. Royer, A.; Yelo, B.A.G.; Laffont, R.; Fernández, M.H. New bioclimatic models for the quaternary palaearctic based on insectivore and rodent communities. *Palaeogeogr. Palaeoclimatol. Palaeoecol.* **2020**, *560*, 110040.
40. Jeannet, M. L'écologie quantifiée. Essai de description de l'environnement continental à l'aide des microvertébrés. *Préhistoires méditerranéennes* **2010**, *1*, 1–26.
41. Gamisch, A. Oscillayers: A dataset for the study of climatic oscillations over Plio-Pleistocene time-scales at high spatial-temporal resolution. *Glob. Ecol. Biogeogr.* **2019**, *28*, 1552–1560.
42. Lebreton, L.; Bailon, S.; Guillaud, E.; Testu, A.; Perrenoud, C. Multi-taxa referential of a modern Eurasian Eagle-Owl (*Bubo bubo*) aerie. *J. Archaeol. Sci. Rep.* **2020**, *32*, 102417.
43. Hanquet, C.; Desclaux, E. Analyse paléoécologique des communautés de micromammifères de la Caune de l'Arago (Tautavel, France) dans le contexte des migrations de faunes en Europe méridionale au cours du Pléistocène moyen. *Quaternaire* **2011**, *22*, 34–45.

44. Guthrie, D.; van Kolfschoten, T. Neither warm and moist, nor cold and arid: The ecology of the Mid Upper Palaeolithic. In *Hunters of the Golden Age*; Roebroeks, W., Mussi, M., Svoboda, J., Fennema, K., Eds.; Leiden University: Leiden, The Netherlands, 2000; pp. 3–20.
45. Bennett, K.D.; Provan, J. What do we mean by 'refugia'? *Quat. Sci. Rev.* **2008**, *27*, 2449–2455.
46. Stewart, J.R.; Lister, A.M.; Barnes, I.; Dalén, L. Refugia revisited: Individualistic responses of species in space and time. *Proc. Royal Soc. B* **2010**, *277*, 661–671.
47. Royer, A.; Montuire, S.; Legendre, S.; Discamps, E.; Jeannet, M.; Lécuyer, C. Investigating the influence of climate changes on rodent communities at a regional-scale (MIS 1-3, Southwestern France). *PLoS ONE* **2016**, *11*, e0145600.
48. Fürsich, F.T.; Aberhan, M. Significance of time-averaging for palaeocommunity analysis. *Lethaia* **1990**, *23*, 143–152.
49. Denys, C. Rodent faunal lists in karstic and open-air sites of Africa: An attempt to evaluate predation and fossilization biases on paleodiversity. *Cuad. De Geol. Ibérica* **1997**, *23*, 73–94.
50. Moigne, A.M.; Grégoire, S.; Lumley, H.D. Les territoires de chasse et d'exploitation des matières premières des hommes préhistoriques de la Caune de l'Arago entre 600 000 ans et 400 000 ans. *Actes Des Congrès Natx. Des Sociétés Hist. Et Sci.* **2005**, *126*, 17–31.
51. Lisiecki, L.E.; Raymo, M.E. A Pliocene-Pleistocene stack of 57 globally distributed benthic δ18O records. *Paleoceanography* **2005**, *20*, 1–17.
52. López-García, J.M.; Fagoaga, A.; Nabais, M.; Póvoas, L.; Zilhão, J. Late Quaternary (MIS 5a-5b) climate and environments of western Iberia inferred from the small-mammal assemblage of Gruta da Oliveira, Torres Novas, Portugal. *Palaeogeogr. Palaeoclimatol. Palaeoecol.* **2022**, *603*, 111194.

quaternary

Article

The Influence of Hand Dimensions on Finger Flexion during Lower Paleolithic Stone Tool Use in a Comfortable Grip

Annapaola Fedato [1,*], María Silva-Gago [2], Marcos Terradillos-Bernal [3], Rodrigo Alonso-Alcalde [4,5] and Emiliano Bruner [6]

1 Centro Nacional de Investigación sobre la Evolución Humana, 09002 Burgos, Spain
2 Incipit CSIC—Consejo Superior de Investigaciones Científicas, 15707 Santiago de Compostela, Spain
3 Facultad de Humanidades y Ciencias Sociales, Universidad Internacional Isabel I de Castilla (Ui1), 09003, Burgos, Spain
4 Área de Didáctica y Dinamización, Museo de la Evolución Humana, Paseo Sierra de Atapuerca n°2, 09002 Burgos, Spain
5 Departamento de Historia, Geografía y Comunicación, Universidad de Burgos, 09001 Burgos, Spain
6 Museo Nacional de Ciencias Naturales—CSIC, 28006 Madrid, Spain
* Correspondence: fedato.anna@gmail.com

Abstract: Considering the biomechanical and cognitive aspects involved in tool manipulation, hand size emerges as a critical factor. Males, on average, exhibit greater grip strength attributed to larger hand dimensions. Beyond mere physical factors, cognitive components tied to visuospatial abilities also influence stone tool use. However, the intricate relationship between hand size, grip strength, and ergonomic patterns necessitates further exploration. Here, we study the ergonomic pattern of phalanx flexion during the manipulation of Lower Paleolithic stone tools (choppers and handaxes) to understand the nuanced interplay between hand dimensions and grasping behaviors in Lower Paleolithic stone tool use. The static hand posture during the comfortable grasping of each tool is measured using a motion capture hand glove. Flexions are measured at the metacarpophalangeal joint, the proximal interphalangeal joint and the distal interphalangeal joint of each finger. Our investigation into Lower Paleolithic stone tool manipulation reveals gender-based differences in phalanx flexion, with hand dimensions showing correlation only in pooled samples. However, these associations diminish when analyzing males and females separately. This study suggests a minimal link between hand size and grasping behavior within our sample, hinting at the influence of cognitive, behavioral, and motor factors. Exploring lifestyle and psychometric profiles could provide further insights. In the context of early human technology, our results prompt considerations on the evolution of the hand-tool interaction system, linking our tool-dependent culture to our phylogenetic history.

Keywords: Oldowan; Acheulean; haptics; ergonomics; hand size

Citation: Fedato, A.; Silva-Gago, M.; Terradillos-Bernal, M.; Alonso-Alcalde, R.; Bruner, E. The Influence of Hand Dimensions on Finger Flexion during Lower Paleolithic Stone Tool Use in a Comfortable Grip. *Quaternary* **2024**, 7, 29. https://doi.org/10.3390/quat7030029

Academic Editor: Ran Barkai

Received: 10 January 2024
Revised: 3 June 2024
Accepted: 17 June 2024
Published: 29 June 2024

1. Introduction

Human hand morphology, intricately shaped by evolutionary pressures associated with tool making and usage, serves as a tangible record of our species' adaptive journey [1]. Specific characteristics, such as the thumb/index length proportion, have garnered attention for their presumed role in conferring enhanced precision during tool handling [2–6]. Napier's seminal classification of grips into "precision" and "power" provides a conceptual foundation for understanding the diverse grasping possibilities exhibited by humans, crucial for both daily activities and tool use [7,8].

While studies in experimental archaeology have shed light on limited grasping patterns associated with Lower Paleolithic stone cutting tools [9], the overarching ergonomic schemes inherent in early and extinct human species remain speculative [9]. The handling and manipulation of tools depends significantly on the tool function, and the results of traceological examinations can be helpful in defining the actions for which tools were used. While traceology is commonly applied to tools from later periods, its principles can still be applied to Lower Paleolithic stone tools to infer their use. Despite the challenges, traceological analysis remains a valuable method for gaining insights into the use and functionality of early stone tools [10].

Biomechanical responses during tool use are not only shaped by human physiology but are also intricately linked to the morphology of the stone tools themselves, as evidenced by the work of Patiño et al. (2017) [11]. Furthermore, recent studies examining the ergonomic aspects of Lower Paleolithic technology have underscored the influence of tool type and metrics on the intricate patterns of phalanx flexion during comfortable tool grasping [12,13].

Beyond the biomechanical considerations, effective tool manipulation involves a cognitive component tethered to an individual's capacity to perceive object affordances [13]. This cognitive aspect, intertwined with sensing capacity, initiates physiological responses associated with attention and arousal [14,15]. The role of the parietal cortex in visuospatial integration is particularly pronounced in the context of stone tool use, aligning the cognitive dimensions with the evolutionary trajectory of our visuospatial abilities [16–18].

Hand size emerges as a pivotal factor in the nuanced landscape of Lower Paleolithic ergonomic patterns. The applied forces and contact area, intricately linked to grip diameter and hand size, underscore the importance of hand dimensions in shaping the efficiency and comfort of tool grasping [19]. Notably, the sexual dimorphism in grip strength, with males generally producing greater force, is a well-established phenomenon [19–25]. The predictive power of hand dimensions, particularly palm width, in estimating grip force has been corroborated by Nicolay and Walker (2005) [25]. However, the complex interplay between sociodemographic predictors and grip strength, as explored by Hart (2018) [26], adds layers of nuance to our understanding.

In the context of stone tool use, the biomechanical aspects of the hand have been found to be intimately related to efficiency in cutting tasks [27]. Notably, manipulative strength stands out as the primary predictor for efficiency in flake tools, while hand size takes precedence for handaxes [9]. Building upon our recent work [13], where we meticulously measured the phalanx flexion of 78 subjects during comfortable stone tool handling for both Oldowan pebble tools and Acheulean handaxes, the present study endeavors to deepen our understanding of the ergonomic hand patterns associated with the grasping of Lower Paleolithic stone tools. Specifically, we aim to assess whether hand dimensions play a role in sexual differences in manipulation patterns, systematically analyzing grasping patterns in males and females under the null hypothesis of no differences and exploring the potential effect of hand dimensions on finger flexion under the null hypothesis of no correlation. This multidimensional approach integrates biomechanical, cognitive, and sociodemographic dimensions, contributing to a more comprehensive understanding of the intricate relationship between human hands and the tools they manipulate.

2. Material and Methods

2.1. Tool Selection and Reproduction

This study focused on two prominent Lower Paleolithic stone tool types—20 choppers and 20 handaxes (Figure 1)—under the null hypothesis that hand size has no effect on their ergonomic grasping patterns, as indicated by finger flexion degrees. To ensure uniformity, an experienced tool maker (MTB) meticulously replicated forty stone tools with standardized dimensions and form, following the methodology outlined in Fedato et al. (2020) [13] (Table 1). The tools were made from Paleozoic quartzite, characterized by a fine grain and consistent structure, and sourced from large, irregular pebbles lacking any major fissures or fractures. Specifically, the tool reproductions were designed to facilitate one-handed manipulation, considering the distinctive grip requirements associated with each tool type. As demonstrated experimentally, handaxes typically involve the whole palm and all five digits, whereas flake tools necessitate controlled pad-to-pad pinching, a feature not considered in our analysis, which is exclusively tailored for a holistic hand tactile experience [9,28].

Figure 1. Replicas of Lower Paleolithic stone tools used in the experiment.

Table 1. Median values of stone tool measurements (in mm).

	Handaxes	Choppers
Max Length	172.7	109.2
Max Width	88.9	113.0
Max Thickness	35.2	53.9
Width at 25	88.1	102.1
Width at 50	78.5	107.9
Width at 75	49.9	95.1
Thickness at 25	34.9	43.8
Thickness at 50	31.6	51.6
Thickness at75	25.3	36.5
Length/Width	1.8	0.9

2.2. Participant Recruitment and Characteristics

Enrolled participants comprised 50 females and 28 males, all right-handed adults aged between 23 and 67 years. The recruitment intentionally excluded individuals with previous archaeology experience to maintain this study's focus on the ergonomic relationship between the hand and the tool, specifically emphasizing hand-tool haptic feedback. Elements such as knowledge of tool functions or tasks were deliberately excluded from participant considerations.

2.3. Hand Dimension Measurement

Utilizing a Canon CanoScan 8800F 2D scanner (resolution 4800 dpi), hand images were acquired and analyzed using ImageJ 1.46r [29]. Hand dimensions measured included hand length (HL), palmar length (PL), and palmar width (PW), following established protocols [30]. HL spanned from the distal flexion crease at the wrist to the tip of the third digit, while PL was calculated from the midpoint of the distal transverse crease of the wrist flexures to the most proximal flexion crease of the third finger. PW, representing the distance between the radial aspect of the second metacarpal and the ulnar aspect of the fifth metacarpal, was also measured (Table 2; Figure 2a).

Table 2. Summary statistics of palmar length (PL), palmar width (PW), and hand length (HL) for females (F) and males (M). Values are expressed in cm.

	F	**M**	**F**	**M**	**F**	**M**
	PL	**PL**	**PW**	**PW**	**HL**	**HL**
N	50	28	50	28	50	28
Median	9.9	11.2	8	8.9	17.8	19.7
25 prcntil	9.7	10.8	7.8	8.8	17.2	19
75 prcntil	10.3	11.6	8.2	9.2	18	20.3
Coeff. var	4.6	5.4	4.5	5.1	4.3	4.8

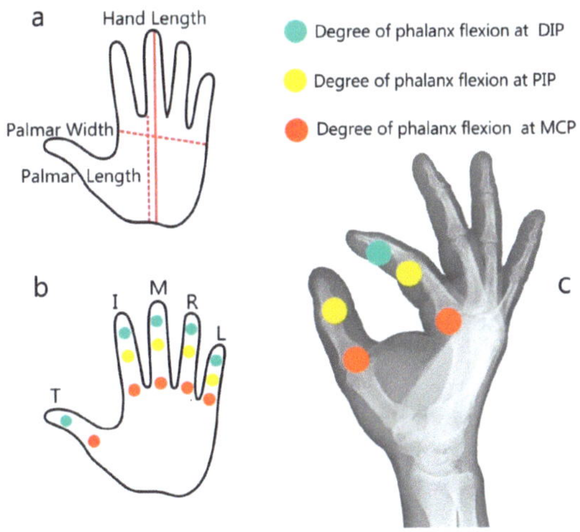

Figure 2. (**a**) The hand dimensions measured in this study are hand length (HL), palmar length (PL) and palmar width (PW). (**b**) Phalanx joints are situated at the metacarpophalangeal joint (MCP—red dots), proximal interphalangeal joint (PIP—yellow dots), and distal interphalangeal joint (DIP—green dots). (**c**) Static hand posture was measured by recording the angular position of the 14 joint angles (**b**) of the fingers and of the thumb, during the comfortable grasping of each tool.

2.4. Data Acquisition

This study focused on finger flexion towards the palm, involving the metacarpophalangeal joint (McP), proximal interphalangeal joint (PiP), and distal interphalangeal joint (DiP) (Figure 2b,c). The static hand posture during comfortable tool grasping was recorded using a VMG 30™ motion capture hand glove (Virtual Motion Labs®). This device, calibrated before each trial, transformed finger motion into real-time sensory data through joint angle measurements [31]. Flexion degrees were measured as the external angle of the phalanx, with labels including McP1–McP5 for the thumb through the little finger and PiP1–PiP5 and DiP2–DiP5 for the proximal and distal interphalangeal joints, respectively (Figure 2).

2.5. Experimental Procedure

In each trial, participants manipulated 20 choppers and 20 handaxes, aiming for a comfortable grip with the right hand only (Figure 3). They were instructed that a comfortable grip should ensure secure handling without slipping, free from tension or unpleasant sensations. Participants were encouraged to choose the most suitable grasp type and inform the experimenter when they reached a final comfortable position with the tool. Flexion values were recorded at this position. All subjects gave their informed consent for inclusion before they participated in this study. This study was conducted in accordance with the Declaration of Helsinki, and the protocol was approved by the Ethics Committee of the University of Burgos.

Figure 3. Experimental procedure. Using a cyber glove, finger flexion during stone tool manipulation was recorded.

2.6. Statistical Analysis

For each participant, median values of phalanx flexion were calculated for the forty Lower Paleolithic stone tools. As these values did not follow a normal distribution (Table 3), non-parametric tests were employed for group comparisons and correlation analyses. A Principal Component Analysis (PCA) was conducted on median phalanx flexions to identify the main patterns associated with comfortable grasping. Correlation analyses explored the relationship between these patterns and hand dimensions. Mann-Whitney tests compared phalanx flexion values between males and females. Statistical analyses were conducted using PAST v. 3.18 [32], SPSS 24, and R 3.5.0.

Table 3. Shapiro-Wilk (W) test for normal distribution of the 14 variables of phalanx flexion. With p values less than 0.05, the data tested are not normally distributed.

	W	p
McP1	0.939	0.001
PiP1	0.954	0.006
McP2	0.981	0.288
PiP2	0.929	<0.001
DiP2	0.928	<0.001
McP3	0.959	0.013
PiP3	0.847	<0.001
DiP3	0.847	<0.001
McP4	0.933	<0.001
PiP4	0.884	<0.001
DiP4	0.871	<0.001
McP5	0.962	0.019
PiP5	0.904	<0.001
DiP5	0.922	<0.001

3. Results

Analyzing the joint angles, the first principal component (PC1) explained 64% of the variance, with PC2 explaining 13%. PC2's eigenvalue, as per the broken stick model, approached the threshold of random variation, urging caution in its interpretation. Subsequent components were deemed insignificant. Figure 2 illustrates the loadings of the variance-covariance matrix for PC1, presenting a violin plot of PC1 scores for males and females. PC1 displayed negative loadings for metacarpophalangeal (McP) joints (indicating increasing flexion) and positive loadings for proximal interphalangeal (PiP) and distal interphalangeal (DiP) joints (indicating decreasing flexion). Loadings for each variable are detailed in Table 4. PC2 was mainly influenced by McP3 flexion, followed by index finger flexion, with other variables showing increasing flexion along PC2, except for PiP5, DiP5, and McP1. Figure 4 also depicts loadings and violin plots for PC2, with loading values in Table 4.

Table 4. Loading values for PC1 and PC2.

	PC 1	PC 2
McP1	0.08	0.04
PiP1	0.27	−0.14
McP2	−0.08	0.39
PiP2	0.36	0.38
DiP2	0.18	0.19
McP3	−0.08	0.7
PiP3	0.41	0.09
DiP3	0.21	0.05
McP4	−0.05	0.34
PiP4	0.37	0.01
DiP4	0.27	0.02
McP5	−0.15	0.09
PiP5	0.48	−0.12
DiP5	0.25	−0.06

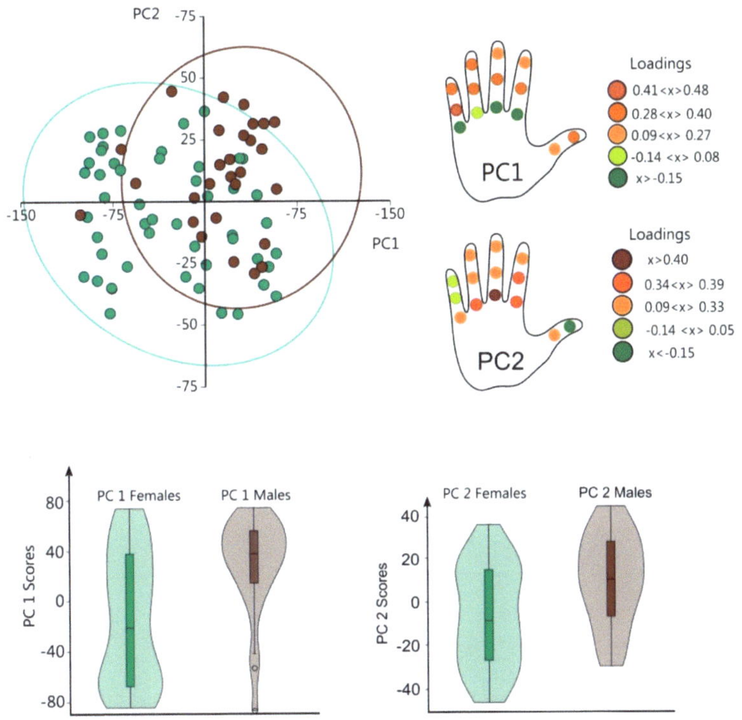

Figure 4. PC1 and PC2 of phalanx flexions: the loading values of the 14 joint angles are shown as colored dots. Colors indicate which range of values the loadings belong to. Positive loadings are increasingly dark red, and negative loadings are increasingly green. PC1 and PC2 scores of males and females are plotted in a violin plot.

Multiple linear regression tested if hand length (HL), palmar length (PL), and palmar width (PW) significantly predicted PC1 and PC2. For PC1, the model was as follows: Finger flexion (PC1) = −334.02 + 12.01(PL) + 14.75(PW) + 4.70(HL). The overall regression was significant (R2 = 0.18, F(3, 74) = 5.33, $p < 0.001$), yet none of the predictors significantly predicted PC1. For PC2, the model was as follows: Finger flexion (PC2) = −164.58 + 0.6(PL) + 9.11(PW) + 4.47(HL). The overall regression was significant (R2 = 0.19, F (3, 74) = 5.97, $p < 0.001$), though no predictors significantly predicted PC2.

Correlations among PC1, PC2, and hand dimensions were significant for the pooled sample but not when analyzed separately for males and females. Correlations were considered significant for $p < 0.017$. Hand dimensions significantly correlated with PC1 (HL: r = 0.40, $p < 0.001$; PL: r = 0.40, $p < 0.001$; PW: r = 0.36, $p = 0.001$) and PC2 (HL: r = 0.41, $p < 0.001$; PL: r = 0.38, $p < 0.001$; PW: r = 0.40, $p < 0.001$) (Figure 5). However, when analyzed separately, correlations lost significance (Table 5). Similar results were obtained when PCA was computed separately for males and females.

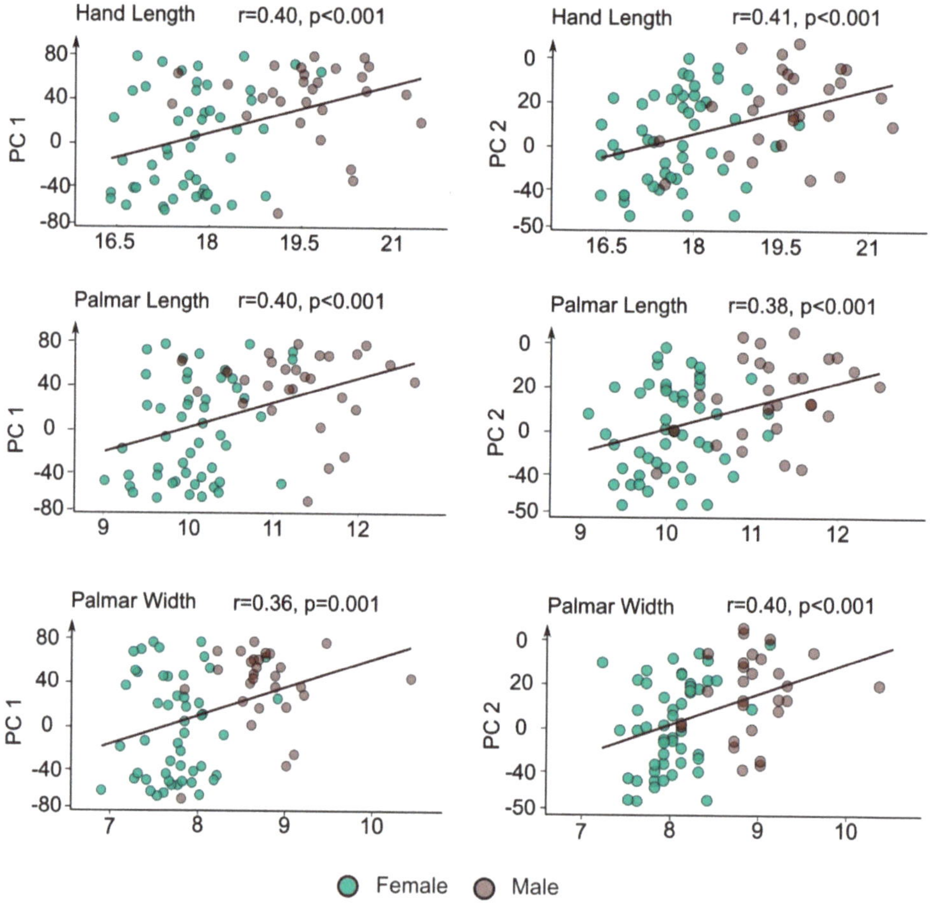

Figure 5. Correlation between hand measures and principal components of phalanx flexion. Hand measures are expressed in centimeters.

Table 5. Correlation between the first two components of phalanx flexion and hand dimensions, for males and females.

	Females		Males	
PC 1	r	*p*	r	*p*
HL	0.27	0.06	−0.04	0.82
PL	0.28	0.05	−0.07	0.72
PW	0.09	0.51	0.11	0.58
PC 2	r	*p*	r	*p*
HL	0.25	0.08	0.15	0.28
PL	0.16	0.26	0.08	0.34
PW	0.32	0.02	0.5	0.13

Correlations between each phalanx flexion variable and hand dimension were considered significant for $p < 0.0036$. However, none of them correlated with hand dimensions (Table 6).

Table 6. Spearman rank correlation test between the 14 variables of phalanx flexion and hand dimensions, for males and females.

	HL		PL		PW	
Females	r	*p*	r	*p*	r	*p*
McP1	0.15	0.29	0.14	0.35	0.05	0.75
PiP1	0.15	0.30	0.26	0.07	−0.02	0.90
McP2	0.17	0.25	0.01	0.96	0.24	0.09
PiP2	0.33	0.02	0.30	0.03	0.18	0.20
DiP2	0.33	0.02	0.30	0.03	0.18	0.20
McP3	0.29	0.04	0.16	0.28	0.31	0.03
PiP3	0.09	0.55	0.17	0.25	0.02	0.88
DiP3	0.09	0.55	0.17	0.25	0.02	0.88
McP4	0.09	0.56	0.01	0.93	0.28	0.05
PiP4	0.15	0.30	0.19	0.18	0.01	0.97
DiP4	0.17	0.24	0.16	0.26	0.02	0.89
McP5	−0.22	0.13	−0.18	0.21	0.03	0.86
PiP5	0.23	0.11	0.19	0.18	0.02	0.88
DiP5	0.24	0.10	0.20	0.17	0.03	0.84
Males	r	*p*	r	*p*	r	*p*
McP1	−0.12	0.55	−0.18	0.35	−0.07	0.72
PiP1	−0.01	0.97	−0.05	0.81	0.19	0.32
McP2	0.12	0.54	0.13	0.52	0.08	0.68
PiP2	0.13	0.52	0.15	0.44	−0.09	0.64
DiP2	0.13	0.50	0.16	0.43	−0.09	0.65
McP3	0.22	0.26	0.26	0.19	0.23	0.23
PiP3	0.08	0.68	0.02	0.91	−0.13	0.50
DiP3	0.08	0.68	0.02	0.91	−0.13	0.50
McP4	0.26	0.18	0.34	0.08	0.17	0.39
PiP4	0.09	0.64	0.01	0.97	−0.08	0.68
DiP4	0.10	0.61	0.01	0.96	−0.07	0.72
McP5	0.03	0.87	0.18	0.37	0.31	0.11
PiP5	−0.04	0.84	−0.10	0.60	−0.12	0.56
DiP5	−0.04	0.84	−0.10	0.60	−0.12	0.56

Considering the distribution along PC1, males displayed higher phalanx flexion values than females (Mann-Whitney Test: U = 373, $p < 0.001$), although female variance was larger (Levene's test for equality of variances: $p = 0.002$). This trend persisted for PC2, with males showing larger phalanx flexion values than females (Mann-Whitney Test: U = 428,

$p < 0.004$). Males and females exhibited several significant differences in finger flexion, with males generally displaying higher flexion (negative Z values), particularly in McP1, Pip5, Dip5, McP5, Pip2, and Dip2, while PiP1, Mcp4, and McP2 showed no significant differences (Table 7, Figure 6).

Table 7. Mann-Whitney U test between males and females. The Holm-Bonferroni sequential method for multiple hypothesis tests is used to correct Type I errors. Statistically significant differences between males and females are displayed in bold.

	U	Z	Sig.
McP1	265	−4.5	**<0.001**
PiP1	527	−1.8	0.072
McP2	553	−1.5	0.126
PiP2	372	−3.4	**0.001**
DiP2	372	−3.4	**0.001**
McP3	438	−2.7	**0.006**
PiP3	418	−2.9	**0.003**
DiP3	418	−2.9	**0.003**
McP4	545	−1.6	0.106
PiP4	416	−3	**0.003**
DiP4	441	−2.7	**0.007**
McP5	357	−3.6	**<0.001**
PiP5	331	−3.8	**<0.001**
DiP5	333	−3.8	**<0.001**

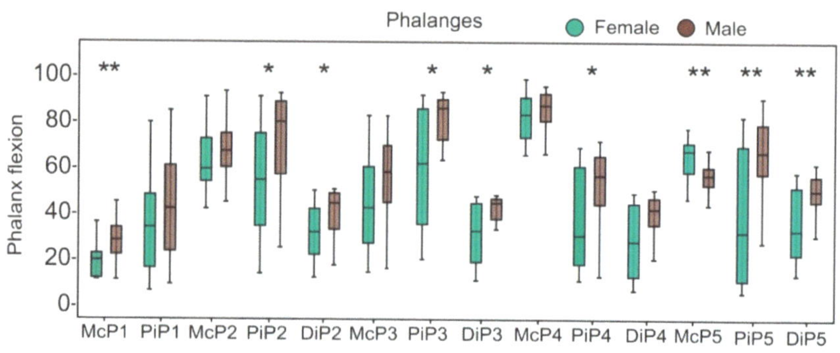

Figure 6. Distribution of the values of degree of phalanx flexion in males and females (median, interquartile, and range). Higher values mean more flexion. Significant differences between males and females are marked with an asterisk.

4. Discussion

In this comprehensive study, we systematically examine comfortable grasping patterns associated with Lower Paleolithic stone tools, particularly choppers and handaxes. Our analysis focuses on quantifying phalanx flexion and considering the potential influence of hand dimensions, including hand length, palmar length, and palmar width, previously correlated with finger length and demonstrating sexual dimorphism [33–35].

The primary grasping pattern, represented by the first principal component, shows a strong association with distal phalanges' flexion, particularly evident in the little and ring fingers. An intriguing biomechanical insight arises from the negative correlation between proximal and distal joint flexion, emphasizing the need to consider potential biomechanical or spatial constraints between finger joints when interpreting tool grasping patterns and ergonomics.

The secondary grasping pattern, represented by the second principal component, is linked to the flexion of the distal phalanges, especially in the index and middle fingers.

Both grasping patterns modestly correlate with hand dimensions (r~0.39), affirming prior experiments suggesting the influence of specific biometric traits on Lower Paleolithic stone-tool-use efficiency. Grip force and contact area, influenced by handle diameter and hand size, emerge as crucial parameters impacting tool manipulation.

An intriguing revelation emerges when examining the distribution of grasping patterns between males and females. On average, males exhibit more flexion, but the correlation between grasping patterns and hand size vanishes when the sexes are analyzed separately. This suggests that differences in hand size alone do not account for the observed sexual disparities in grasping behavior. Despite a potentially spurious correlation in the pooled analysis, dissecting the groups reveals an absence or extreme weakness in their reciprocal influence, pointing to the involvement of other, yet-to-be-explored factors in sexual differences.

The observed variability in the little and ring fingers, as revealed by the main grasping pattern identified through the first principal component, may be indicative of their distinct functional roles compared to the thumb and index finger. The biomechanical considerations suggest that the thumb and index finger, associated with precision grips [3], are more frequently utilized in fine motor tasks requiring delicate manipulation. In contrast, the little and ring fingers, displaying greater variability in our analysis, may be more involved in tasks requiring power or stability grips, where precision is less critical. This aligns with established biomechanical principles, emphasizing the dominant role of the thumb and index finger in precision grasping activities. While these observations are consistent with known functional anatomy, specific references from reputable anatomy or the biomechanics literature can be consulted for further confirmation and to ensure up-to-date accuracy in the scientific context.

Acknowledging the subjective nature of comfortable grasping, influenced by both physical constraints and individual choices, underscores the complexity of grasping behavior involving sensory, motor, and cognitive systems. The intricate interplay among cognitive abilities, visuospatial strategies, and evolutionary aspects may contribute to sex differences in grasping behavior, suggesting the need for future studies to delve into individual differences in visuospatial skills during stone tool manipulation.

It is crucial to recognize the limitations of the analysis, which focuses specifically on finger flexion during ergonomic and comfortable grasping of tools manipulated with the whole hand. By instructing participants to grasp objects in a way that feels comfortable, we attempt to capture the inherent variability in tool manipulation, considering not only the physical constraints imposed by object shape but also individual choices guided by haptic and ergonomic considerations. This approach aligns with the multifaceted nature of tool use in prehistoric contexts, where individuals may have adapted their grasping techniques based on the unique characteristics of the tools and their intended tasks. It is known that choppers, while often classified as tools, served multiple purposes beyond just tool use, such as being used as cores for producing smaller flakes or as percussive instruments for various activities [36]. Additionally, other enigmatic spherical stone objects are present in archaeological Oldowan assemblages and are among the oldest manufactured stone items [37]. Even if their functions and manufacture modes are still poorly known, the remarkable geographical and chronological spread of stone balls warrants attention and calls for further investigation into their potential structural and functional roles in the context of human technological advancement [38].

The exclusion of other haptic and cognitive components involved in grasping behavior and the restriction to larger tools warrant future investigations into the ergonomic patterns associated with manipulating smaller objects and tools with different functional demands.

Ongoing analyses, comparing manipulation patterns between expert and non-expert stone tool users, aim to uncover differences in manipulative strategies, while future surveys should consider grasping strategies related to specific functional tasks.

In exploring comfortable grasping patterns for Lower Paleolithic stone tools—specifically choppers and handaxes—our study offers a unique lens into the behaviors of our prehistoric ancestors. By quantifying the degree of phalanx flexion and considering the potential influence of hand dimensions, we delve into the intimate relationship between human hand morphology and tool manipulation efficiency. This endeavor is not merely an analytical exercise but rather an attempt to unravel the ergonomic intricacies that prehistoric humans may have grappled with during their daily interactions with stone tools.

The insights gained from our study carry implications for understanding the daily lives of prehistoric communities. The consideration of grip force, contact area, and the intricate interplay between hand dimensions and grasping behavior speaks to the challenges and adaptations that our ancestors navigated while utilizing stone tools. Future research could expand to include juveniles, offering a more comprehensive understanding of skill acquisition and the transmission of tool-use knowledge. Additionally, further investigation into the manipulation patterns of expert versus non-expert stone tool users would provide valuable insights into the roles of teaching, learning, and mastery within these ancient societies.

Similarly, studies on stone tool use should take into consideration the morphological differences of the hand among early humans, such as *Homo habilis*, *Homo ergaster*, and modern humans (*Homo sapiens*), particularly noting the more robust and curved finger bones in early humans which suggest powerful grips and climbing adaptations, in contrast to the more gracile and dexterous hands of modern humans that are optimized for fine motor skills and complex tool manipulation [28,39]. However, even though some differences can be appreciated, the comparison between early and modern humans is challenging due to the scarcity and incompleteness of fossil records, which are often not associated with any particular individual, making it difficult to directly quantify intrinsic hand proportions or provide an overall functional interpretation of the hand in these *Homo* species.

To enrich our understanding of these ancestral behaviors, future studies could explore individual differences in visuospatial skills during stone tool manipulation, further connecting cognitive abilities and visuospatial strategies to the nuanced ways in which prehistoric humans interacted with their material culture. By expanding our focus beyond finger flexion to encompass other haptic and cognitive components and by considering tools of varying sizes and functional demands, we aim to contribute to a holistic portrayal of the intricate relationship between human hands and the tools that shaped our evolutionary trajectory.

5. Conclusions

During comfortable manipulation of Lower Paleolithic stone tools, males and females show different patterns of phalanx flexion. Hand dimensions are related to the patterns of phalanx flexion only when the sample is pooled. However, when males and females are considered separately, the correlation no longer stands. Although the sample size is not particularly large, a relevant association between hand size and grasping behavior is very small, negligible, or even absent, at least with regards to our sample.

Considering that the grasping is related with many aspects beside ergonomics, we may suggest that the differences that we found in our study could be due to certain individuals' cognitive, behavioral, or motor aspects.

For example, experience in large-object manipulation, grip strength, sociodemographic and occupational factors could be responsible for the differences [40–44]. If this is the case, information about individuals' lifestyles or psychometric profiles could help us to further understand these results.

To fully understand the Paleolithic techno-complex, it is crucial to consider the broader socio-economic and cultural contexts that influenced lithic production and utilization. Socio-economic factors such as resource distribution, trade networks, and division of labor played significant roles in shaping the production and use of stone tools. Additionally, gender roles in lithic production are essential for understanding past societies, potentially reflecting specific responsibilities and contributions of community members. However, distinguishing gender-specific domains in the archaeological record remains challenging. We must avoid applying preconceived concepts about household structures to archaeological data and critically consider biases in our analogical materials. Addressing these challenges helps us understand the dynamic and fluid nature of gender roles in past societies, avoiding oversimplified interpretations [45].

The continuous use of specific materials for toolmaking over millennia reflects deep familiarity and complex relationships with the environment [46]. Thus, tools crafted by early humans were not merely functional objects but embodied cultural conventions and ontological beliefs [47]. Recognizing these aspects allows us to appreciate the full spectrum of past human behaviors and interactions with their environment. Integrating these perspectives provides a more comprehensive and nuanced interpretation of the prehistoric techno-complex, acknowledging that functional, economic, and spiritual-symbolic meanings are all integral parts of human life.

The results presented here should be interpreted in two different contexts. Firstly, they add to the general study of haptic exploration behavior, bridging perception and cognition associated with hand-tool interaction [17,48–51]. Secondly, taking into consideration that these large-sized tools represent the earliest record of a consistent human technology, we should consider whether or not these behaviors can also provide information regarding the evolution of the hand-tool interaction system. We humans are obligatory tool users [52] and have a tool-dependent culture [53] and technology-based cognition [54]; we should hence consider that our haptic capacities are deeply rooted in our phylogenetic history.

Author Contributions: Conceptualization, A.F., M.S.-G. and E.B.; methodology, A.F., M.S.-G., E.B., M.T.-B. and R.A.-A.; software, A.F. and M.S.-G.; validation, A.F. and M.S.-G.; formal analysis, A.F.; investigation, A.F.; resources, M.T.-B. and R.A.-A.; data curation, A.F. and M.S.-G.; writing—original draft preparation, A.F. and M.S.-G.; writing—review and editing, A.F.; visualization, A.F.; supervision, E.B.; project administration, E.B. funding acquisition, A.F., E.B. and M.S.-G. All authors have read and agreed to the published version of the manuscript.

Funding: This study is funded by the Spanish Government (Atapuerca Project; PGC2018-093925-B-C31/32), by the Junta de Castilla y León (EDU/574/2018) and by the Italian Institute of Anthropology.

Data Availability Statement: The data that support the findings of this study are available from the corresponding author upon reasonable request. The data are not publicly available due to privacy or ethical restrictions.

Acknowledgments: We are grateful to all the volunteers who participated in this survey. We are extremely grateful to three anonymous reviewers, who supplied relevant observations to improve this article.

Conflicts of Interest: One of the authors of this manuscript, Annapaola Fedato, is currently employed by MDPI, but works for a different journal within the company. This manuscript was developed and prepared prior to Annapaola Fedato's employment with MDPI. The research, writing, and submission processes were conducted independently and without any influence from MDPI or its affiliates.

References

1. Marzke, M.W. Tool making, hand morphology and fossil hominins. *Philos. Trans. R. Soc. B Biol. Sci.* **2013**, *368*, 20120414. [CrossRef] [PubMed]
2. Trinkaus, E.; Villemeur, I. Mechanical advantages of the Neandertal thumb in flexion: A test of an hypothesis. *Am. J. Phys. Anthropol.* **1991**, *84*, 249–260. [CrossRef] [PubMed]
3. Feix, T.; Romero, J.; Schmiedmayer, H.B.; Dollar, A.M.; Kragic, D. The grasp taxonomy of human grasp types. *IEEE Trans. Hum.-Mach. Syst.* **2015**, *46*, 66–77. [CrossRef]
4. Key, A.J.; Dunmore, C.J. The evolution of the hominin thumb and the influence exerted by the non-dominant hand during stone tool production. *J. Hum. Evol.* **2015**, *78*, 60–69. [CrossRef] [PubMed]
5. Key, A.J.; Dunmore, C.J.; Marzke, M.W. The unexpected importance of the fifth digit during stone tool production. *Sci. Rep.* **2019**, *9*, 16724. [CrossRef] [PubMed]
6. Karakostis, F.A.; Haeufle, D.; Anastopoulou, I.; Moraitis, K.; Hotz, G.; Tourloukis, V.; Harvati, K. Biomechanics of the human thumb and the evolution of dexterity. *Curr. Biol.* **2021**, *31*, 1317–1325.e8. [CrossRef] [PubMed]
7. Napier, J. The prehensile movements of the human hand. *J. Bone Joint Surg. Br.* **1956**, *38*, 902–913. [CrossRef] [PubMed]

8. Marzke, M.W.; Shackley, M.S. Hominid hand use in the Pliocene and Pleistocene: Evidence from experimental archaeology and comparative morphology. *J. Hum. Evol.* **1986**, *15*, 439–460. [CrossRef]
9. Key, A.J.; Lycett, S.J. Investigating interrelationships between Lower Palaeolithic stone tool effectiveness and tool user biometric variation: Implications for technological and evolutionary changes. *Archaeol. Anthropol. Sci.* **2018**, *10*, 989–1006. [CrossRef]
10. Rots, V. Prehension and Hafting Traces on Flint Tools: A Methodology. *J. Archaeol. Sci.* **2010**, *37*, 1946–1959.
11. Patiño, F.Y.; Luque, M.; Terradillos-Bernal, M.; Martín-Loeches, M. Biomechanics of microlithics manufacture: A preliminary approach to Neanderthal's motor constrains in the frame of embodied cognition. *J. Anthropol Sci.* **2017**, *95*, 1–16.
12. Williams-Hatala, E.M.; Hatala, K.G.; Gordon, M.; Key, A.; Kasper, M.; Kivell, T.L. The manual pressures of stone tool behaviors and their implications for the evolution of the human hand. *J. Hum. Evol.* **2018**, *119*, 14–26. [CrossRef] [PubMed]
13. Fedato, A.; Silva-Gago, M.; Terradillos-Bernal, M.; Alonso-Alcalde, R.; Bruner, E. Hand grasping and finger flexion during Lower Paleolithic stone tool ergonomic exploration. *Archaeol. Anthropol. Sci.* **2020**, *12*, 254. [CrossRef]
14. Sartori, L.; Straulino, E.; Castiello, U. How objects are grasped: The interplay between affordances and end-goals. *PLoS ONE* **2011**, *6*, e25203. [CrossRef]
15. Fedato, A.; Silva-Gago, M.; Terradillos-Bernal, M.; Alonso-Alcalde, R.; Martín-Guerra, E.; Bruner, E. Electrodermal activity during Lower Paleolithic stone tool handling. *Am. J. Hum. Biol.* **2019**, *31*, e23279. [CrossRef] [PubMed]
16. Silva-Gago, M.; Fedato, A.; Hodgson, T.; Terradillos-Bernal, M.; Alonso-Alcalde, R.; Bruner, E. Visual attention reveals affordances during Lower Palaeolithic stone tool exploration. *Archaeol. Anthropol. Sci.* **2021**, *13*, 9. [CrossRef]
17. Bruner, E.; Iriki, A. Extending mind, visuospatial integration, and the evolution of the parietal lobes in the human genus. *Quat. Int.* **2016**, *405*, 98–110. [CrossRef]
18. Bruner, E.; Spinapolice, E.; Burke, A.; Overmann, K.A. Visuospatial integration: Paleoanthropological and archaeological perspectives. In *Evolution of Primate Social Cognition*; Springer: Cham, Switzerland, 2018; pp. 299–326.
19. Seo, N.J.; Armstrong, T.J. Investigation of grip force, normal force, contact area, hand size, and handle size for cylindrical handles. *Hum. Factors* **2008**, *50*, 734–744. [CrossRef]
20. Nevill, A.M.; Holder, R.L. Modelling handgrip strength in the presence of confounding variables: Results from the Allied Dunbar National Fitness Survey. *Ergonomics* **2000**, *43*, 1547–1558. [CrossRef]
21. Peolsson, A.; Hedlund, R.; Öberg, B. Intra-and inter-tester reliability and reference values for hand strength. *J. Rehabil. Med.* **2001**, *33*, 36–41. [CrossRef]
22. Ruiz-Ruiz, J.; Mesa, J.L.; Gutiérrez, A.; Castillo, M.J. Hand size influences optimal grip span in women but not in men. *J. Hand Surg.* **2002**, *27*, 897–901. [CrossRef] [PubMed]
23. Imrhan, S.N. Two-handed static grip strengths in males: The influence of grip width. *Int. J. Ind. Ergon.* **2003**, *31*, 303–311. [CrossRef]
24. Peebles, L.; Norris, B. Filling 'gaps' in strength data for design. *Appl. Ergon.* **2003**, *34*, 73–88. [CrossRef] [PubMed]
25. Nicolay, C.W.; Walker, A.L. Grip strength and endurance: Influences of anthropometric variation, hand dominance, and gender. *Int. J. Ind. Ergon.* **2005**, *35*, 605–618. [CrossRef]
26. Hart, P.D. Multivariate analysis of the sociodemographic predictors of grip strength in US Adults. *Age* **2018**, *1*, 001.
27. Key, A.J.; Lycett, S.J. Technology based evolution? A biometric test of the effects of handsize versus tool form on efficiency in an experimental cutting task. *J. Archaeol. Sci.* **2011**, *38*, 1663–1670. [CrossRef]
28. Marzke, M.W. Precision grips, hand morphology, and tools. *Am. J. Phys. Anthropol. Off. Publ. Am. Assoc. Phys. Anthropol.* **1997**, *102*, 91–110. [CrossRef]
29. Schneider, C.A.; Rasband, W.S.; Eliceiri, K.W. NIH Image to ImageJ: 25 years of image analysis. *Nat. Methods* **2012**, *9*, 671–675. [CrossRef]
30. Kanchan, T.; Krishan, K. Anthropometry of hand in sex determination of dismembered remains-A review of literature. *J. Forensic Leg. Med.* **2011**, *18*, 14–17. [CrossRef]
31. Cobos, S.; Ferre, M.; Ángel Sánchez-Urán, M.; Ortego, J.; Aracil, R. Human hand descriptions and gesture recognition for object manipulation. *Comput. Methods Biomech. Biomed. Eng.* **2010**, *13*, 305–317. [CrossRef]
32. Hammer, Ø.; Harper, D.A.; Ryan, P.D. PAST: Paleontological statistics software package for education and data analysis. *Palaeontol. Electron.* **2001**, *4*, 9.
33. Fedato, A.; Silva-Gago, M.; Terradillos-Bernal, M.; Alonso-Alcalde, R.; Martín-Guerra, E.; Bruner, E. Hand morphometrics, electrodermal activity, and stone tools haptic perception. *Am. J. Hum. Biol.* **2019**, *32*, e23370. [CrossRef] [PubMed]
34. Barut, C.; Dogan, A.; Buyukuysal, M.C. Anthropometric aspects of hand morphology in relation to sex and to body mass in a Turkish population sample. *HOMO* **2014**, *65*, 338–348. [CrossRef] [PubMed]
35. Kanchan, T.; Rastogi, P. Sex determination from hand dimensions of North and South Indians. *J. Forensic Sci.* **2009**, *54*, 546–550. [CrossRef] [PubMed]
36. Isaac, G.L. *Olorgesailie: Archaeological Studies of a Middle Pleistocene Lake Basin in Kenya*; University of Chicago Press: Chicago, IL, USA, 1977.
37. Cabanès, J.; Borel, A.; Preysler, J.B.; Lourdeau, A.; Moncel, M.-H. Palaeolithic polyhedrons, spheroids and bolas over time and space. *PLoS ONE* **2022**, *17*, e0272135. [CrossRef] [PubMed]
38. Assaf, E.; Baena Preysler, J.; Bruner, E. Lower Paleolithic Shaped Stone Balls—What Is Next? Some Cultural–Cognitive Questions. *Quaternary* **2023**, *6*, 51. [CrossRef]

39. Napier, J. Fossil hand bones from Olduvai Gorge. *Nature* **1962**, *196*, 409–411. [CrossRef]
40. Edgren, C.S.; Radwin, R.G.; Irwin, C.B. Grip force vectors for varying handle diameters and hand sizes. *Hum. Factors* **2004**, *46*, 244–251. [CrossRef]
41. Hesse, C.; Deubel, H. Efficient grasping requires attentional resources. *Vis. Res.* **2011**, *51*, 1223–1231. [CrossRef]
42. Murali, A.; Liu, W.; Marino, K.; Chernova, S.; Gupta, A. Same object, different grasps: Data and semantic knowledge for task-oriented grasping. In Proceedings of the 4th Conference on Robot Learning (CoRL 2020), Virtual, 16–18 November 2020.
43. Petersen, P.; Petrick, M.; Connor, H.; Conklin, D. Grip strength and hand dominance: Challenging the 10% rule. *Am. J. Occup. Ther.* **1989**, *43*, 444–447. [CrossRef]
44. Seegelke, C.; Hughes, C.M.; Knoblauch, A.; Schack, T. Grasp posture planning during multi-segment object manipulation tasks—Interaction between cognitive and biomechanical factors. *Acta Psychol.* **2013**, *144*, 513–521. [CrossRef] [PubMed]
45. Scanlan, K. Seeing Women in Stone: A Spatial Analysis of Lithic Technology and Use-Wear to Identify a Norton Tradition Ena on the Kvichak River, Bristol Bay, Alaska. *J. Northwest Anthropol.* **2021**, *55*.
46. Assaf, E.; Romagnoli, F. Beyond tools and function: The selection of materials and the ontology of hunter-gatherers. Ethnographic evidences and implications for Palaeolithic archaeology. *Camb. Archaeol. J.* **2021**, *31*, 281–291. [CrossRef]
47. Malafouris, L. Material engagement and the embodied mind. In *Cognitive Models in Palaeolithic Archaeology*; Oxford University Press: Oxford, UK, 2016; pp. 69–82.
48. Tunik, E.; Rice, N.J.; Hamilton, A.; Grafton, S.T. Beyond grasping: Representation of action in human anterior intraparietal sulcus. *Neuroimage* **2007**, *36*, T77–T86. [CrossRef] [PubMed]
49. Turvey, M.T.; Carello, C. Obtaining information by dynamic (effortful) touching. *Philos. Trans. R. Soc. B Biol. Sci.* **2011**, *366*, 3123–3132. [CrossRef] [PubMed]
50. Ackerley, R.; Kavounoudias, A. The role of tactile afference in shaping motor behaviour and implications for prosthetic innovation. *Neuropsychologia* **2015**, *79*, 192–205. [CrossRef] [PubMed]
51. Bruner, E.; Fedato, A.; Silva-Gago, M.; Alonso-Alcalde, R.; Terradillos-Bernal, M.; Fernández-Durantes, M.Á.; Martín-Guerra, E. Cognitive archeology, body cognition, and hand-tool interaction. *Prog. Brain Res.* **2018**, *238*, 325–345. [PubMed]
52. Shea, J.J. Occasional, obligatory, and habitual stone tool use in hominin evolution. *Evol. Anthropol. Issues News Rev.* **2017**, *26*, 200–217. [CrossRef] [PubMed]
53. Plummer, T. Flaked stones and old bones: Biological and cultural evolution at the dawn of technology. *Am. J. Phys. Anthropol.* **2004**, *125*, 118–164. [CrossRef]
54. Malafouris, L. The brain–artefact interface (BAI): A challenge for archaeology and cultural neuroscience. *Soc. Cogn. Affect. Neurosci.* **2010**, *5*, 264–273. [CrossRef]

quaternary

Article

The Role of Large Mammals as Vitamin C Sources for MIS 3 Hominins

José Luis Guil-Guerrero

Food Technology Division, CeiA3, University of Almería, 04120 Almería, Spain; jlguil@ual.es

Abstract: The acquisition of large prey by hominins living during the Marine Isotope Stage 3, including Neanderthals and Anatomically Modern Humans, had nutritional and bioenergetic implications: these contain high fat amounts, provide a high energy return, and the strategies and skills required to acquire small prey were different from those required to acquire the former. Vitamin C availability at several MIS 3 periods could have had a strong seasonal variability and would have been decisive for hominin groups' survival. During the cold periods of the MIS 3, Paleolithic hominins had variable available amounts of vitamin C-containing plants only in the short summers, and for the remainder of the year, viscera would have been their best source of vitamin C. Meanwhile, the dependence on small mammals could have caused an erratic distribution of viscera to be consumed by such hominins, thus leading to chronic scurvy, and compromising their survival. Then, the hunting of large mammals would have helped to meet the daily vitamin C needs, besides an efficient energy supply. Therefore, the decline of large prey during the MIS 3 could have been critical for hominins survival, and thus the efficient exploitation of alternative vitamin C-rich food resources such as birds and aquatic animals could have favored the evolutionary success of hominin populations.

Keywords: Neanderthals; anatomically modern humans; vitamin C; large fauna; Paleolithic diet; marine isotope stage 3; viscera; scurvy

check for updates

Citation: Guil-Guerrero, J.L. The Role of Large Mammals as Vitamin C Sources for MIS 3 Hominins. *Quaternary* **2023**, *6*, 20. https://doi.org/10.3390/quat6010020

Academic Editor: Ran Barkai

Received: 7 February 2023
Revised: 1 March 2023
Accepted: 7 March 2023
Published: 10 March 2023

1. Introduction

Hominins living during the Marine Isotope Stage 3 (M3H, ~60–20 ka) were Anatomically Modern Humans (AMH), Denisovans, and Neanderthals. The latter inhabited Western Eurasia from approximately ~430 ka [1], and their demise took place after Anatomically Modern Humans' (AMH) dispersal, at ~42 ka [2,3]. Different hypotheses have been formulated for this event: abrupt climate and vegetation change [4,5], competitive exclusion [6], pathogens [7], assimilation [8,9], demographic weakness [10], and volcanic eruptions [11]. The proposed mechanisms are not mutually exclusive, and their relative roles may have also varied regionally and temporally [12]. Mitochondrial DNA nucleotide sequence differences among Neanderthals suggest three different geographical populations occurring in Western Europe, Southern Europe, and West Asia [13], while the earliest AMH in Europe are thought to have appeared around ~46 ka, and thus were directly contemporary with the latest European Neanderthals [14].

The demise of Neanderthals could have been linked to the paleoecology of other mammals. The biogeographical analysis of a database of European mammalian fossils for MIS 3 revealed the ecological conditions of this period, in which Neanderthals appear to belong to the extinct southern grouping of Late Pleistocene Megafaunal elements [15]. The geographical distribution of large mammals (>500 kg, corresponding to Bunn's criteria [16], levels 4–) at MIS 3 confirms a retreat toward the south and west of Neanderthals [15], although it has been proposed that some millennia before their demise, about 59–49 ka, Neanderthals entered southern Siberia [17].

1.1. Controversy in the Diet of M3H

Northern European M3H had to face a medium with a great shortage of edible plants, with seasonal intensity, so they relied mainly on foods of animal origin for most of the year. For instance, Neanderthals seem to have had a predilection for large- and mega-faunal elements, and the woolly mammoth (*Mammuthus primigenius* Blumenbach, 1799) was the more prominent animal belonging to the Megafaunal complex. Its extinction is largely attributed to the worldwide wave of extinction of megafauna, which took place prior to the Holocene [18].

Both lithic and faunal assemblages and stable isotope analysis provide a measure of human diets in the past and support the idea that M3H subsistence was based mainly on the consumption of large herbivores [19–22]. Stable isotope analysis of human skeletons has been used successfully for years to determine the type of food animals consumed by M3H. For instance, the use of Ca isotopes allowed the dietary reconstruction of an OIS 5 Neandertal fossil from Southwest France, pointing out a predominantly carnivorous diet, composed of a mixture of various herbivorous prey [23]. The diet of Neanderthals was also assessed by the strontium–calcium (Sr/Ca) and barium–calcium (Ba/Ca) ratios of bones, which predicted that such a diet was composed of about 97% of meat with a weak contribution of vegetables or fish [24]. Recently, the evaluation of C and N stable isotope ratios in bone collagen allowed the assessment of the Paleolithic diet of Siberia and Eastern European hominins, and it was found that the Neanderthal diet in Siberia was based on the consumption of terrestrial animal protein, while for a Neanderthal/Denisovan hybrid the contribution of aquatic food like freshwater fish was suggested [25]. Overall, Paleolithic AMH in Siberia and Eastern Europe procured mainly terrestrial herbivores, such as reindeer, horses, and bison. Moreover, there is evidence that the oldest AMH individuals supplemented their diet with a certain amount of aquatic food [25]. In this context, isotopic analyses suggest that mammoths and plants were important contributors to the diets of the oldest AMH from far Southeast Europe [26], and that both Neanderthals and AMH relied on the same terrestrial herbivores, whereas mobility strategies indicate considerable differences between Neandertal groups, as well as in comparison to AMH, who exploited their environment to a greater extent than Neanderthals [27].

The use of plants as foods by Neanderthals, whether it was seasonal or occasional, has been documented [28,29]. Such use was inferred from the analyses of dental calculus, which has led to the argument that Neanderthals ate as many plants and used as many species of plants as modern humans do [30]. However, it is likely that a combination of mobility, energetics, and fatty acids requirements prevented high plant use by Neanderthals [31]. It must be taken into account that the consumption of herbivores' stomach contents, which contain high amounts of carbohydrates, should be considered as a possible source of plant foods, including "medicinal" ones, in the archeological and fossil records [32]. Furthermore, the possible cooking of plant foods by Neanderthals is a hot topic, which currently promotes much debate among researchers.

Another means of reconstructing the diet and behavior of Neanderthals is dental macrowear analysis. Based on comparisons with modern hunter–gatherer populations, Neanderthals shows high dietary variability in Mediterranean evergreen habitats but a more restricted diet in upper latitude steppe/coniferous forest environments, suggesting a significant consumption of high-protein meat resources in this environment [33,34].

Overall, all the evidence points to the fact that Neanderthals were adapted to an essentially carnivorous diet, which was more marked for Northern European populations. This fact is supported by some morphological features of Neanderthals, such as their large lower thorax, which may represent an adaptation to a high-protein diet [35]. In addition to terrestrial mammals, various studies prove the use of birds and fish by Neanderthals, especially in the northwestern Mediterranean [36–40], but in all cases the most appreciated part of the prey by hunter–gatherers, given their high-protein diet with little access to carbohydrates, would always have been the fat [41,42].

Coexistence and hybridization between AMH and Neanderthals are known to have taken place in the Near and Middle East [43]. Therefore, it is more than likely that in some areas of the Neanderthal range, the coexistence of AMH and Neanderthals led to a reduction in their resource pool, and the technologies and cognitive abilities of the former allowed them to access a wider range of resources [44].

All these facts suggest a complex food scenario at the time of the replacement of Neanderthals by AMH. Another factor of uncertainty regarding food use in this period derives from the fact that assemblages checked from the most careful excavations are likely to have been affected by the mixing of materials during their very long curatorial histories [45]; therefore, some fossil finds may be biased and lead to inferences regarding M3H's use of food.

1.2. Prevalence of Scurvy in M3H Populations

Ascorbic acid, or vitamin C (Vit C), has antioxidant functions as a free radical scavenger and allows the synthesis of collagen [46,47]. Today, there is ample consensus that scurvy can be prevented with a Vit C intake of 10 mg/day [48,49]. Scurvy shows a great variety of symptoms, such as arthralgias, bruising, or joint swelling, and common signs include pedal edema, bruising, mucosal changes, general fatigue, and myalgias [50]. Scurvy for M3H could have been more frequent than supposed. The most important diagnostic features are related to changes in soft tissues. Since these features are often difficult to diagnose in bone, obtaining archeological evidence for this condition is likely to be problematic [51]. Scurvy occurred in earlier human populations, and the subperiosteal lesions associated with scurvy assist in the differential diagnosis [52]. Scurvy's cranial symptoms consist of porous and hypertrophic lesions of the vault, affecting frontal and parietal bosses, and related signs have been discovered in Paleolithic hominins' bones. However, most cases of scurvy could have gone unnoticed, given that it is very difficult to attribute some doubtful pathological cases to scurvy since lesions of porotic hyperostosis are also linked to some type of anemia, infections, and/or poor nutrition [53]. In any case, pathologies related to scurvy have been unequivocally identified in some M3H skeletal remains, and some cases can be attributed to omnivorous populations, whether climatic features are considered. So, scurvy probably occurred in long winters, during which Vit C supply would have depended exclusively on animal food sources. Thereby, pathologies attributable to Vit C deficiency in several Neanderthal skeletons from Kiik-Koba (Crimea) have been described [54]. The diagnosis of an adult Neanderthal man in La Ferrassie indicated that he may have suffered scurvy in addition to lung disease [55]. Gardner and Smith [56] inventoried all the pathologies found in Neanderthal bones at the Krapina site, and some cases of porotic hyperostosis were found, which were attributed to scurvy. Other findings in human fossil assemblages from the early Upper Paleolithic (UP) in Mladec (Central Europe) were hemorrhagic processes due to scurvy [57]. In another report on three Neanderthal bones found in the French cave of Combe-Grenal, bony lesions were interpreted as the result of a reaction to chronic hemorrhages, as a possible result of scurvy [58].

1.3. Vit C Status of Arctic Populations: Implications for MIS 3 Hominins

Human groups that are supposed to have a diet more or less similar to that of Northern European M3H, such as Inuit people, may offer clues as to their Vit C-dependent nutritional status. This human group constitutes the current exception to the reliance on plant foods. For most of the year, Inuit people depend on mammals and fish for subsistence, and their survival is possible due to some amounts of Vit C found in mammal organs, while the gathering of plants is possible only in the short Arctic summers, although some Inuit groups never practice this [59].

Specifically, in the Arctic environment, it was found that Vit C requirements are similar to those in temperate climates, and both Inuit and Caucasoids can subsist on ascorbic acid intakes of less than 15 mg daily without showing clinical evidence of Vit C deficiency [60]. Some studies have suggested that the Inuit are able to obtain a minimum level of Vit C from a diet of frozen/raw, fermented, and dried animal food [61]. Moreover, the daily needs of Vit C also depend on genetic factors: Inuit are genetically characterized by a high Haptoglobin 1 allele frequency, which provides them with a greater antioxidant capacity, and therefore they are less likely to suffer from scurvy [62].

Adaptation to a low intake of Vit C offers a clear evolutionary advantage. The metabolism of Vit C, including absorption and its uptake by several cell types, is inhibited by increasing glucose concentration due to a glucose–ascorbate antagonism: due to their molecular similarity, glucose hinders the entry of Vit C into cells [63]. Thus, Arctic populations that are largely dependent on foods of animal origin, whose diets are low in carbohydrates, would need significantly lower amounts of Vit C than are recommended for Western societies. Then, an evolutionarily adapted human diet based on meat, fat and offal would provide enough Vit C to cover physiological needs and to ward off diseases associated with Vit C deficiency [63]. However, according to Høygaard and Rasmusson [64], who performed studies on the nutrition and physiopathology of Eskimos, analyses of Vit C in their blood reflected hypovitaminosis C and scurvy levels for 47% and 18% of their population, respectively. Other researchers indicated that it is highly probable that Inuit with a traditional nutrition live on the edge of scurvy [65]. It has been stated that a reliance on some animals, such as reindeer (*Rangifer tarandus* L.), which has been widely cited for both Inuit and several North European M3H populations, could have caused the onset of scurvy. In this regard, Geraci and Smith [59] described the occurrence of subacute scurvy in several Inuit children when consuming frozen reindeer for several days. Consistently, Vit C availability could have been a bottleneck for the survival of M3H, especially in the long winters, when there is a notable shortage of food plants.

Long ago, Levine [66] and Stefansson [67] reported that by following an exclusive animal food diet, it is possible to prevent the onset of scurvy. Interestingly, in all reported cases, such diet included viscera or meat from marine mammals or birds, such as penguins. In this regard, it was argued that the antiscorbutic properties of the meat in the diet of Arctic explorers and inhabitants of the polar regions lie in the fact that the meat was obtained almost entirely from seals and bears, and included not only muscle tissue but other tissues, such as liver [68]. As discussed below, marine mammals and birds are adequate Vit C sources, able to prevent the appearance of scurvy. However, marine mammals were marginally available for M3H, and there are only reports on their exploitation in Gibraltar [69].

1.4. The Bottleneck of Vitamin C for the Survival of M3H

The reliance of hominins on the acquisition of large mammals during most of the Pleistocene had nutritional and bioenergetic implications: large prey contain high fat amounts, provide a high energy return, and the strategies and skills required to acquire small prey are different from those required to acquire larger prey. Thus, the increased use of plant foods and technological changes that appear in the UP could be explained as adaptations to the decline in large prey in the context of late Quaternary faunal extinction, and the resulting need to acquire smaller prey efficiently and to process higher quantities of plant foods [22,41,70–72].

As for Inuit, the availability of Vit C for M3H could have had a strong seasonal component. However, unlike M3H, Inuit people have regularly available fish all year round, and in some cases, seals, and cetaceans for hunting [59], which are notable Vit C sources, so Inuit people would have rarely suffered deficiency symptoms of this essential nutrient. The problem for the survival of M3H in Northern Europe was that under a very cold climate most of the year, there is an absence of plant foods to obtain Vit C. In this scenario, viscera have been identified as sources of this vitamin, which could have supplied the daily amounts necessary to survive [73]. However, the regular consumption of viscera would have had many added problems. First, this resource is limited because the highest proportion of the edible weight of any hunted mammal corresponds to meat and fat, and thus the dependence on animals supplying minor amounts of viscera, i.e., small prey, could have caused an erratic distribution of viscera to be consumed by the hominin groups. Secondly, viscera contain large amounts of water and bacteria, and therefore are subject to rapid spoilage, so their consumption is a priority after any hunting episode.

This work clarifies how mammal's food resources could have supplied to M3H the daily amount of Vit C needed to maintain good health. I hypothesize that the acquisition of viscera from large- and megafaunal elements by M3H had, besides other physiological roles, the function of totally or partially supplying the daily amount of Vit C needed for hominin groups' survival, and that obtaining this essential nutrient from small prey would not have been as effective as from the previous ones. This fact could have played a crucial role in the size and survival of hominin populations during several Paleolithic periods.

2. Material and Methods

Data were acquired from databases such as Scopus, Google Scholar, and similar resources.

The body weight for M3H was set at 65 kg, considering average values given for M3H both male and female individuals [73]. The total energy expenditure for hominins was computed following previous calculations for both male and female M3H adults, which was set at ~4500 kcal/day, following Guil-Guerrero's calculations [73], which agrees with the other values recently calculated fir this [74]. However, the influence of daily energy requirements on Vit C status will be discussed for a 3000–4500 kcal/day range, according to recent calculations for Neanderthals' energy expenditure [75].

The average protein content of meat from free range mammals was set at ~22 g/100 g, and the protein content of fatty tissues was ~7 g/100 g, which is the average value for several food mammals previously calculated [73]; thus, estimates from both tissues have been considered to calculate daily protein intake. The daily intakes of both meat and fatty tissues are conditioned by protein intake, for which the safe range was between 52 and 162 g daily, following Paddon-Jones and Rasmussen's [76] and Bilsborough and Mann's calculations [77]. Then, the maximum daily protein intake was set at 2.5 g/kg·day. This value corresponds to the maximum safe intake of meat, which also entails the highest intake of polyunsaturated fatty acids (PUFA) and Vit C [73]. Considering a muskox, for optimal nutrition, the maximum consumption of meat supplemented with fatty tissues was set at ~585 g of meat and ~485 of fatty tissues (e.g., subcutaneous fat), for a consumption of 2.5 g of protein per kg and day, and such proportions provide ~4500 kcal [73]. For the remaining animals, these amounts of meats and fatty tissues had small variations, given slight differences in energy provided by the meat and fatty tissues among the various mammals. Furthermore, the influences of possible adaptations to higher amounts of protein intake on Vit C status have also been considered for calculations.

The ascorbic acid content of the main organs of grass-fed mammals usually consumed by Paleolithic hominins is detailed in Table 1, and the ascorbic acid contents of farmed food animals are shown in Table 2. The energy (kcal) and ascorbic acid content of the subcutaneous fat, meat, and viscera of the animals that are evaluated in the present work were taken from previous works, as detailed in Table 3 and the Supporting Tables S1–S6. The average energy content and protein of viscera from the various mammals were considered to be similar to those of meat, as previously reported [73]. For extinct animals, such as the giant deer, woolly rhinoceros, and woolly mammoth, the data for the energy of the different organs and body proportions have been taken from their closest relatives, as indicated in the corresponding supporting tables.

Energy derived from carbohydrates was considered negligible as compared with that from fat and meat, assuming a consumption of ~50 g daily (~187 kcal, considering 3.75 kcal/g), as noted in the traditional Inuit diet [78]. For an adult, such energy represents ~4% of the daily energy, and thus this amount has not been considered in energy calculations.

The Vit C contents of the main organs of free-range mammals usually consumed by M3H are summarized in Table 1. Data on the Vit C content of farmed mammals are detailed in Table 2. Data on the Vit C content of muskox organs were calculated following data on

Vit C concentrations and body organs' weights, as detailed in Table 3. The Vit C contents of aquatic resources available to M3H are displayed in Table 4. Supporting Tables S1–S6 summarize data for organ weights, Vit C, and energy content of free-range mammals commonly consumed by M3H.

The size data of the M3H group were taken from previous calculations [79–81]. The size of the hunter group was taken from Smith [82]. The energy/meat/fatty tissues necessary to fulfill the daily needs of the hominin group was computed using paleodemographic data for M3H populations [83,84], and weighting food amounts for the different age groups proportionally, following the Dietary Guidelines Advisory Committee [85], as detailed in Supporting Table S7.

Table 1. Ascorbic acid contents of the main organs of grass-fed mammals usually consumed by Paleolithic hominins.

Species	mg/100 g Fresh wt	Method of Vitamin C Determination	Reference
Bone marrow			
Caribou/Reindeer (*Rangifer tarandus* L.)	0.00	Microfluorometric	[86]
Brain			
Musk ox (*Ovibos moschatus* Zimmermann)	12.80	Tritation with 2,6-dichlorophenolindophenol	[87]
Depot fat			
Caribou/Reindeer (*R. tarandus* L.)	1.80	Spectrophotometric with 2,4-dinitrophenylhydrazine	[59]
Caribou/Reindeer (*R. tarandus* L.)	0.00	Microfluorometric	[86]
Hearth			
Caribou/Reindeer (*R. tarandus* L.)	2.60	HPLC-ED	[88]
Muskox (*O. moschatus* Zimmermann)	1.50	Tritation with 2,6-dichlorophenolindophenol	[87]
Kidney			
Caribou/Reindeer (*R. tarandus* L.)	8.88	HPLC-ED	[88]
Muskox (*O. moschatus* Zimmermann)	5.90	Tritation with 2,6-dichlorophenolindophenol	[87]
Liver			
Caribou/Reindeer (*R. tarandus* L.)	23.76	HPLC-ED	[88]
Caribou/Reindeer (*R. tarandus* L.)	11.88	Microfluorometric	[86]
Mean for caribou/reindeer liver	*17.82 ± 8.40*		
Muskox (*O. moschatus* Zimmermann)	10.40	Tritation with 2,6-dichlorophenolindophenol	[87]
Meat			
Bear, polar (*Ursus maritimus* Phipps)	1.00	Spectrophotometric with 2,4-dinitrophenylhydrazine	[59]
Bear, polar (*U. maritimus* Phipps)	2.00	-	[89] FDC ID: 169794
Mean for polar bear meat	*1.5 ± 0.7*		
Bear, brown (*Ursus arctos* L.)	0.00	-	[89] FDC ID: 173845
Beaver (*Castor* spp. L.)	2.00	-	[89] FDC ID: 175294
Bison (*Bison bison* L.)	0.00	HPLC-FD	[90]
Bison (*B. bison* L.)	0.00	Tritation with 2,6-dichlorophenolindophenol	[91]
Bison (*B. bison* L.)	0.00	-	[89] FDC ID: 175293
Mean for bison meat	*0.00*		
Caribou/Reindeer (*R. tarandus* L.)	0.86	HPLC-ED	[88]
Caribou/Reindeer (*R. tarandus* L.)	1.40	Spectrophotometric with 2,4-dinitrophenylhydrazine	[59]
Caribou/Reindeer (*R. tarandus* L.)	0.00	Microfluorometric	[86]
Caribou/Reindeer (*R. tarandus* L.)	0.00	-	[89] FDC ID: 173853
Mean for caribou meat	*0.56 ± 0.69*		
Horse (*Equus ferus caballus* L.)	0.00	Tritation with 2,6-dichlorophenolindophenol	[92]
Horse (*E. ferus caballus* L.)	1.00	-	[89] FDC ID: 175086
Mean for horse meat	*0.50 ± 0.71*		
Moose (*Alces alces* L.)	4.00	-	[89] FDC ID: 174344
Muskox (*O. moschatus* Zimmermann)	1.50	Spectrophotometric with 2,4-dinitrophenylhydrazine	[59]
Muskox (*O. moschatus* Zimmermann)	0.80	Tritation with 2,6-dichlorophenolindophenol	[87]
Mean for muskox meat	*1.15 ± 0.49*		
Red deer (*Cervus elaphus* L.)	0.00	-	[89] FDC ID: 173855
Wild boar (*Sus scrofa* L.)	0.00	-	[89] FDC ID: 175297
Stomach			
Caribou/Reindeer (*Rangifer tarandus* L.)	0.45	HPLC-ED	[88]

Table 2. Ascorbic acid contents of farmed food animals.

Organ/Animal	mg/100 g	Reference
Red bone marrow		
Beef	1.0–3.7	[93]
Brain		
Beef	19.20	[94]
Lamb/sheep/mutton	19.40	[94]
Pork	13.50	[94]
Mean for brain	17.37 ± 3.35	
Heart		
Beef	7.00	[94]
Lamb/sheep/mutton	7.30	[94]

Table 2. *Cont.*

Organ/Animal	mg/100 g	Reference
Pork	0.60	[95]
Pork	2.54	[96]
Pork	5.00	[94]
Mean for hearth	4.49 ± 2.89	
Kidney		
Beef	10.40	[94]
Lamb/sheep/mutton	12.90	[94]
Pork	2.52	[95]
Pork	14.20	[94]
Mean for kidney	10.01 ± 5.23	
Liver		
Beef	22.40	[94]
Beef	21.80	[97]
Beef	25.00	[98]
Beef	32.00	[99]
Lamb/sheep/mutton	25.00	[94]
Lamb/sheep/mutton	46.00	[99]
Pork	8.35	[95]
Pork	14.66	[96]
Pork	21.60	[94]
Pork	12.00	[99]
Mean for liver	22.88 ± 10.68	
Lung		
Beef	38.50	[94]
Lamb/sheep/mutton	31.40	[94]
Lamb/sheep/mutton	24.50	[100]
Pork	13.10	[94]
Pork	27.00	[100]
Rabbit	29.00	[100]
Mean for lung	27.25 ± 8.42	
Meat		
Beef	0.00	[101]
Beef	2.53	[102]
Beef	0.00	[103]
Beef	1.60	[98]
Buffalo	0.00	[104]
Camel	0.00	[103]
Lamb/sheep/mutton	0.60	[105]
Mutton	0.00	[103]
Pork	0.27	[95]
Pork	0.90	[96]
Mean for meat	0.66 ± 0.89	
Pancreas		
Lamb/sheep/mutton	17.50	[94]
Pork	15.30	[94]
Mean for pancreas	16.40 ± 1.56	
Spleen		
Beef	45.50	[94]
Lamb/sheep/mutton	23.20	[94]
Pork	19.51	[95]
Pork	27.22	[96]
Pork	30.00	[94]
Mean for spleen	29.09 ± 10.00	
Tonge		
Beef	3.30	[94]
Lamb/sheep/mutton	6.80	[94]
Pork	4.40	[94]
Mean for tonge	4.83 ± 1.79	
Tripe/intestines		
Beef	3.40	[94]
Lamb/sheep/mutton	6.60	[94]
Mean for tripe/intestines	5.00 ± 2.26	
Subcutaneous fat		
Beef raw	0.00	[89] FDC ID: 173091
Lamb raw	0.00	[89] FDC ID: 174435
Pork	0.00	[95]
Pork	0.00	[89] FDC ID: 167813
Mean for subcutaneous fat	0.00 ± 0.00	
Thymus		
Beef	34.00	[94]

Table 3. Organ weights, energy, and ascorbic acid contents of a medium-sized muskox (285 kg) [a].

	Body%	g	Energy kcal/g	Total kcal	Vit C mg/g	Total Vit C in Organ, mg	Reference for Vit C	Reference for Organ wt.%	Reference for Energy
Viscera									
Adrenal	0.0209	59.5	0.680	40	0.392	23.3	[87] [b]	[106] [c]	Estimated as testes
Bile	0.0522	148.7	0.045	7	0.01	1.5	[87] [b]	[107] [c]	Derived from [108] [d]
Blood	3.4052	9704.7	0.700	6793	0.0007	6.8	[109] [e]	[110] [e]	[111] [e]
Brain	0.2210	629.8	1.234	777	0.128	80.6	[87] [b]	[106] [f]	[73] [g]
Digestive tract	5.3000	15,105.0	0.960	14,501	0.074	1117.8	[86] [b]	[106] [e]	[112] [f]
Epididymis	0.1098	312.9	0.680	213	0.042	13.1	[87] [b]	[113] [i]	Estimated as testes
Eyes	0.0393	112.0	3.260	365	0.007	0.8	[87] [b]	[106] [c]	[89] FDC ID: 169801[f]
Heart	0.4400	1254.0	1.120	1404	0.015	18.8	[87] [b]	[114] [b]	[112] [f], [115] [f]
Kidney	0.1855	528.7	0.917	485	0.059	31.2	[87] [b]	[114] [b]	[73] [g]
Liver	1.0382	2958.9	1.240	3669	0.104	307.7	[87] [b]	[114] [b]	[115] [f]
Lung	1.3528	3855.5	1.090	4202	0.081	312.3	[87] [b]	[106] [c]	[112] [f]
Pancreas	0.1405	400.4	2.350	941	0.039	15.6	[87] [b]	[116] [e]	[89] FDC ID: 169452 [h]
Spleen	0.2284	650.9	0.900	586	0.06	39.1	[87] [b]	[106] [c]	[73] [g]
Testes	0.5160	1470.6	0.680	1000	0.182	267.6	[87] [b]	[107] [c]	[89] FDC ID: 172619 [c]
Tongue	0.1891	538.9	2.60	1401	0.01	5.4	[87] [b]	[107] [c]	[89] FDC ID: 168985 [f]
Thyroid	0.0196	55.9	0.680	38	0.038	2.1	[87] [b]	[107] [c]	Estimated as testes
Total viscera	13.2584	37,786.4	-	36,423	-	2243.8	-	-	-
Bones	14.2748	40,683.2	0.000	-	-	-	-	[114] [b]	-
Bone marrow	2.5000	7125.0	4.837	34,464	0.000	0.000	[86] [f]	[117] [f]	[73] [g]
Dissectible fat	17.7099	50,473.2	7.894	398,436	0.000	0.000	[87] [b]	[114] [b]	[73] [b]
Total fat		57,715.5	-	432,899	-	-	-	-	-
Muscles	41.9084	119,438.9	1.132	135,205	0.0115	1373.5	[87] [b]; [59] [b]	[114] [b]	[73] [b]
Skin	9.0043	25,662.3	5.440	139,603	0.0050	128.3	[89] FDC ID: 234130 [j]	[110] [e]	[89] FDC ID: 2341308 [j]
Rest [k]	3.6675	10,452.4	1.000	10,452	0.0010	10.5	Estimated	Estimated	Estimated

[a] Body weight was taken from Knott et al. [118]; [b] muskox sample; [c] sheep sample; [d] based on 0.51% fats in bile; [e] goat sample; [f] *Rangifer* sample; [g] mean for several food animals; [h] beef sample; [i] macaco sample; [j] pork skin rinds; [k] tendon, blood vessels, etc.

Table 4. Vitamin C contents of birds and aquatic food resources available to Paleolithic hominins.

	Scientific Name	Organ	mg/100 g Fresh wt	Reference
Birds				
Duck	*Anas platyrhynchos*	Flesh	6.2	[89] FDC ID: 174469
Goose	*Anser anser*	Flesh	7.2	[89] FDC ID: 172413
Pigeon	*Columba livia*	Flesh	7.2	[89] FDC ID: 171079
Quail	*Perdix perdix*	Flesh	7.2	[89] FDC ID: 172419
Seaweeds				
Dulse	*Palmaria palmata*	Thallus	0.61	[119]
Irishmoss	*Chondrus crispus*	Thallus	3	[89] FDC ID: 168456
Kelp	Laminariales	Thallus	3	[89] FDC ID: 168457
Kombu	*Laminaria* sp.	Thallus	1.34	[119]
Laver	*Porphyra umbilicalis*	Thallus	39	[89] FDC ID: 168458
Laver	*P. umbilicalis*	Thallus	33.29	[119]
Sea spaghetti	*Himanthalia elongata*	Thallus	46.66	[119]
Wakame	*Undaria pinnatifida*	Thallus	3	[89] FDC ID: 170496
Invertebrates				
Abalone	Haliotidae	Edible portion	2	[89] FDC ID: 174212
Crab	*Clibanarius vittatus*	Whole body	1.8	[120]
Eastern Oysters	*Crassostrea virginica*	Edible portion	3	[121]
Eastern Oysters	*Crassostrea virginica*	Edible portion	4.7	[89] FDC ID: 175172
Mussels	*Mytilus edulis*	Edible portion	8	[89] FDC ID: 174216
Octopus, common	*Octopus vulgaris*	Edible portion	5	[89] FDC ID: 174218
Polychaete	*Neanthes virens*	Whole body	1.56	[122]
Sea urchin	*Paracentrotus lividus*	Gonad	26.57	[123]
Squid, mixed species	Cephalopods	Edible portion	4.7	[89] FDC ID: 174223
Shrimp	*Palaemonetes pugio*	Edible portion	1.64	[122]
Snail	Polinices duplicatus	Kidney	2.34	[120]
Whelk, unspecified	Mollusk	Edible portion	4	[89] FDC ID: 171983
Freshwater fish				
Arctic char	*Salvelinys alpinus*	Whole fish	5.8	[59]
Arctic char	*Salvelinys alpinus*	Flesh	1.8	[59]
Arctic char	*Salvelinusalpinus*	Flesh	1.3	[88]
Bass, fresh water, mixed species	Perciformes	Flesh	2	[89] FDC ID: 174184
Broad whitefish	*Coregonos nasus*	Flesh	2.83	[88]
Carp	Cyprinidae	Flesh	1.6	[89] FDC ID: 171952
Carp	*Cyprinus carpio*	Flesh	1.14	[124]
Cisco	*Coregonos nasus*	Eggs	49.6	[87]
European catfish	*Silurus glanis*	Flesh	2.15	[124]
Pike, northern, raw	*Esox lucius*	Flesh	3.8	[89] FDC ID: 173680
Pike perch	*Sander lucioperca*	Flesh	1.91	[124]
Whitefish	*Coregonus clupeaformis*	Eggs	12	[89] FDC ID: 167643
Salmon, chinook	*Oncorhynchus tshawytscha*	Flesh	4.0	[89] FDC ID: 173688
Sculpin	*Myoxocephalus* spp.	Flesh	1.05	[88]
Trout, rainbow	*Oncorhynchus mykiss*	Flesh	2.4	[89] FDC ID: 175154
Trout, rainbow	*Oncorhynchus mykiss*	Flesh	1.8	[121]
Marine fish				
Mullet	Fam. Mugilidae	Flesh	1.2	[89] FDC ID: 175123
Needlefish	Fam. Belonidae	Brain	29.5	[122]
		Gill	3.88	[122]
		Kidney	2.93	[122]
		Flesh	1.41	[122]
		Liver	2.13	[122]
Snapper, mixed species	-	Flesh	1.6	[89] FDC ID: 173698
Marine mammals				
Beluga	*Delphinapterus leucas*	Mattak	36.02	[88]
Narwhal	*Monodon monoceros*	Mattak	31.51	[88]
Ringed seal	*Phoca hispida*	Brain	14.86	[88]
		Liver	23.8	[88]
		Liver	35	[59]
		Meat	1.55	[88]
		Meat	3	[59]

3. Results and Discussion

3.1. Sources of Vitamin C for M3H

The Vit C contents of several organs of free-range mammals usually consumed by M3H are summarized in Table 1. Notice that most muscles contain negligible amounts of, or lack, Vit C. The best Vit C-supplying meats are those of moose and beaver (4 and 2 mg/100 g of fresh weight, [89]). However, no analytical data have been found in the literature supporting such relatively high figures. In any case, moose and beaver bone remains are not very frequent in Paleolithic faunal assemblages (e.g., [125,126]; thus, such meats can be considered marginal in M3H diets. Among meats, that of muskox contains small amounts of Vit C, ~1 mg/100 g, which would need to be taken in large amounts (>1 kg) to reach the 10 mg daily amount needed to prevent scurvy [48,49]

As for Vit C data in Table 1, those corresponding to Geraci and Smith [59] should be considered with caution. This is because the method used for Vit C determination was the spectrophotometric approach, using 2,4-dinitrophenylhydrazine, and this usually produces an overestimation of total Vit C in the analyzed samples [127].

Unlike meat, viscera constitute an appropriate source of this vitamin (Table 1). For instance, ~110 g of kidney from caribou could have fulfilled the daily needs of Vit C for M3H. Although nutritional data on viscera of free-range mammals are very scarce, data on viscera and meats of farm animals are better known, and those on commonly consumed mammals are shown in Table 2. Notice that Vit C contents in the viscera of these animals do not differ from those of wild ones. Interestingly, the Vit C contents of each organ are quite similar among the various species.

Nowadays, there are discrepancies in global Vit C recommendations for optimal health, ranging from 40 to 220 mg Vit C per day [128]. Such differences are due to the variety of recommendations; for instance, to prevent scurvy or for optimizing health. Scurvy can be effectively prevented with ~10 mg Vit C/day [48,49,129]. However, subclinical Vit C deficiency may be much more common than overt disease, and it is difficult to recognize, because the early symptoms of deficiency are unremarkable and nonspecific [130]. Examination of the data in Tables 1 and 2 suggests that viscera, not meat, would have allowed M3H to survive on a carnivorous diet. Moreover, for obtaining the daily needs of ~4500 kcal at such times only from lean meat, almost ~4 kg would have been necessary. The toxicity of nitrogen inherent to high levels of meat consumption would have made this option unfeasible and, given an absence of Vit C in most meats, in no case would the daily amount of Vit C required to avoid the deficiency status have been obtained. The better strategy to achieve the daily energy needs for M3H survival would have involved consuming the higher toxicologically safe amount of meat, supplemented with fatty tissues [73]. This would have allowed for obtaining the maximum amounts of essential omega-3 PUFA and in some cases minor amounts of Vit C. However, the only way to acquire the Vit C required for survival would have been supplementing the diet with viscera from mammals or, when available, by eating plant foods, fish, or birds.

Although a strictly carnivorous diet has been cited for human groups, the consumption of some vegetable organs during the winter by M3H could have been marginally practiced, perhaps as famine foods. In this regard, some plant foods, particularly underground storage organ (USOs), have been proposed as valuable resources to foragers in the late winter and early spring, and it was speculated that such organs would have been carbohydrates and Vit C suppliers during the winter for Paleolithic hominins [131]. For instance, chicory (*Cichorium intybus* L.) roots contain 5 mg/100 g of Vit C ([89] FDC ID: 169993); thus, USOs could have acted as antiscorbutic foods. However, it is difficult to consider the consumption of USOs in winter by M3H: most of them are cryptic in winter; thus, efficient harvesting is unlikely as this activity would have had high energy costs. This way, given the low energy content of USOs (e.g., 72 kcal/100 g for chicory ([89] FDC ID 169993), the energy return derived from their collecting would have been very low. Another possibility would have been to freeze plant organs to be consumed in winter. Although the preservation of frozen foods has been reported for some traditional Arctic societies, the possibility that this was done by M3H for the long-term preservation of foods seems remote, as there is no archeological evidence to support it. Other plant foods resistant to spoilage and therefore storable during the winter are seeds, such as those of poppies, oak, cereals, and flax. However, all of these contain negligible amounts of Vit C [89].

Recently, pine and juniper needles have been proposed as winter sources of ascorbic acid for Pleistocene hominins in northern Eurasia [132]. This possibility seems to be highly unlikely. Even though, like most plant organs, they contain Vit C, their potential toxicity makes their consumption on a constant basis unfeasible, although these could have been drunk occasionally as a therapeutic remedy [66]. However, most of these vegetable organs contain turpentine and other toxic compounds that can cause various health problems, including hepatotoxicity [133]. Moreover, in pregnant women, both juniper and pine needles can pose risks to the developing fetus as the toxic substances present in needles can cross the placenta and potentially cause harm, while laboratory and animal studies have suggested that *Juniperus* and *Pinus* species share similar chemistries and have antifertility effects in both males and females [134]. Therefore, the presence of charcoal derived from pine or other coniferous wood and phytoliths in hominin sites is probably a reflection of the suitability of this material as a rapid fire-starter due to its high resin content, instead of a proof of pine tea-making.

3.2. M3H Viscera, a Valuable Food Resource Hard to Equally Distribute among M3H

Most viscera are nutritionally advantageous over lean meat. For example, caribou liver in addition to Vit C contains at least twice as much fat as meat from the same animal. As muscles, viscera contain high moisture amounts, although the latter ones have a slightly lower protein and higher fat contents than meat [135]. The consumption of viscera has been widely practiced by past and present peoples. Brain, kidneys, liver, heart, and lungs, in addition to Vit C, provide valuable amounts of essential PUFA [73], folacin, and vitamin A [136]. Usually, most viscera are immediately consumed after hunting. This is because these contain large amounts of water and nutrients, which favor the rapid development of microorganisms coming from the digestive tract. For instance, buffalo liver has been evaluated as microbiologically and organoleptically acceptable up to the third day at 4 °C storage [137]. Priority given to the consumption of viscera has been reported for all hunter groups belonging to native cultures, and for chimpanzees. As the animal corpse is completely corrupted in a very short time if not disburdened of its viscera, hunters have no choice but to start by consuming these [138]. This is universally described. For instance, at the Howard Goodhue site, central Iowa, late prehistoric exploitation by Plains Indians of bison, elk, and deer was described by McHugh [139]: "As they worked, most of the participants snacked on raw morsels taken still warm from the slain buffalo, including livers, kidneys, tongues, eyes, testicles, belly fat, parts of the stomach, gristle from snouts, marrow from leg bones, and the hoofs of tiny unborn calves as well as tissue from their sacs."

A more recent case has been exhaustively described. Inuit people from Kivalina, the northwestern coast of Alaska, develop a well-established protocol for the consumption of caribou viscera. After killing, the stomachs and kidneys were immediately eaten raw, while kidneys were sometimes saved, though they were mostly eaten raw. In the same way, the heart and liver were also eaten as soon as they become available [140].

Several authors reported evisceration activities through direct ethnographic observations for understanding carcass acquisition and butchery by M3H. Evisceration processes have been detected through cutmarks occurring on the ventral surface of the ribs [141,142]. It is commonly reported that viscera were the first animal organs eaten in the consumption sequence, while a great number of authors have reported that viscera are highly valued organs by all current and past hunting peoples (e.g., [143]). Previously, through a nutritional model, it was calculated that the daily meal for Paleolithic hunters should have consisted of ~1250 g of animal food, including both meat/viscera and fatty tissues [73]. This value agrees with previous observations about the daily amount of food consumed by Inuit peoples [59], which reaches ~1200 g daily for males and somewhat less for women, corresponding to the ingestion of a mix of flesh, marrow, and selected viscera from several animals.

Considering the high content of Vit C in viscera, priority in consumption would also have bene determined by this vitamin. However, viscera provide a low fraction of the total caloric content of the prey, and whether small- or medium-sized animals were hunted, it would have been difficult to make viscera distribution equitable among the whole hominin group.

3.3. Paleolithic Mammals as Vitamin C Suppliers

Table 3 and Supporting Tables S1–S6 detail the organ weights, Vit C and energy content of most representative mammals whose remains are usually found in Paleolithic sites. Data for Vit C content of the various mammal organs are well known for muskox [87], while for other hunted mammals, data are incomplete. However, as deduced from Tables 1 and 2, the Vit C contents in the organs of all mammalian species are quite similar, and, in the absence of specific data, figures for the Vit C of extinct species could be considered very similar to those of their current relatives.

Bone marrow in this work has been computed considering the energy contribution of its fat content. Many authors have cited the predilection of M3H for this nutritional resource (e.g., [144]), and there is evidence of the storage and delayed consumption of bone marrow at the Middle Pleistocene in Qesem Cave, Israel [145]. The nutritional content of bone marrow is highly variable depending on the age of the animal, the bone considered and the target tissue (red marrow vs. yellow marrow). The limited data available regarding the bone marrow of grass-fed Paleolithic mammals indicate negligible amounts of Vit C contained therein (Table 1). Although this topic is worthy of detailed studies, for now, it is assumed that the fat content of the bones was the target of M3H.

The nutritional and energy needs of hominins should be computed as a group rather than at the individual level, since a small-sized group of Paleolithic hunters would have survived whether the group was able to respond to the nutritional needs of each of its individuals. Calculations for Vit C intake have been undertaken for a 22-individual hominin group, as detailed in the Material and Methods Section. Estimates of Vit C intake are also performed for bigger groups, considering a 20–50-individual range for M3H groups [80], and for both higher daily protein and lower energy intakes.

Muskox (*Ovibos moschatus* Zimmermann, 1780) is probably the best-known animal in relation to the Vit C content in its various organs [87]. Given that this mammal was also studied in detail regarding its body and organ weights [114], both sets of data allowed for computing the total content of Vit C in the different organs of a real animal (Table 3). Some gaps in the data for the Vit C content of minor organs have been filled in by establishing analogies with similar free-range mammals, as detailed in this table. The main organs for food purposes are viscera, muscles, and fatty tissues (subcutaneous fat, marrow, and visceral fat), while other organs of the animal would be wasted and thus not regularly consumed (skin, some bones, blood vessels, etc.).

Modeling the consumption of muskox will help us to understand the features of the distribution of Vit C among an M3H group. A case of a 285 kg hunted animal was selected from the literature (Table 3). The whole viscera would have provided about ~36,000 kcal and ~2244 mg of Vit C, the muscles would yield ~135,000 kcal and ~1400 mg of Vit C, and the fatty tissues ~433,000 kcal and no Vit C.

Considering a 22-individual group, according to paleodemographic data for Paleolithic populations [83,84], the energy needs are ~76,000 kcal (Supporting Table S7), and a minimum of 220 mg of Vit C daily would also have been needed [48,49,129] (this last value is for short-term survival). As implied, the optimal consumption of each individual would have been that which allowed the ingestion of the maximum safe amount of proteinaceous and visceral tissues (sources of essential PUFA and Vit C).

Obviously, the equitable distribution of viscera would have been of crucial importance since the consumption of extra amounts of Vit C-rich organs by some individuals would have had deleterious effects on the remaining individuals. Whether viscera of muskox were equally distributed among the 22-people group on the hunting day, the group would still have had a deficit of ~40,000 kcal, which should have been fulfilled by variable amounts of meat and fatty tissues. Such a deficit could have been covered by ~5.2 kg of meat

(~5900 kcal) and the remainder would have been fulfilled with ~4.3 kg of fatty tissues (~34,000 kcal). Then, on the hunting day, ~2300 mg of Vit C would have been available for the 22-hominin group, and an intake of ~105 mg of Vit C for each individual was expected. From the second day, the remaining energy content (~129,000 kcal from meat and ~400,000 kcal from fatty tissues) of the muskox carcass would have allowed enough energy for about 12 days (considering meat) and 6 days (considering fatty tissues), according to calculations detailed in Supporting Table S7. Thus, it is likely that an absence of fatty tissues in the diet would have forced a new hunting episode, given a surplus of meat over fatty tissues during the 6-day period. Within such a viscera-free period, Vit C intake would have been ~6 mg daily, all of which would have been obtained from the animal's muscles. Obviously, for larger hominin groups, the amount of Vit C consumed by each individual on the hunting day would be less, although probably sufficient for the whole group, and after a Vit C-deficiency period it is likely that a new hunting episode would occur earlier.

Considering daily protein intake higher than the 2.5 g safe value used for previous calculations, i.e., 3.5 g/kg·day, as indicated for well-adapted athletes [146], and similar energy needs, ~440 g of fatty tissues and ~895 g of meat would have been needed to achieve 4500 kcal for adults. This situation would have allowed a longer period for consuming the whole carcass, which would have delayed the start of a new hunting event and thus scurvy would have been more likely. As for muskox, the consumption of its meat would allow for obtaining ~10 mg of Vit C daily, which is on the threshold of a satisfactory Vit C intake. However, it is necessary to consider that the meat of this mammal is the richest source of Vit C among all considered free-range mammals, and such Vit C intake would be unattainable by consuming meat from other mammals. Then, at the protein level of 3.5 g/kg·day, the amount of Vit C ingested daily from the meats of reindeer, horse, red deer, giant deer, and bison, would have been about 4, 4, 0.0, 0.0, and 0.0 mg, respectively, according to calculations made with data detailed in Table 1. Thus, except for muskox, the amounts obtainable of Vit C are lower than needed. Clearly, all meats from free-range mammals, except those of moose and beaver, contain Vit C amounts unable to provide 10 mg daily, the above-indicated threshold necessary to prevent the onset of scurvy. Then, higher protein intakes, including values up to 5.0 mg/kg·day, would marginally modify the above-described situation.

Concerning a lower daily energy expenditure than previously set, as suggested by Venner [75], it is likely that the daily intake of fatty tissues would progressively decrease to reduce energy expenditure, but protein intake would probably remain at a similar level. Thus, the effect of this situation on the daily intake of Vit C would be negligible. Furthermore, in the event that meat intake was reduced, Vit C status would be negative, accentuating the previously described deficit situations.

Unlike the previous case, the most realistic form of prey consumption would have been similar to the one that is detailed today for native hunter peoples; that is, viscera would have been consumed only by some preferential individuals. On the hunting day, the caloric content of muskox viscera would have been enough to feed around nine individuals—the group of hunters, probably adults. In the case of the exclusive consumption of viscera, these individuals would have obtained ~250 mg of Vit C each. The remaining individuals would have consumed muscles and fatty tissues, and each of them would have just fulfilled their daily needs of Vit C by consuming meat, as previously explained. Obviously, there are intermediate situations between those previously discussed, corresponding to cases where the viscera would have been partially transported to the base camp; however, all of them are unsatisfactory regarding Vit C status for all free-range mammals except for the muskox case. For all the remaining animal cases (see supporting Tables S1–S6), in the event of an insufficient viscera intake, all of them would have provided unsatisfactory nutritional results regarding the daily supply of Vit C, which would have led to chronic subacute scurvy, and thus to increased mortality.

Of course, several prey could have been captured in the same hunting episode, and this would have prevented the previously discussed situation of a Vit C deficit. In late

Mousterian sites, communal hunting involved deferred consumption in anticipation of future needs. For instance, at the Les Pradelles site, the preferential introduction of nutritionally rich elements onto the site is consistent with the slaughter of several reindeer during the same hunting event [147]. Assuming this scenario, considering the low energy contents of the viscera of all prey, the hunters would have needed to catch several animals to ensure the equitable distribution of viscera and therefore an adequate intake of Vit C for the whole hominin group. However, considering the surplus of meat and fat once the viscera was consumed, it is likely that such food availability would have retarded a new hunting episode, and that latent scurvy occurrence would have been more than frequent.

For instance, when a medium-sized prey, such as a reindeer (Supporting Table S1), had been caught, about ~76,000 kcal from its viscera would have been needed to meet the daily caloric needs of the 22-individual group (Supporting Table S7), and considering that a reindeer specimen of about 135 kg provides ~19,500 kcal from its viscera (Supporting Table S1), more than three prey would be necessary to this end. This would have involved the hunting and complete transport of at least four animals from the hunting site to the base camp. However, as previously exposed, not all viscera from medium-sized animals are transported to the base camp, and thus the actual amount of viscera consumed would have been less than the actual content in the prey. Moreover, considering the surplus of meat and fatty tissues available for the entire hominin group, these organs would have been used to meet the daily energy needs for several days, instead of Vit C-rich visceral organs. In any case, although not totally satisfactory, this last possibility seems the most feasible regarding the supply of Vit C to M3H groups, since the amount of viscera available would have been greater and it therefore would have been better distributed.

3.4. The Importance of Large Prey for Paleolithic Hominins

A predilection for large mammals by Paleolithic hominins has been widely reported. Driver and Maxwell [148] demonstrated that Old World Paleolithic hunters who lived in markedly seasonal environments conducted the killing of animals with high fat levels, such as bison. The acquisition of large prey was vital for M3H since they provided large amounts of fat, which constituted an energy reserve for the hominin group [71,149]. Considering large mammals, it is more than likely that big visceral organs were transported to the base camp, in addition to muscles and fat, since the hunting group would not have been able to consume all viscera completely. This would have favored an equitable distribution of visceral organs among all members of the hominin group for short-term consumption. Moreover, other parts of the carcass would have been transported from the hunting camp to the residential camp, i.e., fatty tissues, meat and large bones, and other high-utility elements of the animals, such as hides, to make clothes and other useful objects [150]. This implies high-energy expenditure and delayed consumption [151].

The survival of a 22-individual M3H group (days) facilitated by the caloric content of viscera, meat, and fatty tissues, and by the Vit C content of the viscera and meat from the typically hunted different-sized Paleolithic mammals, is plotted in Figure 1.

As for energy needs, meat and fatty tissues provide longer survival to the hominin group, but these organs effectively lack Vit C. Concerning viscera, only those of giant deer and bigger animals contain enough energy for more than one day of survival, thus simultaneously fulfilling the daily needs of Vit C for the whole M3H group. On the other hand, the viscera of medium- and small-sized animals—for instance, horses and reindeer—can fulfill the needs for Vit C of the M3H group for more than one day, although without covering its energy needs. This causes a significant nutritional imbalance for the hominin group because it can survive for many days using the energy content of fatty tissues and meat from medium-size animals, although without obtaining enough Vit C for preserving good nutritional status. Furthermore, as previously described, the surplus of carcasses would have discouraged hunters from obtaining new prey, and therefore the M3H group would be doomed to consume foods lacking significant amounts of Vit C for several days.

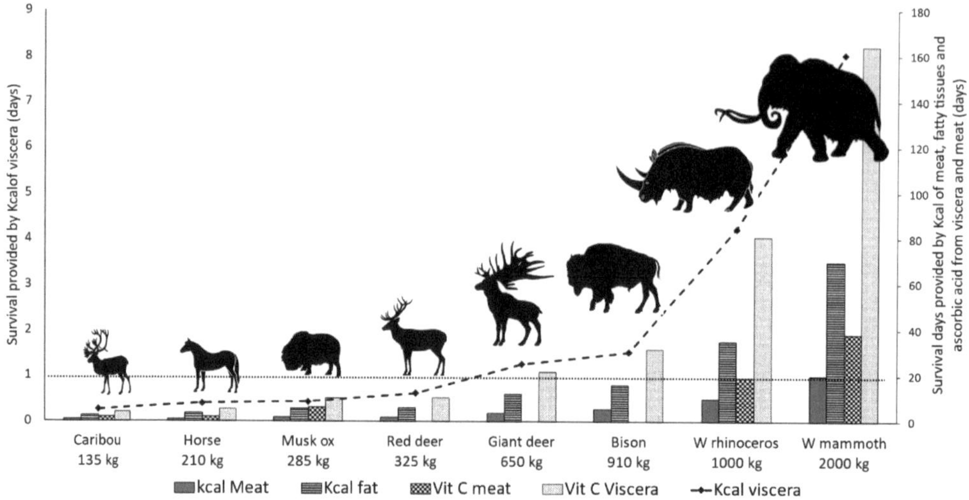

Figure 1. Survival (days) of a 22-individual hominin group provided by typically hunted Paleolithic mammals, assessed by the energy content of viscera, meat, and fatty tissues, and by Vit C from viscera and meat. Daily needs were set at 76,000 kcal and 220 mg of Vit C. Data were obtained from Table 3 and Supporting Tables S1–S6. Data for giant deer were scaled from those of red deer. Note that the energy content of viscera provides at least one day of survival to the hominin group only when the size of the animal is approximately that of the giant deer.

Conversely, the hunting of a single large mammal, e.g., a mammoth, would have provided viscera to the M3H group more effectively than that of a large number of medium- and small-sized animals. From this mammal, just the liver (~65 kg) would have provided the ~9.890 kg daily requirement of a proteinaceous organ for ~7 days to a 22-individual M3H group, while providing energy-rich fatty tissues, and it would also have been enough to meet the energy needs of larger hominin groups for some days (see Supporting Table S6). Besides the daily needs of Vit C, this organ would have fulfilled the daily requirements for several essential nutrients, such as omega-3 and omega-6 PUFA and vitamin A, for the whole M3H group. Considering the presence of other large viscera in mammoths, such as kidneys (~11 kg) and brain (~3.6 kg), which contain high amounts of Vit C and omega-3 PUFA, good-quality nutrients for the whole hominin groups are expected from a single specimen. Viscera would have been preserved below 4 °C (for example, stored under ice), to be consumed in short- or medium-periods, probably in moderate amounts and supplemented with meat and fatty tissues to avoid possible toxicities due to excesses of vitamin A. So, the advantages of hunting large animals over small ones are clear, not only considering energy acquisition. Concerning viscera, the hunting of a medium-size mammoth would have been equivalent to that of approximately 30 reindeers, and viscera, not meat, would have contributed to fulfilling the daily needs of Vit C for the M3H group for several days. Additionally, when large prey was caught, there could have been a surplus of meat and fat after the consumption of the viscera. However, it is more than likely that not all parts of the carcasses from such large prey would have been transported to the base camp. Logically, the hunter group would have transported the optimal amount to subsist on based on the longevity of the meat and viscera, leaving the rest at the hunting site. The transport of elements of large carcasses would have also been conditioned by the number of carriers, damage sustained when the animal was wounded, and by the nutritional status of the individual animal [152]. In this regard, in studies on walrus exploitation during the Paleoeskimo period at the Tayara site, Arctic Canada, it has been described that in the process of walrus butchering, which animals reach 675 kg on average, all the abdominal viscera were recovered and transported in parts or consumed locally [153].

Considering small-sized animals, meeting Vit C needs would have required intense hunting activity, beyond covering the energy needs of the hominin group. Furthermore, food acquisition would not have focused on viscera, and once they had achieved the required amount of meat and fatty tissues for several days, the hunting activity would have stopped, thus depriving the M3H group of the Vit C needed to survive during the extreme winters of the cold events characteristic of the MIS 3.

3.5. The Decline of in Mammals during the Marine Isotope Stage 3: Differential Consequences for the Vit C Status of M3H

MIS 3 was an interstadial stage, a relatively warm climatic period that developed roughly between 60 and 50 and 30 ka. Several very cold periods, known as Heinrich (H) events, developed during the MIS 3 as a result of a partial collapse of the North American ice sheet margins [154]. The early UP is included in this period, which corresponds to an episode of significant climatic deterioration associated with a reduction in mammalian and plant species diversity. Then, the rapid cooling characteristic of H events could have produced drastic periodic reductions in food resources [155]. All the previous suggests the extreme difficulty of obtaining Vit C within the H events from the progressively declining plant foods [21], and from large-size animals, as a consequence of their decline [22,156], which could have led to the gradual onset of chronic scurvy-induced malnutrition. Specifically, mammoth decline is confirmed by the more frequent use of ivory at Châtelperronian sites than at the later Aurignacian ones [157]. In this regard, it has been demonstrated that hominins extirpated megafauna throughout the Pleistocene, and when the largest species were depleted, the next-largest were targeted [70]. Eeuwes [158] collected spatiotemporal data on all archeological and paleontological finds of megafauna weighing above 2000 kg and hominin species between 71.5 ka and the present. Among others, several mammoth species and woolly rhinoceros (*Coelodonta antiquitatis* Blumenbach) were selected. Interestingly, he found fewer megaherbivore species per time frame before 40 ka, and after that, an increasing number of AMH sites were found. However, both woolly rhinoceros and woolly mammoth recovered some millennia after this decrease, and their records increased in numbers, particularly for the woolly mammoth. Later, the gradual increase in the records of AMH coincides with a decrease in the records of these megafaunal elements till around 20 ka. Regardless of what caused such a decrease in large fauna (overkilling, climate change, or other causes), the underrepresentation of large fauna before 40 ka coincided with the disappearance of Neanderthals in Eurasia.

In Eastern Europe, climate change has been held responsible for the demise of Neanderthals. Cooling events in the late Paleolithic contributed to shifts from a periglacial moderate cold and wet climate to a colder and drier climate, and the opening of the landscape, involving shifts in seasonality, plant phenology, reindeer body size and animal population densities across Central and Western Europe. Overall, abrupt climate oscillations during the MIS 3 contributed to the decline of Neanderthal populations, notably because of habitat fragmentation [159]. M3H, especially Northern European animal food-adapted populations, could have faced difficulty adapting to environments with less big game. For instance, faunal data from the Saint-Césaire site, France, suggest that reliance on reindeer during this period would also have promoted declines in human population densities, and the MP/Upper Paleolithic (M/UP) transition represented for hominins a period of significant niche contraction in Western Europe [160]. In the open grasslands resulting from the cooling events characteristic of late Paleolithic, the Neanderthal diet was mainly based on meat from terrestrial animals, whereas Aurignacian's AMH, whose dispersal in Europe took place during the MIS 3, exploited plants and aquatic foods more frequently than the former [19,161]. Furthermore, resource competition between both hominins may have accelerated the disappearance of the Neanderthal populations [12,162].

Recent studies of human subsistence strategies from the lower Paleolithic onwards have shown that hunter–gatherers in the western Mediterranean area had broader diets than those from adjacent, colder regions [39]. Thus, reliance on large mammals to obtain the necessary daily amount of Vit C could have been less marked for Neanderthals inhabiting the southernmost regions, as they would have regularly obtained this vitamin from wild edible plants. However, the cold steppe that southern Europe became during several H events, such as the H4 one, would have caused the decline in most of the plant foods sources of Vit C. In this regard, the cooling of a geographical area, the reduction in plant species, and this fact could have been decisive in the decline of Southern European Neanderthals. For instance, during the H4 event, an expansion of aridity over the Iberian Peninsula and the desertification of the south demonstrate that such events led to drastic changes over southern Iberia [163].

The situation of the decline in large prey during the OIS 3 period and the consequent effects on the survival of Neanderthals were preceded by similar ones. For instance, Earth's climate experienced major warming during the Eemian, OIS 5e, 128–114 ka. The rapid climate change in this period altered ecosystems and the human population collapsed, maintaining itself only in a few regions. Evidence from a well-preserved archeological layer at Baume cave (Moula-Guercy, southeastern France) shows environmental upheavals, including the depletion of prey biomass at the beginning of the Upper Pleistocene, which contributed to the rise of cannibalistic behavior in Neanderthals [164].

3.6. Vitamin C-Rich Food Resources Available to M3H

In addition to the use of viscera, especially from large prey, M3H may have exploited other Vit C-rich foods. The use of aquatic food resources by Neanderthals and AMH is controversial. AMH in South Africa around 164 ka had the ability to fish and hunt marine mammals, thus it is likely that such knowledge traveled North into Europe with the AMH as they moved [44]. It has been argued that the diet breadth at MP and early UP should be assessed by C and N stable isotope ratios in fossil bone collagen when establishing the intake of aquatic foods, and by zooarcheological evidence too. This is because the interpretation of a human's isotopic signature does not necessarily imply a significant proportion of aquatic-derived protein in the diet [165]. In this regard, an intensification in freshwater fish exploitation by AMH has been verified, in comparison with Neanderthals, which is supported by isotopic analyses of human bone (e.g., [166]).

The assemblages of fish bone provide evidence for a shift to new resources in connection with the colonization of most of Europe by AMH and the local extinction of Neanderthals. For instance, in Geißenklösterle (southwestern Germany), five species of fish from the Danube were identified, all of them belonging to the MP-Gravettian transition. Excavators identified fish bones and fish scales in the Aurignacian and Gravettian deposits [167]. This, however, was not the case in the MP deposits, which contained only a few fish remains [168], and this discovery associated with human sites took place after the occupation of Europe by AMH [169]. Moreover, the analysis of the fauna of the Swabian Jura (southwestern Germany), which contains some of the oldest and richest Aurignacian assemblages in Europe, show an increase in the exploitation of birds and hares, starting with the Aurignacian [45]. This fact could have been due to an incipient shortage of large mammals and to the necessity to hunt smaller and smaller game, which was possible thanks to modifications in hunting technology. Overall, this situation led to a diversification of the diet by AMH [70].

Marine resources have been part of the subsistence of Paleolithic hominins since very ancient times. In the Mediterranean Basin, the use and consumption of littoral resources (mollusks, fish, and marine mammals) has been documented from the MP [170], although the record of marine food remains has increased progressively in later times [171]. This fact proves the early diversification of the diet from the early UP onwards. Overall, the consumption of coastal resources from MP to UP contributes to a progressive change in the consideration of these resources from "marginal" to "usual" in the hunter–gatherers' economy. The use of coastal seaweed resources leaves no remains, although these could have been used by M3H. Costal algae could have been harvested in parallel with mollusks, echinoderms, and fish, and would have contributed to balancing the diet, not just in terms of Vit C. However, evidence of this activity has been found mostly on the European Mediterranean coasts. Algae, echinoderms, mollusks, fish, and marine mammals contain Vit C in concentrations much higher than those found in terrestrial mammalians (Table 4).

Many algae contain Vit C in concentrations similar to fruits. Furthermore, invertebrates such as mussels, seashells, crabs, squid, and urchins are Vit C-rich sources. Table 4 summarizes the Vit C contributions of aquatic resources, determined by maximum daily protein intake. Notice that, unlike meat and fatty tissues, most of these resources can fulfill the daily needs of Vit C, as discussed above.

As for the fish from rivers and lakes, these would have been available to both AMH and Neanderthals. However, as previously discussed, the last hominins hardly left any traces of their consumption. Likewise, fish appear to have been more restricted to Southern European hominins. As noted in Table 4, the Vit C content of this resource is much higher than that of meat, and considering the daily amounts that were safe, whole fish consumption would have fulfilled the daily needs of Vit C for hominins. For example, consuming just 250 g of Atlantic salmon, which contains 4 mg of Vit C per 100 g of edible tissue, would have provided the daily amount of Vit C needed to prevent the development of scurvy.

In Western Europe, archeozoological analyses of Mousterian collections have shown that Neandertals regularly obtained large mammals and, to a lesser extent, small fast game such as rabbits [39], while in Iberia this hominin exploited birds [4]. The use of marine mammals would have provided significant benefits in terms of nutritional status. These, in addition to viscera, contain exceptionally Vit C-rich large organs—the skin and subcutaneous tissues. For instance, the concentration of Vit C in raw mattak from beluga (*Delphinapterus leucas* Pallas) is 36.02 mg/100 g [88]. Another important fact to be considered is that Vit C-rich organs of the terrestrial animals, such as the liver and kidney, represent a modest weight percentage of the total live weight of any mammal, thus, small- and medium-sized land mammals could have only offered a limited amount of Vit C to M3H. This is not the case of marine mammals; for instance, up to a quarter of the live weight of beluga is mattak [172]. However, as previously stated, there are only reports on the exploitation of marine mammals in Gibraltar [69].

The capture of fish and birds, whose meats are rich in Vit C (Table 4) and additionally contain omega-3 PUFA, requires strategies and skills different from those required to acquire large prey. Access to small quick game is limited by technical skills, but this skill is usually learned in late childhood [173]. Although it is documented that Neanderthals made use of such resources, AMH were more adapted to this type of complex hunting, thanks to newer and more effective learning strategies [174]. In this regard, AMH have more than 200 unique non-protein-coding genes regulating the co-expression of many more protein-coding genes in coordinated networks that underlie their capacities for self-awareness, creativity, prosocial behavior, and healthy longevity, which are not found in chimpanzees or Neanderthals [175]. Such cognitive capacities could have helped the domestication of dogs. There is evidence that the dogs or incipient dog remains found at some sites were from sub-Arctic and Arctic environments. In fact, the domestication of dogs may have predated what is traditionally accepted [176]. Considering that food animals contain much more meat than fat, AMH in Eurasia would have used the previously discussed surplus of meat from carcasses to share with incipient dogs [177]. This strategy would have favored the survival of AMH populations as compared to Neanderthal ones, since dogs would have helped the former in hunting and other well-known related tasks that still remain prevalent in current societies.

4. Conclusions

The results of this study underscore the importance of the viscera of large mammals as a source of vitamin C for Paleolithic hominins. Until now, the exploitation of large mammals by Paleolithic hunters was considered exclusively as a consequence of the high energetic return derived from their hunting. As demonstrated in this work, the meat and fatty tissues of terrestrial mammals could hardly have provided the daily amount of vitamin C that Paleolithic hominins needed to survive. Conversely, viscera from food animals could have supplied the Vit C needed for optimal health, while the regular intake of such a source of Vit C had key features. This food resource is limited because the highest proportion of the edible weight of any hunted mammal corresponds to meat and fatty tissues, and thus, reliance on small- and medium-sized animals supplying minor amounts of viscera could have caused an erratic intake of Vit C within the whole hominin group.

Paleolithic hominins experienced abrupt, large-magnitude changes in climate during the cold stages, known as Heinrich events, that occurred during the MIS 3 period, which caused changes in the biomes from which such hominins obtained plant and animal foods. In this context, the progressive decline of large mammals detected during such a period would have deprived Paleolithic hominins of a considerable proportion of the Vit C needed for survival, so they would have experienced a progressive deterioration in their nutritional status. In this context, there are proofs that Anatomically Modern Humans exploited birds and aquatic resources more successfully than Neanderthals, which allowed the former to develop a better Vit C-dependent nutritional status than Neanderthals during Heinrich events. However, in light of the current knowledge, more studies are needed to delve into the use of differential food resources between both types of hominins. This would help us to discern to what extent the decline in large mammals could have contributed to the demise of Neanderthals and to the evolutionary success of Anatomically Modern Humans.

Supplementary Materials: The following supporting information can be downloaded at: https://www.mdpi.com/article/10.3390/quat6010020/s1, File S1: Supporting Table S1. Organ weights, kcal and ascorbic acid content of a medium-sized reindeer (135 kg); File S2: Supporting Table S2. Organ weights, kcal, and ascorbic acid content of a medium-sized horse (210 kg); File S3: Supporting Table S3. Organ weights, kcal, and ascorbic acid content of a medium-sized reed deer (325 kg); File S4: Supporting Table S4. Organ weights, kcal, and ascorbic acid content of a medium-sized bison (910 kg); File S5: Supporting Table S5. Organ weights, kcal, and ascorbic acid content of a medium-sized woolly rhinoceros (2000 kg); File S6: Supporting Table S6. Organ weights, kcal, and ascorbic acid content of a medium-sized woolly mammoth (4000 kg); File S7: Supporting Table S7. Daily needs of energy, meat, and fatty tissues for a 2.5 g protein/kg·day muskox-fed 22-individualss M3H group.

Funding: The authors acknowledge the financial support of Junta de Andalucía (Proyect P20_00806), Campus de Excelencia Internacional Agroalimentario (ceiA3), and Centro de Investigación en Agrosistemas Intensivos Mediterráneos y biotecnología Agroalimentaria (CIAMBITAL).

Data Availability Statement: The data presented in this study are available in the article and Supplementary Material.

Conflicts of Interest: The sponsors had no role in the design, execution, interpretation, or writing of the study.

References

1. Arsuaga, J.L.; Martínez, I.; Arnold, L.J.; Aranburu, A.; Gracia-Téllez, A.; Sharp, W.D.; Quam, R.M.; Falguères, C.; Pantoja-Pérez, A.; Bischoff, J.; et al. Neandertal roots: Cranial and chronological evidence from Sima de los Huesos. *Science* **2014**, *344*, 1358–1363. [CrossRef] [PubMed]
2. Devièse, T.; Abrams, G.; Hajdinjak, M.; Pirson, S.; De Groote, I.; Di Modica, K.; Toussaint, M.; Fischer, V.; Comeskey, D.; Spindler, L.; et al. Reevaluating the timing of Neanderthal disappearance in Northwest Europe. *Proc. Natl. Acad. Sci. USA* **2021**, *118*, e2022466118. [CrossRef] [PubMed]
3. Guil-Guerrero, J.L.; Manzano-Agugliaro, F. Worldwide research trends on Neanderthals. *J. Quat. Sci.* **2022**, *38*, 208–220. [CrossRef]
4. Finlayson, C.; Carrión, J.S. Rapid ecological turnover and its impact on Neanderthal and other human populations. *Trends Ecol. Evol.* **2007**, *22*, 213–222. [CrossRef] [PubMed]
5. Staubwasser, M.; Drăgușin, V.; Onac, B.P.; Assonov, S.; Ersek, V.; Hoffmann, D.L.; Veres, D. Impact of climate change on the transition of Neanderthals to modern humans in Europe. *Proc. Natl. Acad. Sci. USA* **2018**, *115*, 9116–9121. [CrossRef]
6. Banks, W.E.; d'Errico, F.; Peterson, A.T.; Kageyama, M.; Sima, A.; Sánchez-Goñi, M.F. Neanderthal extinction by competitive exclusion. *PLoS ONE* **2008**, *3*, e3972. [CrossRef]
7. Houldcroft, C.; Underdown, S.J. Neanderthal genomics suggests a pleistocene time frame for the first epidemiologic transition. *Am. J. Phys. Anthropol.* **2016**, *160*, 379–388. [CrossRef]

8. Smith, F.H.; Janković, I.; Karavanić, I. The assimilation model, modern human origins in Europe, and the extinction of Neandertals. *Quat. Int.* **2005**, *137*, 7–19. [CrossRef]
9. Lalueza-Fox, C. Neanderthal assimilation? *Nat. Ecol. Evol.* **2021**, *5*, 711–712. [CrossRef]
10. Degioanni, A.; Bonenfant, C.; Cabut, S.; Condemi, S. Living on the edge: Was demographic weakness the cause of Neanderthal demise? *PLoS ONE* **2019**, *14*, e0216742. [CrossRef]
11. Fitzsimmons, K.E.; Hambach, U.; Veres, D.; Iovita, R. The Campanian Ignimbrite eruption: New data on volcanic ash dispersal and its potential impact on human evolution. *PLoS ONE* **2013**, *8*, e65839. [CrossRef] [PubMed]
12. Timmermann, A. Quantifying the potential causes of Neanderthal extinction: Abrupt climate change versus competition and interbreeding. *Quat. Sci. Rev.* **2020**, *238*, 106331. [CrossRef]
13. Fabre, V.; Condemi, S.; Degioanni, A. Genetic evidence of geographical groups among Neanderthals. *PLoS ONE* **2009**, *4*, e5151. [CrossRef] [PubMed]
14. Haws, J.A.; Benedetti, M.M.; Talamo, S.; Bicho, N.; Cascalheira, J.; Ellis, M.G.; Carvalho, M.M.; Friedl, L.; Pereira, T.; Zinsious, B.K. The early Aurignacian dispersal of modern humans into westernmost Eurasia. *Proc. Natl. Acad. Sci. USA* **2020**, *117*, 25414–25422. [CrossRef] [PubMed]
15. Stewart, J.R. Neanderthal extinction as part of the faunal change in Europe during Marine Isotope Stage 3. *Acta Zool. Crac.-Ser. A Vertebr.* **2007**, *50*, 93–124. [CrossRef]
16. Bunn, H.T., III. *Meat-Eating and Human Evolution: Studies on the Diet and Subsistence Patterns of Plio-Pleistocene Hominids in East Africa*; University of California: Berkeley, CA, USA, 1982.
17. Kolobova, K.A.; Roberts, R.G.; Chabai, V.P.; Jacobs, Z.; Krajcarz, M.T.; Shalagina, A.V.; Krivoshapkin, A.I.; Li, B.; Uthmeier, T.; Markin, S.V.; et al. Archaeological evidence for two separate dispersals of Neanderthals into southern Siberia. *Proc. Natl. Acad. Sci. USA* **2020**, *117*, 2879–2885. [CrossRef]
18. Stuart, A.J. Late Quaternary megafaunal extinctions on the continents: A short review. *Geol. J.* **2015**, *50*, 338–363. [CrossRef]
19. Richards, M.P.; Trinkaus, E. Out of Africa: Modern human origins special feature: Isotopic evidence for the diets of European Neanderthals and early modern humans. *Proc. Natl. Acad. Sci. USA* **2009**, *106*, 16034-9. [CrossRef]
20. Wißing, C.; Rougier, H.; Crevecoeur, I.; Germonpré, M.; Naito, Y.I.; Semal, P.; Bocherens, H. Isotopic evidence for dietary ecology of late Neandertals in North-Western Europe. *Quat. Int.* **2016**, *411*, 327–345. [CrossRef]
21. Ben-Dor, M.; Barkai, R. *Supersize Does Matter: The Importance of Large Prey in Paleolithic Subsistence and a Method for Measurement of Its Significance in Zooarchaeological Assemblages, in Human–Elephant Interactions: From Past to Present*; Konidaris, G., Barkai, R., Tourloukis, V., Eds.; Tübingen University Press: Tübingen, Germany, 2021.
22. Ben-Dor, M.; Barkai, R. Prey Size Decline as a Unifying Ecological Selecting Agent in Pleistocene Human Evolution. *Quaternary* **2021**, *4*, 7. [CrossRef]
23. Dodat, P.J.; Tacail, T.; Albalat, E.; Gómez-Olivencia, A.; Couture-Veschambre, C.; Holliday, T.; Madelaine, S.; Martin, J.E.; Rmoutilova, R.; Maureille, B.; et al. Isotopic calcium biogeochemistry of MIS 5 fossil vertebrate bones: Application to the study of the dietary reconstruction of Regourdou 1 Neandertal fossil. *J. Hum. Evol.* **2021**, *151*, 102925. [CrossRef] [PubMed]
24. Balter, V.; Person, A.; Labourdette, N.; Drucker, D.; Renard, M.; Vandermeersch, B. Les Néandertaliens étaient-ils essentiellement carnivores? Résultats préliminaires sur les teneurs en Sr et en Ba de la paléobiocénose mammalienne de Saint-Césaire. *Comptes Rendus L'acad. Sci.-Ser. IIA-Earth Planet. Sci.* **2001**, *332*, 59–65. [CrossRef]
25. Kuzmin, Y.V.; Bondarev, A.A.; Kosintsev, P.A.; Zazovskaya, E.P. The Paleolithic diet of Siberia and Eastern Europe: Evidence based on stable isotopes (δ13C and δ15N) in hominin and animal bone collagen. *Archaeol. Anthropol. Sci.* **2021**, *13*, 179. [CrossRef]
26. Drucker, D.G.; Naito, Y.I.; Péan, S.; Prat, S.; Crépin, L.; Chikaraishi, Y.; Ohkouchi, N.; Puaud, S.; Lázničková-Galetová, M.; Patou-Mathis, M.; et al. Isotopic analyses suggest mammoth and plant in the diet of the oldest anatomically modern humans from far southeast Europe. *Sci. Rep.* **2017**, *7*, 6833. [CrossRef]
27. Wißing, C.; Rougier, H.; Baumann, C.; Comeyne, A.; Crevecoeur, I.; Drucker, D.G.; Gaudzinski-Windheuser, S.; Germonpré, M.; Gómez-Olivencia, A.; Krause, J.; et al. Stable isotopes reveal patterns of diet and mobility in the last Neandertals and first modern humans in Europe. *Sci. Rep.* **2019**, *9*, 4433. [CrossRef]
28. Miras, Y.; Barbier-Pain, D.; Ejarque, A.; Allain, E.; Allué, E.; Marín, J.; Vettese, D.; Hardy, B.; Puaud, S.; Llach, J.M.; et al. Neanderthal plant use and stone tool function investigated through non-pollen palynomorphs analyses and pollen washes in the Abri du Maras, South-East France. *J. Archaeol. Sci. Rep.* **2020**, *33*, 102569. [CrossRef]
29. Hardy, K.; Bocherens, H.; Miller, J.B.; Copeland, L. Reconstructing Neanderthal diet: The case for carbohydrates. *J. Hum. Evol.* **2022**, *162*, 103105. [CrossRef]
30. Henry, A.G.; Brooks, A.S.; Piperno, D.R. Plant foods and the dietary ecology of Neanderthals and early modern humans. *J. Hum. Evol.* **2014**, *69*, 44–54. [CrossRef]
31. Churchill, S.E. *Thin on the Ground: Neandertal Biology, Archeology, and Ecology*; John Wiley & Sons: Hoboken, NJ, USA, 2014.
32. Buck, L.T.; Stringer, C.B. Having the stomach for it: A contribution to Neanderthal diets? *Quat. Sci. Rev.* **2014**, *96*, 161–167. [CrossRef]
33. Fiorenza, L.; Benazzi, S.; Tausch, J.; Kullmer, O.; Bromage, T.G.; Schrenk, F. Molar macrowear reveals Neanderthal eco-geographic dietary variation. *PLoS ONE* **2011**, *6*, e14769. [CrossRef]

34. Droke, J.L.; Schmidt, C.W.; Williams, F.L.; Karriger, W.M.; Smith, F.H.; Becam, G.; de Lumley, M.-A. Regional variability in diet between Northern European and Mediterranean Neandertals: Evidence from dental microwear texture analysis. In *Dental Wear in Evolutionary and Biocultural Contexts*; Academic Press: Cambridge, MA, USA, 2020; pp. 225–241. [CrossRef]
35. Ben-Dor, M.; Gopher, A.; Barkai, R. Neandertals' large lower thorax may represent adaptation to high protein diet. *Am. J. Phys. Anthropol.* **2016**, *160*, 367–378. [CrossRef] [PubMed]
36. Hardy, B.L.; Moncel, M.-H. Neanderthal use of fish, mammals, birds, starchy plants and wood 125–250,000 years ago. *PLoS ONE* **2011**, *6*, e23768. [CrossRef] [PubMed]
37. Fiore, I.; Gala, M.; Romandini, M.; Cocca, E.; Tagliacozzo, A.; Peresani, M. From feathers to food: Reconstructing the complete exploitation of avifaunal resources by Neanderthals at Fumane cave, unit A9. *Quat. Int.* **2016**, *421*, 134–153. [CrossRef]
38. Finlayson, S.; Finlayson, G.; Guzman, F.J.G.; Finlayson, C. Neanderthals and the cult of the Sun Bird. *Quat. Sci. Rev.* **2019**, *217*, 217–224. [CrossRef]
39. Morin, E.; Meier, J.; El Guennouni, K.; Moigne, A.-M.; Lebreton, L.; Rusch, L.; Valensi, P.; Conolly, J.; Cochard, D. New evidence of broader diets for archaic Homo populations in the northwestern Mediterranean. *Sci. Adv.* **2019**, *5*, eaav9106. [CrossRef]
40. Guillaud, E.; Béarez, P.; Daujeard, C.; Defleur, A.R.; Desclaux, E.; Roselló-Izquierdo, E.; Morales-Muñiz, A.; Moncel, M.-H. Neanderthal foraging in freshwater ecosystems: A reappraisal of the Middle Paleolithic archaeological fish record from continental Western Europe. *Quat. Sci. Rev.* **2021**, *252*, 106731. [CrossRef]
41. Tanner, A. *An embarrassment of riches: The ontological aspect of meat and fat harvesting among subarctic hunters. In Human–Elephant Interactions: From Past to Present*; Konidaris, G., Barkai, R., Tourloukis, V., Eds.; Tübingen University Press: Tübingen, Germany, 2021; Available online: https://publikationen.uni-tuebingen.de/xmlui/handle/10900/114229 (accessed on 19 November 2022).
42. Ben-Dor, M.; Gopher, A.; Hershkovitz, I.; Barkai, R. Man the fat hunter: The demise of Homo erectus and the emergence of a new hominin lineage in the Middle Pleistocene (ca. 400 kyr) Levant. *PLoS ONE* **2011**, *6*, e28689. [CrossRef]
43. Navazo Ruiz, M.; Benito-Calvo, A.; Alonso-Alcalde, R.; Alonso, P.; de la Fuente, H.; Santamaría, M.; Santamaría, C.; Álvarez-Vena, A.; Arnold, L.J.; Iriarte-Chiapusso, M.J.; et al. Late Neanderthal subsistence strategies and cultural traditions in the northern Iberia Peninsula: Insights from Prado Vargas, Burgos, Spain. *Quat. Sci. Rev.* **2021**, *254*, 106795. [CrossRef]
44. McLeod, H. Plenty of Fish in the Sea? A Comparison of Marine Resource Use in Early Hominins. *Ethnograph* **2018**, *4*, 36–55.
45. Bertacchi, A.; Starkovich, B.M.; Conard, N.J. The Zooarchaeology of Sirgenstein Cave: A Middle and Upper Paleolithic site in the Swabian Jura, SW Germany. *J. Paleolit. Archaeol.* **2021**, *4*, 7. [CrossRef]
46. Englard, S.; Seifter, S. The biochemical functions of ascorbic acid. *Annu. Rev. Nutr.* **1986**, *6*, 365–406. [CrossRef]
47. Padh, H. Cellular functions of ascorbic acid. *Biochem. Cell Biol.* **1990**, *68*, 1166–1173. [CrossRef]
48. Baker, E.M.; Hodges, R.E.; Hood, J.; Sauberlich, H.E.; March, S.C.; Canham, J.E. Metabolism of ^{14}C and ^{3}H-labeled L-ascorbic acid in human scurvy. *Am. J. Clin. Nutr.* **1971**, *24*, 444–454. [CrossRef]
49. Burri, B.J.; Jacob, R.A. Human metabolism and the requirement for vitamin C. In *Vitamin C in Health and Disease*; Packer, L., Fuchs, J., Eds.; Marcel Dekker Inc.: New York, NY, USA, 1997; pp. 341–366.
50. Olmedo, J.M.; Yiannias, J.A.; Windgassen, E.B.; Gornet, M.K. Scurvy: A disease almost forgotten. *Int. J. Dermatol.* **2006**, *45*, 909–913. [CrossRef] [PubMed]
51. Huss-Ashmore, R.; Goodman, A.H.; Armelagos, G.J. Nutritional inference from paleopathology. In *Advances in Archaeological Method and Theory*; Academic Press: Cambridge, MA, USA, 1982; pp. 395–474. Available online: https://www.jstor.org/stable/20210060 (accessed on 12 January 2023).
52. Ortner, D.J.; Ericksen, M.F. Bone changes in the human skull probably resulting from scurvy in infancy and childhood. *Int. J. Osteoarchaeol.* **1997**, *7*, 212–220. [CrossRef]
53. Eddie, D.M. Examination of Trauma in a Neandertal Ulna. Master's Thesis, University of Kansas, Lawrence, KS, USA, 2013. Available online: https://kuscholarworks.ku.edu/handle/1808/12956 (accessed on 27 October 2022).
54. Mednikova, M.B. Bioarcheology of Neanderthal burials from the territory of France and Crimea. *Brief Commun. Inst. Archeol.* **2015**, *238*, 243–261.
55. Fennell, K.J.; Trinkaus, E. Bilateral femoral and tibial periostitis in the La Ferrassie 1 Neanderthal. *J. Archaeol. Sci.* **1997**, *24*, 985–995. [CrossRef]
56. Gardner, J.C.; Smith, F.H. The paleopathology of the Krapina Neandertals. *Period. Biol.* **2006**, *108*, 471.
57. Teschler-Nicola, M.; Czerny, C.; Oliva, M.; Schamall, D.; Schultz, M. Pathological alterations and traumas in the human skeletal remains from Mladeč. In *Early Modern Humans at the Moravian Gate*; Springer: Vienna, Austria, 2006; pp. 473–489.
58. Garralda, M.D.; Vandermeersch, B. ¿Escorbuto en los neandertales? Posibles casos en Combe-Grenal (Domme, Francia). *Veleia* **2008**, *24–25*, 385–395.
59. Geraci, J.R.; Smith, T.G. Ascorbic acid in the diet of Inuit hunters from Holman, Northwest Territories. *Arctic* **1979**, *32*, 135–139. [CrossRef]
60. Rodahl, K. Nutritional requirements in cold climates. *J. Nutr.* **1954**, *53*, 575–588. [CrossRef] [PubMed]
61. Fediuk, K. Vitamin C in the Inuit Diet: Past and Present. School of Dietetics & Human Nutrition. Master's Thesis, McGill University, Montreal, QC, Canada, 2000.
62. Delanghe, J.R.; Langlois, M.R.; De Buyzere, M.L.; Na, N.; Ouyang, J.; Speeckaert, M.M.; Torck, M.A. Vitamin C deficiency: More than just a nutritional disorder. *Genes Nutr.* **2011**, *6*, 341–346. [CrossRef]

63. Clemens, Z.; Tóth, C. Vitamin C and disease: Insights from the evolutionary perspective. *J. Evol. Health Jt. Publ. Ancestral Health Soc. Soc. Evol. Med. Health* **2013**, *1*, 13. [CrossRef]
64. Høygaard, A.; Rasmussen, H.W. Vitamin C sources in eskimo food. *Nature* **1939**, *3631*, 943. [CrossRef]
65. Mullie, P.; Deliens, T.; Clarys, P. Vitamin C in East-Greenland traditional nutrition: A reanalysis of the Høygaard nutritional data (1936–1937). *Int. J. Circumpolar Health* **2021**, *80*, 1951471. [CrossRef]
66. Levine, V.E. The value of meat as an antiscorbutic. *Am. J. Dig. Dis.* **1941**, *8*, 454–463. [CrossRef]
67. Stefansson, V. The Dilemma in Vitamins. *Science* **1939**, *89*, 484. [CrossRef]
68. Pierson, E.M. The Antiscorbutic Properties of Some Common Food Materials. Ph.D. Thesis, The University of Minnesota, Minneapolis, MN, USA, 1922.
69. Stringer, C.B.; Finlayson, J.C.; Barton, R.N.E.; Fernández-Jalvo, Y.; Cáceres, I.; Sabin, R.C.; Rhodes, E.J.; Currant, A.P.; Rodríguez-Vidal, J.; Giles-Pacheco, F.; et al. Neanderthal exploitation of marine mammals in Gibraltar. *Proc. Natl. Acad. Sci. USA* **2008**, *105*, 14319–14324. [CrossRef]
70. Dembitzer, J.; Barkai, R.; Ben-Dor, M.; Meiri, S. Levantine overkill: 1.5 million years of hunting down the body size distribution. *Quat. Sci. Rev.* **2022**, *276*, 107316. [CrossRef]
71. Ben-Dor, M.; Barkai, R. The importance of large prey animals during the Pleistocene and the implications of their extinction on the use of dietary ethnographic analogies. *J. Anthropol. Archaeol.* **2020**, *59*, 101192. [CrossRef]
72. Ben-Dor, M.; Sirtoli, R.; Barkai, R. The evolution of the human trophic level during the Pleistocene. *Am. J. Phys. Anthropol.* **2021**, *175*, 27–56. [CrossRef] [PubMed]
73. Guil-Guerrero, J.L. Evidence for chronic omega-3 fatty acids and ascorbic acid deficiency in Palaeolithic hominins in Europe at the emergence of cannibalism. *Quat. Sci. Rev.* **2017**, *157*, 176–187. [CrossRef]
74. Goldfield, A.E.; Booton, R.; Marston, J.M. Modeling the role of fire and cooking in the competitive exclusion of Neanderthals. *J. Hum. Evol.* **2018**, *124*, 91–104. [CrossRef] [PubMed]
75. Venner, S.J. *A New Estimate for Neanderthal Energy Expenditure*; CUNY Hunter College: New York, NY, USA, 2018.
76. Paddon-Jones, D.; Rasmussen, B. Dietary protein recommendations and the prevention of sarcopenia: Protein, amino acid metabolism and therapy. *Curr. Opin. Clin. Nutr. Metab. Care* **2009**, *12*, 86–90. [CrossRef]
77. Bilsborough, S.; Mann, N. A review of issues of dietary protein intake in humans. *Int. J. Sport. Nutr. Exerc. Metab.* **2006**, *16*, 129. [CrossRef]
78. Shaffer, P.A. Antiketogenesis. II. The ketogenic antiketogenic balance in man. *J. Biol. Chem.* **1921**, *47*, 463–473.
79. Hassan, F.A. Demographic archaeology. In *Advances in Archaeological Method and Theory*; Academic Press: Cambridge, MA, USA, 1981; pp. 225–279.
80. Kelly, R.L. *The Lifeways of Hunter-Gatherers: The Foraging Spectrum*, 2nd ed.; Cambridge University Press: Cambridge, CA, USA, 2013.
81. Goren-Inbar, N.; Belfer-Cohen, A. Reappraisal of hominin group size in the Lower Paleolithic: An introduction to the special issue. *J. Hum. Evol.* **2020**, *144*, 102821. [CrossRef]
82. Smith, E.A. The application of optimal foraging theory to the analysis of hunter-gatherer group size. *Hunt.-Gatherer Strateg.* **1981**, *137*, 36–65.
83. Trinkaus, E. Neanderthal mortality patterns. *J. Archaeol. Sci.* **1995**, *22*, 121–142. [CrossRef]
84. Trinkaus, E. Late Pleistocene adult mortality patterns and modern human establishment. *Proc. Natl. Acad. Sci. USA* **2011**, *108*, 1267–1271. [CrossRef]
85. Dietary Guidelines for Americans. *United States Department of Agriculture and United States Department of Health and Human Services*, 7th ed.; U.S. Government Printing Office: Washington, DC, USA, 2010. Available online: http://www.dietaryguidelines.gov (accessed on 25 December 2022).
86. Hassan, A.A.; Sandanger, T.M.; Brustad, M. Level of selected nutrients in meat, liver, tallow and bone marrow from semi-domesticated reindeer (*Rangifer t. tarandus* L.). *Int. J. Circumpolar Health* **2012**, *71*, 17997. [CrossRef] [PubMed]
87. Rodahl, K. Vitamin sources in Arctic regions. *Nor. Polarinst. Skr.* **1949**, *91*, 1–47.
88. Fediuk, K.; Hidiroglou, N.; Madère, R.; Kuhnlein, H.V. Ascorbic acid in Inuit traditional food and women's diets. *J. Food Compos. Anal.* **2002**, *15*, 221–235. [CrossRef]
89. USDA National Nutrient Database for Standard Reference Legacy Release. Available online: https://ndb.nal.usda.gov (accessed on 25 June 2022).
90. Driskell, J.A.; Marchello, M.J.; Giraud, D.W.; Sulaeman, A. Vitamin and selenium content of ribeye cuts from grass-and grain-finished bison of the same herd. *J. Food Qual.* **2004**, *27*, 388–398. [CrossRef]
91. Marchello, M.J.; Driskell, J.A. Nutrient composition of grass-and grain-finished bison. *Great Plains Res.* **2001**, *11*, 65–82. [CrossRef]
92. Lee, C.-E.; Seong, P.-N.; Oh, W.-Y.; Ko, M.-S.; Kim, K.-I.; Jeong, J.-H. Nutritional characteristics of horsemeat in comparison with those of beef and pork. *Nutr. Res. Pract.* **2007**, *1*, 70–73. [CrossRef] [PubMed]
93. Serum, D.K.; Field, R.A.; Miller, G.J. Contribution of bone marrow to the vitamin content of mechanically separated meat. *Meat Sci.* **1986**, *17*, 73–77. [CrossRef]
94. Kizlaitis, L.; Steinfeld, M.I.; Siedler, A.J. Nutrient Content of Variety Meats. I. Vitamin A, Vitamin C, Iron, and Proximate Composition a, b. *J. Food Sci.* **1962**, *27*, 459–462. [CrossRef]

95. Eichenberger, B.; Pfirter, H.P.; Wenk, C.; Gebert, S. Influence of dietary Vitamin E and C supplementation on Vitamin E and C content and thiobarbituric acid reactive substances (TBARS) in different tissues of growing pigs. *Arch. Anim. Nutr.* **2004**, *58*, 195–208. [CrossRef]
96. Gebert, S.; Eichenberger, B.; Pfirter, H.P.; Wenk, C. Influence of different dietary vitamin C levels on vitamin E and C content and oxidative stability in various tissues and stored m. *longissimus dorsi of growing pigs. Meat Sci.* **2006**, *73*, 362–367. [CrossRef]
97. Mori, K.; Kidawara, M.; Iseki, M.; Umegaki, C.; Kishi, T. A simple fluorometric determination of vitamin C. *Chem. Pharm. Bull.* **1998**, *46*, 1474–1476. [CrossRef] [PubMed]
98. Phillips, P.H.; Stare, U.F. The distribution of a reducing substance (vitamin C) in the tissues of fluorine-fed cows. *J. Biol. Chem.* **1934**, *104*, 351–358. [CrossRef]
99. Svirbely, J.L.; Szent-Györgyi, A. The chemical nature of Vitamin C. *Biochem. J.* **1933**, *27*, 279–285. [CrossRef] [PubMed]
100. Slade, R.; Stead, A.G.; Graham, J.A.; Hatch, G.E. Comparison of lung antioxidant levels in humans and laboratory animals. *Am. Rev. Respir. Dis.* **1985**, *131*, 742–746.
101. Chant, J.L.; Day, L.; Field, R.A.; Kruggel, W.G.; Chang, Y. Composition and palatability of mechanically deboned meat and mechanically separated tissue. *J. Food Sci.* **1977**, *42*, 306–309. [CrossRef]
102. Descalzo, A.M.; Rossetti, L.; Grigioni, G.; Irurueta, M.; Sancho, A.M.; Carrete, J.; Pensel, N.A. Antioxidant status and odour profile in fresh beef from pasture or grain-fed cattle. *Meat Sci.* **2007**, *75*, 299–307. [CrossRef] [PubMed]
103. Mohammed, H.H.H.; Jin, G.; Ma, M.; Khalifa, I.; Shukat, R.; Elkhedir, A.E.; Zeng, Q.; Noman, A.E. Comparative characterization of proximate nutritional compositions, microbial quality and safety of camel meat in relation to mutton, beef, and chicken. *LWT* **2020**, *118*, 108714. [CrossRef]
104. Tamburrano, A.; Tavazzi, B.; Callà, C.A.M.; Amorini, A.M.; Lazzarino, G.; Vincenti, S.; Zottola, T.; Campagna, M.C.; Moscato, U.; Laurenti, P. Biochemical and nutritional characteristics of buffalo meat and potential implications on human health for a personalized nutrition. *Ital. J. Food Saf.* **2019**, *8*, 8317. [CrossRef]
105. Young, O.A.; Lane, G.A.; Podmore, C.; Fraser, K.; Agnew, M.; Cummings, T.L.; Cox, N.R. Changes in composition and quality characteristics of ovine meat and fat from castrates and rams aged to 2 years. *N. Z. J. Agric. Res.* **2006**, *49*, 419–430. [CrossRef]
106. Crile, G.; Quiring, D.P. A record of the body weight and certain organ and gland weights of 3690 animals. *Ohio J. Sci.* **1940**, *15*, 219–259.
107. Kirton, A.H.; Fourie, P.D.; Jury, K.E. Growth and development of sheep: III. Growth of the carcass and non-carcass components of the Southdown and Romney and their cross and some relationships with composition. *N. Z. J. Agric. Res.* **1972**, *15*, 214–227. [CrossRef]
108. Barrett, K.E. *Ganong's Review of Medical Physiology*, 24th ed.; McGraw-Hill Medical: New York, NY, USA, 2012.
109. Richmond, M.S.; Satterfield, G.H.; Grinnells, C.D.; Dann, W.J. Ascorbic acid Content of Goat's Milk and Blood: Influence of Ascorbic Acid Injection and Diet: Three Figures. *J. Nutr.* **1940**, *20*, 99–108. [CrossRef]
110. Awuk, A.; Tamir, B. Feed Intake, Weight Gain and Carcass Yield Characteristics of Intact Hararghe Highland Male Goats Fed on Different Hay to Concentrate Ratios. *East Afr. J. Sci.* **2007**, *1*, 45–54. [CrossRef]
111. Nollet, L.M.; Toldrá, F. (Eds.) *Handbook of Analysis of Edible Animal by-Products*; CRC Press: Boca Raton, FL, USA, 2011.
112. Kuhnleini, H.V.; Soueida, R. Use and nutrient composition of traditional Baffin Inuit foods. *J. Food Compos. Anal.* **1992**, *5*, 112–126. [CrossRef]
113. Bercovitch, F.B.; Rodriguez, J.F. Testis size, epididymis weight, and sperm competition in rhesus macaques. *Am. J. Primatol.* **1993**, *30*, 163–168. [CrossRef]
114. Adamczewski, J.Z.; Flood, P.F.; Gunn, A. Seasonal Patterns in Body Composition and Reproduction of Female Muskoxen (*Ovibos moschatus*). *J. Zool.* **1997**, *241*, 245–269. [CrossRef]
115. Kuhnlein, H.; Chan, H.; Leggee, D.; Barthet, V. Macronutrient, mineral and fatty acid composition of Canadian Arctic traditional food. *J. Food Compos. Anal.* **2002**, *15*, 545–566. [CrossRef]
116. Sebsibe, A.; Casey, N.; Van Niekerk, W.; Tegegne, A.; Coertze, R. Growth performance and carcass characteristics of three Ethiopian goat breeds fed grainless diets varying in concentrate to roughage ratios. *S. Afr. J. Anim. Sci.* **2007**, *37*, 221–232. [CrossRef]
117. Nieminen, M.; Laitinen, M. Bone marrow and kidney fat as indicators of condition in reindeer. *Rangifer* **1986**, *6*, 219–226. [CrossRef]
118. Knott, K.K.; Barboza, P.S.; Bowyer, R.T. Growth in arctic ungulates: Postnatal development and organ maturation in *Rangifer tarandus* and *Ovibos moschatus*. *J. Mammal.* **2005**, *86*, 121–130. [CrossRef]
119. Ferraces-Casais, P.; Lage-Yusty, M.A.; De Quirós, A.R.-B.; López-Hernández, J. Evaluation of bioactive compounds in fresh edible seaweeds. *Food Anal. Methods* **2012**, *5*, 828–834. [CrossRef]
120. Carr, R.S.; Neff, J.M. Determination of ascorbic acid in tissues of marine animals by liquid chromatography with electrochemical detection. *Anal. Chem.* **1980**, *52*, 2428–2430. [CrossRef] [PubMed]
121. Nettleton, J.A.; Exler, J. Nutrients in wild and farmed fish and shellfish. *J. Food Sci.* **1992**, *57*, 257–260. [CrossRef]
122. Carr, R.S.; Bally, M.B.; Thomas, P.; Neff, J.M. Comparison of methods for determination of ascorbic acid in animal tissues. *Anal. Chem.* **1983**, *55*, 1229–1232. [CrossRef]
123. de Quirós, A.R.-B.; Lopez-Hernandez, J.; Simal-Lozano, J. Determination of ascorbic acid in sea urchin: Comparison of two HPLC methods. *Chromatographia* **2001**, *53*, S246–S249. [CrossRef]

124. Özyurt, G.; Polat, A.; Loker, G.B. Vitamin and mineral content of pike perch (*Sander lucioperca*), common carp (*Cyprinus carpio*), and European catfish (*Silurus glanis*). *Turk. J. Vet. Anim. Sci.* **2009**, *33*, 351–356. [CrossRef]
125. Daujeard, C.; Abrams, G.; Germonpré, M.; Le Pape, J.-M.; Wampach, A.; Di Modica, K.; Moncel, M.-H. Neanderthal and animal karstic occupations from southern Belgium and south-eastern France: Regional or common features? *Quat. Int.* **2016**, *411*, 179–197. [CrossRef]
126. Lebreton, L.; Moigne, A.-M.; Filoux, A.; Perrenoud, C. A specific small game exploitation for Lower Paleolithic: The beaver (*Castor fiber*) exploitation at the Caune de l'Arago (Pyrénées-Orientales, France). *J. Archaeol. Sci. Rep.* **2017**, *11*, 53–58. [CrossRef]
127. Collie, J.T.; Greaves, R.F.; Jones, O.A.; Eastwood, G.; Bellomo, R. Ascorbic acid measurement in critical illness: Challenges, methodologies and quality improvements. *Clin. Chem. Lab. Med. (CCLM)* **2020**, *58*, 460–470. [CrossRef]
128. Carr, A.C.; Lykkesfeldt, J. Discrepancies in global vitamin C recommendations: A review of RDA criteria and underlying health perspectives. *Crit. Rev. Food Sci. Nutr.* **2021**, *61*, 742–755. [CrossRef]
129. Hodges, R.E.; Hood, J.; Canham, J.E.; Sauberlich, H.E.; Baker, E.M. Clinical manifestations of ascorbic acid deficiency in man. *Am. J. Clin. Nutr.* **1971**, *24*, 432–443. [CrossRef] [PubMed]
130. Masci, D.; Rubino, C.; Basile, M.; Indolfi, G.; Trapani, S. When the limp has a dietary cause: A retrospective study on scurvy in a tertiary Italian pediatric hospital. *Front. Pediatr.* **2022**, *10*, 1581. [CrossRef] [PubMed]
131. Henry, A.G.; van Amerongen, Y. Tubers in Winter. Seventh Annual Meeting of the European Society for the Study of Human Evolution. *Evol. Anthropol. Issues News Rev.* **2017**, *27*, 87.
132. Schwarcz, H.P. Winter sources of ascorbic acid for Pleistocene hominins in northern Eurasia. *Res. Sq.* **2023**, preprint. [CrossRef]
133. Ng, W.Y.; Hung, L.Y.; Lam, Y.H.; Chan, S.S.; Pang, K.S.; Chong, Y.K.; Ching, C.K.; Mak, T.W. Poisoning by toxic plants in Hong Kong: A 15-year review. *Hong Kong Med. J.* **2019**, *25*, 102. [CrossRef]
134. Applequist, W.L.; Bridges, M.C.; Moerman, D.E. North American Fertility–Regulating Botanicals: A Review. *Econ. Bot.* **2022**, *76*, 84–113. [CrossRef]
135. Ali, M.; Hossain, M.; Akhter, S.; Islam; Hashem, M. Effect of age on slaughterhouse by-products of indigenous cattle of Bangladesh. *Bangladesh J. Anim. Sci.* **2013**, *42*, 62–66. [CrossRef]
136. Bunn, H.T.; Ezzo, J.A. Hunting and scavenging by Plio-Pleistocene hominids: Nutritional constraints, archaeological patterns, and behavioural implications. *J. Archaeol. Sci.* **1993**, *20*, 365–398. [CrossRef]
137. Devatkal, S.; Mendiratta, S. Evaluation of shelf life and spoilage pattern of aerobically stored buffalo liver. *J. Muscle Foods* **2007**, *18*, 276–284. [CrossRef]
138. Tennie, C.; O'Malley, R.C.; Gilby, I.C. Why do chimpanzees hunt? Considering the benefits and costs of acquiring and consuming vertebrate versus invertebrate prey. *J. Hum. Evol.* **2014**, *71*, 38–45. [CrossRef]
139. McHugh, T. *The Time of the Buffalo*; University of Nebraska Press: Lincoln, NE, USA, 1979; p. 85.
140. Saario, D.J. *Human Ecological Investigations at Kivalina, Alaska. Final Report. No. PNE-425*; University of Alaska and Oak Ridge: Fairbanks, AK, USA, 1962.
141. Blumenschine, R.J.; Caro, T.M. Unit flesh weights of some East African bovids. *Afr. J. Ecol.* **1986**, *24*, 273–286. [CrossRef]
142. Nilssen, P.J. An actualistic Butchery Study in South Africa and Its Implications for Reconstructing Hominid Strategies of Carcass Acquisition and Butchery in the Upper Pleistocene and Plio-Pleistocene. Ph.D. Thesis, University of Cape Town, Cape Town, South Africa, 2000.
143. Blumenschine, R.J. Carcass consumption sequences and the archaeological distinction of scavenging and hunting. *J. Hum. Evol.* **1986**, *15*, 639–659. [CrossRef]
144. Morin, E.; Ready, E. Foraging Goals and Transport Decisions in Western Europe During the Paleolithic and Early Holocene. In *Zooarchaeology and Modern Human Origins: Human Hunting Behavior During the Later Pleistocene*; Clark, J., Speth, J., Eds.; Springer: New York, NY, USA, 2013; pp. 227–269.
145. Blasco, R.; Rosell, J.; Arilla, M.; Margalida, A.; Villalba, D.; Gopher, A.; Barkai, R. Bone marrow storage and delayed consumption at Middle Pleistocene Qesem Cave, Israel (420 to 200 ka). *Sci. Adv.* **2019**, *5*, eaav9822. [CrossRef]
146. Wu, G. Dietary protein intake and human health. *Food Funct.* **2016**, *7*, 1251–1265. [CrossRef]
147. Rendu, W.; Costamagno, S.; Meignen, L.; Soulier, M.-C. Monospecific faunal spectra in Mousterian contexts: Implications for social behavior. *Quat. Int.* **2012**, *247*, 50–58. [CrossRef]
148. Driver, J.C.; Maxwell, D. Bison death assemblages and the interpretation of human hunting behaviour. *Quat. Int.* **2013**, *297*, 100–109. [CrossRef]
149. Dusseldorp, G.L. Studying prehistoric hunting proficiency: Applying optimal foraging theory to the Middle Palaeolithic and Middle Stone Age. *Quat. Int.* **2012**, *252*, 3–15. [CrossRef]
150. Rodríguez-Hidalgo, A.; Saladié, P.; Ollé, A.; Arsuaga, J.L.; de Castro, J.M.B.; Carbonell, E. Human predatory behavior and the social implications of communal hunting based on evidence from the TD10. 2 bison bone bed at Gran Dolina (Atapuerca, Spain). *J. Hum. Evol.* **2017**, *105*, 89–122. [CrossRef] [PubMed]
151. Saladié, P.; Huguet, R.; Díez, C.; Rodríguez-Hidalgo, A.; Cáceres, I.; Vallverdú, J.; Rosell, J.; de Castro, J.M.B.; Carbonell, E. Carcass transport decisions in *Homo antecessor* subsistence strategies. *J. Hum. Evol.* **2011**, *61*, 425–446. [CrossRef]
152. O'Connell, J.F.; Hawkes, K.; Jones, N.B. Reanalysis of large mammal body part transport among the Hadza. *J. Archaeol. Sci.* **1990**, *17*, 301–316. [CrossRef]

153. Monchot, H.; Houmard, C.; Dionne, M.M.; Desrosiers, P.M.; Gendron, D. The modus operandi of walrus exploitation during the Palaeoeskimo period at the Tayara site, Arctic Canada. *Anthropozoologica* **2013**, *48*, 15–36. [CrossRef]
154. Rabassa, J.; Ponce, J.F. The Heinrich and Dansgaard-Oeschger climatic events during Marine Isotopic Stage 3: Searching for appropriate times for human colonization of the Americas. *Quat. Int.* **2013**, *299*, 94–105. [CrossRef]
155. Clarke, A.; Gaston, K.J. Climate, energy and diversity. *Proc. R. Soc. B Biol. Sci.* **2006**, *273*, 2257–2266. [CrossRef] [PubMed]
156. Nogués-Bravo, D.; Rodríguez, J.; Hortal, J.; Batra, P.; Araújo, M.B. Climate change, humans, and the extinction of the woolly mammoth. *PLoS Biol.* **2008**, *6*, e79. [CrossRef]
157. D'Errico, F.; Zilhão, J.; Julien, M.; Baffier, D.; Pelegrin, J. Neanderthal acculturation in Western Europe? A critical review of the evidence and its interpretation. *Curr. Anthropol.* **1998**, *39*, S1–S44. [CrossRef]
158. Eeuwes, D.D.M. Megafauna Extinction Caused by Homo Sapiens? Reports Environmental Science No. 589. Master's Thesis, Radboud University Nijmegen, Nijmegen, The Netherlands, 2018.
159. Magniez, P.; Boulbes, N. Environment during the Middle to Late Palaeolithic transition in southern France: The archaeological sequence of Tournal Cave (Bize-Minervois, France). *Quat. Int.* **2014**, *337*, 43–63. [CrossRef]
160. Morin, E. Evidence for declines in human population densities during the early Upper Paleolithic in western Europe. *Proc. Natl. Acad. Sci. USA* **2008**, *105*, 48–53. [CrossRef]
161. El Zaatari, S.; Grine, F.E.; Ungar, P.S.; Hublin, J.-J. Neandertal versus modern human dietary responses to climatic fluctuations. *PLoS ONE* **2016**, *11*, e0153277. [CrossRef]
162. Fabre, V.; Condemi, S.; Degioanni, A.; Herrscher, E. Neanderthals versus modern humans: Evidence for resource competition from isotopic modelling. *Int. J. Evol. Biol.* **2011**, *2011*, 689315. [CrossRef]
163. Sepulchre, P.; Ramstein, G.; Kageyama, M.; Vanhaeren, M.; Krinner, G.; Sánchez-Goñi, M.F.; d'Errico, F. H4 abrupt event and late Neanderthal presence in Iberia. *Earth Planet. Sci. Lett.* **2007**, *258*, 283–292. [CrossRef]
164. Defleur, A.R.; Desclaux, E. Impact of the last interglacial climate change on ecosystems and Neanderthals behavior at Baume Moula-Guercy, Ardèche, France. *J. Archaeol. Sci.* **2019**, *104*, 114–124. [CrossRef]
165. Drucker, D.; Bocherens, H. Carbon and nitrogen stable isotopes as tracers of change in diet breadth during Middle and Upper Palaeolithic in Europe. *Int. J. Osteoarchaeol.* **2004**, *14*, 162–177. [CrossRef]
166. Richards, M.P.; Pettitt, P.B.; Stiner, M.C.; Trinkaus, E. Stable isotope evidence for increasing dietary breadth in the European mid-Upper Paleolithic. *Proc. Natl. Acad. Sci. USA* **2001**, *98*, 6528–6532. [CrossRef] [PubMed]
167. Torke, W. *Fischreste als Quellen der Okologie und Okonomie in der Steinzeit Siidwest-Deutschlands;* Urgeschichtliche Materialhefte Nr. 4; Verlag Archaeologica Venatoria; Institut für Urgeschichte der Universität: Tübingen, Germany, 1981.
168. Conard, N.J.; Kitagawa, K.; Krönneck, P.; Böhme, M.; Münzel, S.C. The importance of fish, fowl and small mammals in the Paleolithic diet of the Swabian Jura, southwestern Germany. In *Zooarchaeology and Modern Human Origins;* Clark, J., Speth, J., Eds.; Springer: Berlin/Heidelberg, Germany, 2013; pp. 173–190. [CrossRef]
169. Van Neer, W.; Wouters, W.; Germonpré, M. Fish remains from three Upper Palaeolithic cave deposits in southern Belgium. *Anthropol. Praehist.* **2007**, *118*, 5–22.
170. Tortosa, J.; Pardo, J.F.J.; Álvarez-Fernández, E.; Ripoll, M.P.; Aristu, B.A.; Morales-Pérez, J.V.; García, M.J.R.; Marlasca, R.; Alcover, J.A.; Jardón, P.; et al. Palaeolithic-epipalaeolithic Sea People of the southern Iberian coast (Spain): An overview. In *Archaeology of Maritime Hunter-Gatherers. From Settlement Function to the Organization of the Coastal Zone;* Pôle editorial: Paris, France, 2016; pp. 69–92.
171. Álvarez-Fernández, E. Humans and Marine Resource Interaction Reappraised: Archaeofauna Remains During the Late Pleistocene and Holocene in Cantabrian Spain. *J. Anthropol. Archaeol.* **2011**, *30*, 327–343. [CrossRef]
172. The Royal Canadian Geographical Society. *Indigenous Peoples Atlas of Canada;* Canadian Geographic: Ottawa, ON, USA, 2018.
173. Stiner, M.C. Small game use and expanding diet breadth. Petits animaux et sociétés humaines du complément alimentaire aux ressources utilitaires. In *Actes des XXIVe Rencontres Internationales D'archéologie et D'histoire d'Antibes;* APDCA: Antibes, France, 2004.
174. Nishiaki, Y.; Jöris, O. Learning Behaviors Among Neanderthals and Palaeolithic Modern Humans: An Introduction. In *Learning Among Neanderthals and Palaeolithic Modern Humans;* Springer: Singapore, 2019; pp. 1–6.
175. Zwir, I.; Del-Val, C.; Hintsanen, M.; Cloninger, K.M.; Romero-Zaliz, R.; Mesa, A.; Arnedo, J.; Salas, R.; Poblete, G.F.; Raitoharju, E.; et al. Evolution of genetic networks for human creativity. *Mol. Psychiatry* **2021**, *27*, 354–376. [CrossRef]
176. Germonpré, M.; Fedorov, S.; Danilov, P.; Galeta, P.; Jimenez, E.-L.; Sablin, M.; Losey, R.J. Palaeolithic and prehistoric dogs and Pleistocene wolves from Yakutia: Identification of isolated skulls. *J. Archaeol. Sci.* **2017**, *78*, 1–19. [CrossRef]
177. Lahtinen, M.; Clinnick, D.; Mannermaa, K.; Salonen, J.S.; Viranta, S. Excess protein enabled dog domestication during severe Ice Age winters. *Sci. Rep.* **2021**, *11*, 7. [CrossRef]

quaternary

MDPI

Article

Neanderthal Hunting Weapons Re-Assessed: A Tip Cross-Sectional Area Analysis of Middle Palaeolithic Point Assemblages from South Eastern France

Marlize Lombard [1,*] and Marie-Hélène Moncel [2]

1 Palaeo-Research Institute, University of Johannesburg, Auckland Park, P.O. Box 524, Johannesburg 2006, South Africa
2 UMR 7194 CNRS-Museum National d'Histoire Naturelle, Institut de Paleontologie Humaine, 1 rue René Panhard, 75013 Paris, France
* Correspondence: mlombard@uj.ac.za

Abstract: There are many opinions and arguments about the types of weapons that Neanderthals may have used. We list five assumptions about Neanderthal weapon-assisted hunting and suggest that the tip cross-sectional area (TCSA) approach may be used to assess these, and to provide a hypothetical overview of stone-tipped weaponry used in south eastern France between MIS 7 and MIS 3. We analysed stone points from Abri du Maras, Saint-Marcel, Grotte du Figuier, and Payre, and discuss possible hunting strategies in tandem with faunal evidence. Our results suggest that early Neanderthals may have hunted with bimanual thrusting spears in combination with one-handed stabbing spears, but that later groups possibly introduced javelin hunting to the Neanderthal arsenal. Stone-tipped assegais or one-handed stabbing spears, however, stand out as the Neanderthal weapons of choice throughout the Middle Palaeolithic, and we discuss the adaptive advantages of hunting with these weapons. Comparative results that include assemblages from elsewhere in the Old World suggest similarities in hunting with stone-tipped weapons between Neanderthal and *H. sapiens* before MIS 5, marked differences during MIS 5–4, but similarities again during MIS 3. We suggest that caution is needed when attributing MIS 3 point assemblages to Neanderthals based on age and/or technology only.

Keywords: stone points; thrusting spears; stabbing spears; javelins; ballistic features; visuo-spatial integration

check for updates

Citation: Lombard, M.; Moncel, M.-H. Neanderthal Hunting Weapons Re-Assessed: A Tip Cross-Sectional Area Analysis of Middle Palaeolithic Point Assemblages from South Eastern France. *Quaternary* **2023**, 6, 17. https://doi.org/10.3390/quat6010017

Academic Editor: Ran Barkai

Received: 29 January 2023
Revised: 10 February 2023
Accepted: 20 February 2023
Published: 1 March 2023

1. Introduction

Similar to *Homo sapiens* foraging groups, Neanderthals gathered plant foods and probably also collected other morsels such as eggs, snails, crabs and lizards—to name but a few [1–4]. Remains of such veld-foods seldom preserve in the archaeological record. Meat consumption, on the other hand, is well-documented in the many faunal assemblages, showing that Neanderthals hunted prey, scavenged from predators and probably harvested weakened, trapped or recently dead animals. They pursued and killed large prey such as *Elephas* with simple wooden spears [5], and herding angulates such as *Equus* with stone-tipped weapons [6,7]. The multi-functionality of Middle Palaeolithic convergent stone tools—and their possible disconnect with hunting technologies—remains a topic of debate [8,9]. This discourse may be a remnant of thinking about Neanderthals and other early human hunters as incapable of effective hunting or making and using composite (multi-part) weapons [10–14]. In much of the discussion there is also a skewed understanding about which methods are appropriate for generating micro-scale evidence of hunting, as opposed to those designed for setting up wide-ranging, testable hypotheses about ancient hunting strategies (see discussion in [15]).

The method we apply aims to demonstrate the ballistic suitability of the tools studied to form part of different types of hunting weapons. It cannot directly assess whether each of the artefacts were used as weapon tips, and some of the elements could have been used for other activities. Use trace evidence for Neanderthal hunting with stone-tipped weapons is, however, mounting [16,17], and together with archaeozoological and isotopic studies [18–22], we may accept that at least some convergent lithics were hafted to wooden shafts and used as weapon tips. Lazuén [16] suggested that hunting with stone-tipped weapons was pervasive throughout southern Europe as part of organised Middle Palaeolithic subsistence strategies. It would also seem that Neanderthals coordinated their hunting behaviours to exploit game that congregated in large herds [14]. Reconstructions of the palaeo-landscape and animal ethology by White et al. [23] show that Neanderthals did not necessarily pre-select and isolate individual animals from a herd to pursue and kill—in the way many current hunter-gatherers do. Instead, they ambushed whole herds, slaughtering at random amongst the animals. When it came to processing the carcasses for transport and consumption, however, they were highly selective. It thus appears that Neanderthals were "excellent tacticians, casual executioners and discerning diners" [23] (p. 10).

Mithen [24] (p. 198) wrote the following about Neanderthal hunting:

"It is widely thought that their main hunting weapons appear to have been short thrusting spears, tipped with stone points. The points themselves required considerable skill to make, but the weapons themselves remained quite simple in design. With such weapons, the Neanderthals appear to have frequently used a confrontational hunting technique. This seems to have been a major cause of the high frequency of bone fractures found on Neanderthal skeletal remains and the high mortality of young adults. Why Neanderthals did not invent more effective hunting weapons and methods to avoid such injuries remains unclear. We have no evidence for throwing spears, for bows and arrows, for spear-throwers, for pits, traps and snares".

With this contribution, we analyse seven stone point assemblages from sites around the confluence of the Rhône and Ardèche-Payre tributaries on the south eastern margins of the Massif Central Mountains in France, (Figure 1). The assemblages include stone points from Abri du Maras (MIS 3 and 5), Saint-Marcel (MIS 3 and 5), Grotte du Figuier (MIS 3–4), and Payre (MIS 6 and 7) (Figure 1). We use the revised tip cross-section area (TCSA) method [15] as testable hypothesis-building tool to assess the following assumptions about Neanderthal hunting weapons:

1. Neanderthal stone-tipped weaponry was limited to double-handed thrusting spears [25,26].
2. Neanderthals used stone-tipped spears for both thrusting and short-distance throwing [27,28].
3. Neanderthals did not make or use mechanically projected technologies such as the spear-thrower and dart or the bow and arrow for killing at a distance [29,30].
4. Neanderthals used string-thrown darts [31].
5. Neanderthal weapons were similar to those used by broadly contemporaneous *Homo sapiens* hunters in Africa [32].

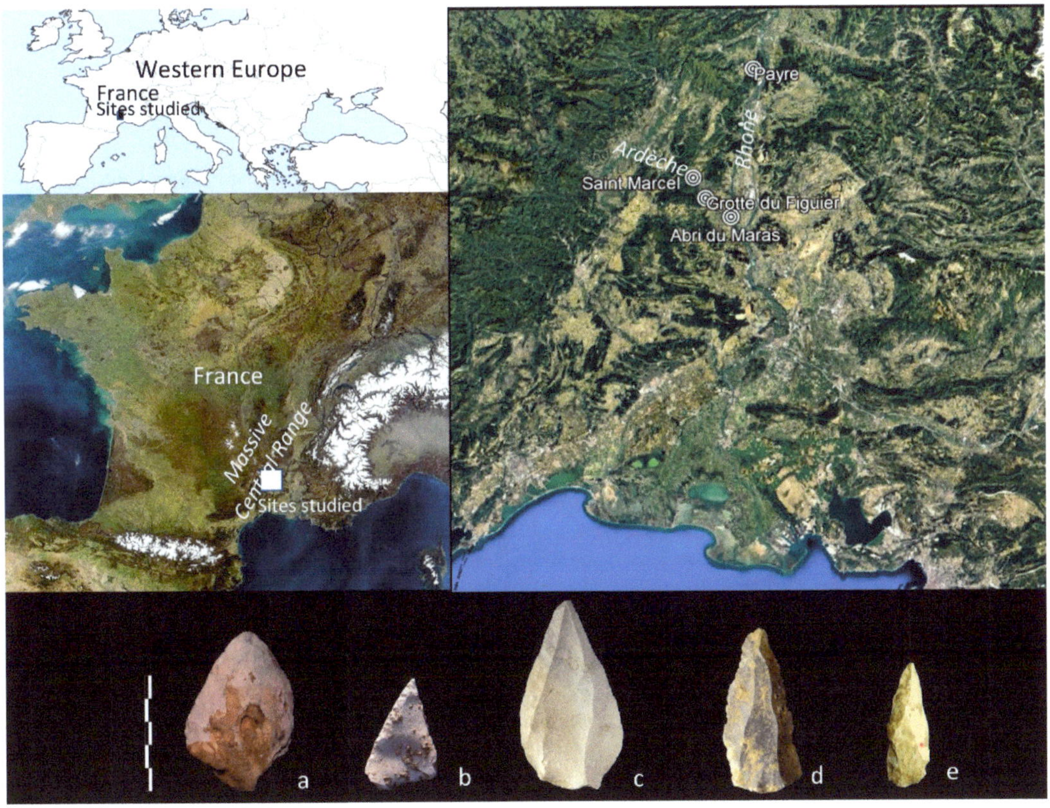

Figure 1. Location of the sites from which assemblages were analysed for this study, and points from the oldest Payre unit Ga, MIS 7 (**a**,**b**) and youngest Abri du Maras units 4.1 and 4.2, MIS 3 (**c**–**e**) assemblages in our study. Maps by M Lombard and artefact photographs by M-H. Moncel.

2. Our Approach

Whereas the design of weapon components may vary in terms of style, weapon tips must articulate effectively with the physiological and cognitive constraints of the hunters as well as with their hunting ecology [30]. In addition, weapons must suit the chosen subsistence, personal risk management, and mobility strategies of the hunters who use them. Design aspects that impact their ballistic success or failure will therefore co-vary with the different selective forces [33–37]. The TCSA statistic is ballistically relevant, because: (a) it is key to the success of most hunting weapons by representing a tip's ability to cut a hole through the hide of a prey animal [15,38]; and (b) it predicts the penetration depth of weapons—with smaller TCSA values reaching higher velocities and penetrating deeper than larger, heavier tips under similar launching conditions [39]. TCSA values ($0.5 \times$ maximum width $\times$ maximum thickness) cannot determine how any individual artefact was used. However, the method enables the assessment of large samples for best-fit ballistic probability over a range of weapon types [36,38,40], generating directly comparable quantitative data with little room for calculation or explanatory error [41].

The method, its interpretative potential, its shortcomings and improvements have been synthesised recently [15], so that we do not repeat that discussion here. Instead, we directly apply the calibrated TCSA values with statistical integrity for artefacts used to tip poisoned arrows, un-poisoned arrows, javelins and single-handed stabbing spears (Table 1) to our samples from south eastern France. The artefact sample setting the TCSA standard for bimanual-thrusting spears is still too small to be robust, for example, the TCSA range for four iron-tipped thrusting spears housed at the KwaZulu-Natal Museum is 180–360 mm^2, and that of eight European Pleistocene wooden spears amounts to 408–878 mm^2 [15]. For this study, we therefore use TCSA values of >200 mm^2 as a working standard for bimanual thrusting spears.

Table 1. Newly calibrated TCSA standards for six weapon types used by recent hunter-gatherers [15].

TCSA Standard	N	Mean mm^2	SD mm^2	Range mm^2	Median mm^2
Poisoned arrow tips (tested/calibrated)	565	11	7	4–18	8
Un-poisoned arrow tips (tested/calibrated)	338	32	15	17–47	31
Dart tips (working standard)	40	58	17	41–75	60
Javelin tips (tested/calibrated)	270	66	24	42–90	63
Stabbing spear tips (tested/calibrated)	141	140	60	80–200	128
Thrusting spear tips (working standard)	12	519	266	>200	465

With the introduction of a discrete TCSA category for javelin tips—a weapon system still in use by hunter-gatherers across the globe and therefore not to be dismissed in the context of deep-time hunting—the dart-tip category lost its statistical integrity [36,37]. What is more, the original dart-tip category does not consider the possibility of two distinct dart-tip TCSA values wherein large terrestrial spear-throwers were used for hunting sheep-sized and larger animals, and small Arctic spear-throwers to hunt marine mammals and birds [42,43]. We cannot resolve these issues here, and although the samples for other weapon-tip types have been augmented and their TCSA values calibrated [15], we still use the Thomas [44] and Shott [45] samples, and their TCSA range calculated by Shea [40], as representative of dart hunting. Using the scheme in Table 1, results in an almost complete overlap between the probability of dart use vs. the probability of javelin use. Thus, whilst relatively robust for hypothesising about the other weapon-delivery systems, TCSA results on their own are currently less able to highlight a distinction between javelin hunting and dart hunting. The most parsimonious approach to deal with this quandary is to accept the most conservative interpretation for overlapping TCSA values, namely javelin hunting, unless there is corroborating evidence for the use of spear-throwers.

We also assess hypothetical Neanderthal weapon-assisted hunting strategies on a wider scale. First, by comparing our results for south eastern France with those previously published for concurrent Neanderthal or probable/assumed Neanderthal assemblages and second, by looking at synchronous *H. sapiens* assemblages. We suggest that the groundwork laid in the resulting framework can serve to stimulate discussion and further research until data from many more assemblages are added to shape a robust framework for the evolution of weapon-assisted hunting during the Old World Pleistocene.

3. Our Assemblages

For the assemblages below we calculated the TCSA values for all the material with two convergent edges, regardless of the technical axis. *Déjétée* points and triangular flakes were also selected. The difference between a point (retouched or not) and a convergent scraper, for example, was made based on the value of the tip angle (with a graph to isolate groups if visible).

3.1. Payre

The two oldest assemblages in our study are from Payre, a collapsed cave site first excavated in the 1960s by Combier [46] and more recently by Moncel [47]. Today it is an open-air site of ≥80 m^2 with a 5m-thick stratigraphic sequence with seven units (from bottom to top: G, F, E, D, C, B and A). Four of these units feature occupation layers (G, F, E and D). The site is known for providing Middle Palaeolithic layers with Neanderthal remains, bracketed between flowstone [48–50]. U-series on flowstone, ESR/U-series on teeth and thermoluminescence (TL) on flint suggest a chronology ranging from ~300 ka to ~140 ka [51,52]. Occupations at the base of the sequence (units G and F) are concurrent with the transition between MIS 8 and MIS 7, and the top (units E and D) with the end of MIS 6. Some of the ages obtained from the base flowstone are, however, younger or fall within the same range as the TL ages for overlying layer G [53], and these ages must be used with caution as minimum estimates. The archaeological units are attributed to the early Middle Palaeolithic [54–56]. We analysed the point assemblages from layer G (n = 152) associated with MIS 7, and layer D associated with MIS 6 (n = 141).

3.2. Saint-Marcel

This cave site is located on the left side of the Ardèche River, in Saint-Marcel d'Ardèche, not far from the Abri du Maras. Under the entrance, excavations by Gilles from 1974 to 1988 revealed a 6.50m-thick infill, corresponding to detrital cones built up as a result of the collapse of the entrance [57]. Stratigraphic and sedimentological studies identified about forty layers, grouped into two complexes (upper and lower). The upper complex (layers c to u) contains several recurrent Mousterian occupations showing no behavioural change despite sedimentary breaks, except for level u at the bottom. The uppermost part of the upper complex (layers c to f) belongs to the Late Middle Palaeolithic, deposited during a temperate and wet period (MIS 4/beginning of MIS 3). In layer f, red deer bones with cut marks produced dates of 37,850 ± 550 BP (OxA-19623), 37,850 ± 600 BP (OxA-19624) and 41,300 ± 1700 BP (OxA-19625), which together with sedimentological evidence, indicate a Late Mousterian occupation [58]. Levels k to u (from the bottom to the upper complex) correspond to MIS 5e-a according to biochronology [59]. We analysed points from layers i and j of the MIS 3 occupation (n = 25), and from layer u (n = 8) corresponding to MIS 5e.

3.3. Abri du Maras

Located in the Ardèche gorge, Abri du Maras is a rock shelter that was initially excavated by Gilles and Combier [46]. Since the 2000s, excavations have been conducted by M-H. Moncel. The new stratigraphy consists of layers numbered from 1 at the top to 6 at the bottom, with most archaeological material recovered from layers 5 and 4. Layer 4 is subdivided into levels 4.1 and 4.2, representing the two main occupation phases. Previous dating suggests a chronology ranging from MIS 5 to MIS 3 obtained from U-series ages of bones from the bottom of layer 4 and the top of layer 5 ranging from ~90 ka to ~70 ka [60]. Combined ESR/U-series dating yielded ages ranging from 90 ± 9 ka (layer 5) to 40 ± 3 ka (layer 4.1) [61]. The new ESR/U-series ages obtained for layer 4.2 are similar to those previously obtained for layers 4.1 and 4.2, ranging from 40 ± 3 to 55 ± 2 ka. Together with IRSL ages, they confirm the attribution of this part of the sequence to MIS 3, perhaps representing some of the latest Neanderthal occupations in the area [58,61]. We analysed the points from layers 4.1 and 4.2 as a single assemblage (n = 95) associated with MIS 3, and those from the underlaying layer 5 (n = 62) as associated with the MIS 5a-b occupation of the shelter.

3.4. Grotte du Figuier

The Figuier Cave with its three rooms is located on the bank of the Ardèche River. The 8m-high entrance is oriented to the south and opens into room one with a ~150 m^2 surface area that was excavated by Chiron and Raymond during the end of the 19th and beginning of the 20th centuries. The Upper Palaeolithic at the site spans from the Aurignacian to the Magdalenian with *Homo sapiens* in rooms one and three, and Solutrean engravings [62]. The stratigraphy proposed by Combier [46] also includes two Middle Palaeolithic layers, one attributed to the Quina Mousterian, separated from the Upper Palaeolithic layers by a sterile sediment. More recent excavations were carried out in 2007–2008 by M-H. Moncel.

Correlations between stratigraphic sequences in the three rooms allow for the identification of five layers on top of a sterile, micaceous sand base [62,63]. Previous dating of layer 4 (room one) to 52 ± 9 ka [61] suggests occupation during MIS 4–3. Two U-series ages were obtained for rooms two and three on a bone and a tooth of 34 ± 7 and 13 ± 1 ka, provide a minimum estimate only [53]. We analysed points from the room two Middle Palaeolithic levels (*n* = 21).

4. TCSA Results and Contextual Interpretation for the Assemblages from South Eastern France

4.1. Payre MIS 7

The oldest point assemblage, Payre MIS 7, has a median TCSA value of 197 mm^2 (Table 2), indicating that if stone points were used to tip hunting weapons the trend was to produce tips that would be most effective as large single-handed stabbing spears or relatively small bimanual thrusting spears. Of all our samples, this assemblage shows the highest TCSA frequencies for stabbing spears (64.1%) and thrusting spears (44.7%) (Table 2). The frequencies for all other weapon-tip categories in the assemblage fall below the 15% threshold [41], so that such weapons were probably not part of stone-tipped hunting technologies if these were used during MIS 7 at Payre. Stabbing spears can also be thrown over short distances, and thrusting spears are ethno-historically associated with groups of hunters bringing down large and/or dangerous animals [15], both weapons are used during ambush hunting. We hypothesise that stabbing spears would have been effective for hunting the medium- and small-sized deer, whilst thrusting spears would have been useful for hunting the large (>300 kg) horses and aurochs. Using a combination of the two weapon-delivery types, perhaps with some members of the group using stabbing spears thrown over short distances (5–10 m) and others using thrusting spears for killing injured large animals at close quarters, would have had adaptive advantages for Neanderthal hunters—as opposed to using thrusting spears only.

Table 2. TCSA statistics (values in mm^2) and percentage frequencies of potential weapon-delivery systems based on our analysis.

Assemblage	TCSA Statistics in mm^2			% Potential Weapon-Delivery System					
	Mean	SD	Median	Poisoned Arrow	Arrow	Dart	Javelin	Stabbing Spear	Thrusting Spear
Abri du Maras, MIS 3 (*n* = 95)	117	99	94	6.3	20	21	30.5	47.4	12.6
Saint-Marcel, MIS 3 (*n* = 25)	113	63	100	0	12	36	36	48	12
Grotte du Figuier, MIS 3–4 (*n* = 21)	103	42	108	0	9.5	27.8	28.6	57.1	0
Abri du Maras, MIS 5a-b (*n* = 61)	142	103	108	1.6	9.8	24.2	32.3	46.8	22.5
Saint-Marcel, MIS 5e (*n* = 8)	224	195	146	0	0	12.5	25	50	37.5
Payre, MIS 6 (*n* = 141)	210	165	150	0	1.4	0.7	17	46.1	37.6
Payre MIS 7 (*n* = 152)	237	152	197	0.7	1.3	7.2	9.9	64.1	44.7

For this assemblage it is, however, worth noting that the core technologies are not devoted to symmetrical point production. Instead, discoidal cores provided mostly elongated flakes and *dejetée* triangular flakes or points. Thus far, use trace results have provided evidence of mammal processing, but no conclusive evidence for convergent artefacts being hafted or used to tip hunting weapons [2,64]. Faunal data, however, indicate that during MIS 7 hunting or meat harvesting activities at Payre centred around wild horses (*Equus ferus*), red and roe deer (*Cervus elaphus* and *Capreolus capreolus*) and aurochs (*Bos primigenius*). The age profiles of the animals indicate that year-around hunting took place, but that horses were probably not hunted in summer and deer and aurochs probably not in winter [65]. If this was indeed the case, we may wonder what the people of Payre used as hunting weapons during MIS 7—if not stone-tipped spears, wooden spears are a possibility and may not have been preserved.

4.2. Payre MIS 6

During this phase, the TCSA median value for the point assemblage decreases relative to the MIS 7 phase. Now being 150 mm^2, it is more consistent with a preferred production and use of tips most suited for single-handed stabbing spears. The TCSA frequency indicating the possible use of stabbing spears, however, decreases to 46.1% and for thrusting spears to 37.6% (Table 2). At the same time, the possibility of hunting with light-weight javelins becomes a minor possibility with a frequency of 17%, just over the 15% threshold for probable use. People still hunted wild horses and aurochs, but based on the animals' age estimates it seems that hunting now mostly took place from summer to autumn, no longer during winter and spring [65]. Although the frequency is low, we may see the first tentative use of light-weight javelins by Neanderthal hunters in south eastern France, should they have used stone-tipped weaponry. Perhaps, being free from the constraints of heavy winter clothing and possible snowy or marshy winter-spring landscapes, running after wounded prey and throwing weapons over longer distances became increasingly attractive. For this point sample, corroborating use trace analyses are lacking.

4.3. Saint-Marcel MIS 5e

This sample is too small ($n = 8$) to draw any robust inferences from, but it continues the trend of probable point manufacture for one-handed stabbing spears with a TCSA median value of 146 mm^2 (Table 1). Half of the points in this assemblage conform to ballistic expectations for use in this manner, 37.5% of the pieces could have functioned effectively as tips for bimanual thrusting spears, and the possible javelin-tip component rises to 25% (Table 2). Throughout the sequence, the faunal assemblage is dominated by cervids. In layer u (MIS 5), European fallow deer (*Dama dama*) are most prolific.

4.4. Abri du Maras MIS 5a-b

The median TCSA value of 108 mm^2 is consistent with a focus on producing stone points most effective for use as tips for stabbing spears (Table 1). Compared to the Saint-Marcel sample, however, the decrease in probable thrusting-spear tip production (22.5%), and the increase in the production of tips suitable for javelins (32.3%), are more pronounced (Table 2). For the first time there is a possibility that tips suitable for spear-thrower darts may have been in the mix at 24.2%, but with an ~98% overlap with the javelin-tip category parsimony dictates that additional strands of evidence are required before dart hunting is inferred.

The MIS 5 faunal assemblage at Abri du Maras is dominated by medium sized reindeer (*Rangifer tarandus*) at >65%. Other species such as medium-sized red deer (*Cervus elaphus*), and large wild horse (*Equus ferus*) and steppe bison (*Bison priscus*) were also sometimes targeted during summer when Neanderthal groups used the shelter as a long-term, seasonal campsite [66]. Reindeer are alert, highly mobile and well dispersed so that they are fairly difficult to hunt. The most effective way to hunt them is to intercept migrations, kill in excess, and preserve the surplus meat and fat [67]. The specialisation in summer-time reindeer hunting may have stimulated Neanderthal groups to increase their use of javelins. Such a strategy would mean that they could initiate hunting from a greater distance to compensate for the alert mobility of reindeer herds. Once the herd became confused, and some animals wounded, hunters could come in for the kill at a shorter range with stabbing and thrusting spears to finish off their preferred targets. Detailed use trace analyses on this point assemblage have not been conducted, and corroborating evidence for dart hunting is absent. We therefore suggest that, based on the TCSA results, the most likely hunting scenario associated with this point assemblage is the use of stabbing spears, supported by both javelins and thrusting spears, to optimise reindeer ambush hunting.

4.5. Grotte du Figuier MIS 3–4

The median TCSA value of 108 mm^2 for this assemblage is the same as for the Abri du Maras MIS 5a-b assemblage (Table 2). There is a clear preference for the knapping of points most suitable for tipping single-handed stabbing spears (57.1%), but, different from all the preceding assemblages, none of the points were suitable for tipping bimanual thrusting spears. Instead, 28.6% of the pieces would be most suited for javelin hunting when used to tip hunting weapons, with a 27.8% probability of dart hunting. Again, given the massive overlap in TCSA values for these two weapon systems, and lacking evidence of spear-throwers, the conservative interpretation is that whilst stabbing spears were the weapons of choice, the people who used Grotte du Figuier during the MIS 3–4 phase also hunted with lightweight javelins.

Bison and wild horses were still hunted, but now Alpine ibex (*Capra ibex*) were pursued almost as frequently as reindeer, and chamois (*Rupicapra rupicapra*) were also targeted by the people who used the site as a regular short-term camp [62,68]. Ibex herds are relatively small ($\leq$20 animals) compared to horse or reindeer herds that may number hundreds of animals (swelling into the thousands during seasonal mass migrations), and the chamois is a small goat-antelope (<50 kg). Expanding hunting breadth to include these animals that may have been pursued individually rather than ambushed in herds, may have further stimulated the use of javelins as a hunting strategy.

4.6. Saint-Marcel MIS 3

The TCSA median value of 100 mm^2 for this assemblage is also most consistent with points being knapped for tipping stabbing spears (Table 1)—should they have been intended for use in a composite hunting technology. Points best suited ballistically to tip stabbing spears comprise 48% of the assemblage (Table 2). Frequencies for potential javelin hunting and/or dart hunting is the same at (36%), the highest for their relative weapon-delivery types amongst all our assemblages. Medium-sized red deer is the most abundant prey animal, followed by roe deer, fallow deer (*Dama dama*), ibex (in the form of *Capra caucasica*, west Caucasian tur), extinct giant deer (*Megaloceros giganteus*), horse (*Equus germanicus*), aurochs, European ass (*Equus hydruntinus*) and wild boar (*Sus scrofa*) [69]. Seasonality indexes show that red deer were hunted all year round with most slaughters occurring during the autumn, mainly with young animals and adults killed in herds [68,70]. It seems that hunting with stone-tipped weapons from medium (javelin) to short distances (stabbing spears with short throwing capacity) was the preferred hunting strategy for this context, instead of hunting from closer quarters with stabbing and thrusting spears.

4.7. Abri du Maras MIS 3

For our last sample, the TCSA median value of 94 mm^2 is also most consistent with tips used for hunting with stabbing spears, but now it is skewed towards the smaller range of these weapons (Table 1). The frequency for possible javelin tips is 30.5% and for possible dart tips 21% (Table 2). What stands out for this assemblage is the 20% possibility for bow hunting with un-poisoned stone-tipped arrows. The mammal assemblage is even more dominated by reindeer (88%), compared to the MIS 5a-b occupation of this site, with minor occurrences of horse, bison, giant deer, red deer and ibex, but now lagomorphs are also present [71]. Although a full-scale use trace analysis of all the points and point fragments has not been completed, some of the artefacts have impact traces and polishes consistent with their use as weapon tips [31]. Hardy et al. [31] interpreted the presence of twisted fibres on a few tools, together with their relatively small size, as possible evidence for string-thrown darts in a Neanderthal hunting context. Currently, there is no corroborating evidence for the use of either spear-throwers or bows at the site, and there are no definitive human remains, so that a direct association with Neanderthals remains elusive. Once again, we have to take the most cautious route, and interpret possible hunting with stone-tipped weapons at Abri du Maras during MIS 3 as a preference for the dual-use of stabbing spears and lightweight javelins. The presence of small game such as lagomorphs, and relatively good potential for both dart and arrow hunting, however, indicate a potentially interesting scenario for this context that requires further exploration.

5. The TCSA Results of Our Analysis in the Greater Context

Below we present the hypothetical reconstruction for the use of stone-tipped hunting weapons based on TCSA values. This broad interpretation represents only an initial framework to be continuously fleshed out, tested, strengthened and constrained with additional data and lines of evidence. We start with the earliest MIS stages working through time to MIS 3 and introduce the associated fossil hominins or assumed populations responsible for the production and use of the assemblages.

5.1. Marine Isotope Stage 7 and Older

Of the three assemblages directly associated with *Homo neanderthalensis* fossil remains, the Payre assemblage has the most pieces best suited for use as bimanual thrusting spears, but several tips would also have been suited for tipping single-handed stabbing spears (Figure 2). At Biache-Saint-Vaast in north-western France [72], it seems that Neanderthals were more geared towards hunting with stabbing spears only, with a similar trend represented at Tabun in Israel [40,41]. The older wooden spears from Schöningen in Germany and Clacton in England, presumably made by pre-Neanderthals, have relatively massive TCSA values that are most consistent with bimanual thrusting spears [15,41], even though they may also have been hurled over short distances [73,74].

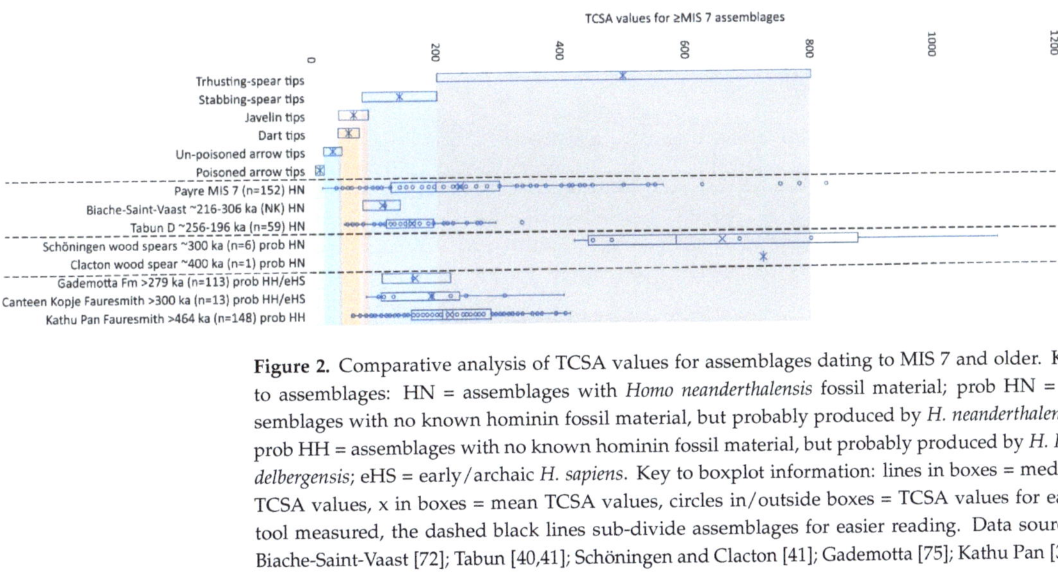

Figure 2. Comparative analysis of TCSA values for assemblages dating to MIS 7 and older. Key to assemblages: HN = assemblages with *Homo neanderthalensis* fossil material; prob HN = assemblages with no known hominin fossil material, but probably produced by *H. neanderthalensis*; prob HH = assemblages with no known hominin fossil material, but probably produced by *H. Heidelbergensis*; eHS = early/archaic *H. sapiens*. Key to boxplot information: lines in boxes = median TCSA values, x in boxes = mean TCSA values, circles in/outside boxes = TCSA values for each tool measured, the dashed black lines sub-divide assemblages for easier reading. Data sources: Biache-Saint-Vaast [72]; Tabun [40,41]; Schöningen and Clacton [41]; Gademotta [75]; Kathu Pan [36].

Of the African assemblages, the TCSA value distribution for points from Kathu Pan in the Northern Cape of South Africa is most similar to that of Payre (Figure 2). Although this assemblage is not directly associated with hominin fossil material, and if we accept the association of the artefacts with the age estimate of >464 ka [76], it is reasonable to suggest that these artefacts were knapped by *Homo heidelbergensis* [77]. *H. heidelbergensis* fossils were recorded from Elandsfontein and Hoedjiespunt in South Africa during MIS 7 [78]. The Canteen Kopje assemblage from South Africa dating to ~300 ka [79] could have been made by either *H. heidelbergensis* or early *H. sapiens*. The latter is represented at Florisbad, South Africa, by ~259 ka [80,81]. Most of these pointed artefacts could serve well as tips for stabbing spears, with some also suitable for tipping bimanual thrusting weapons. The small-bodied, small-brained *H. naledi* may have shared the South African landscape with the larger humans between ~335–236 ka [82], but thus far there is no evidence that they hunted with stone-tipped weapons [83]. The Ethiopian assemblage from the Gademotta Formation—probably also made by either *H. heidelbergensis* or early *H. sapiens*—falls mostly in the TCSA range of stabbing spear tips.

5.2. Marine Isotope Stage 6

By MIS 6 the Pyre point assemblage produced by *H. neanderthalensis* still spans both stabbing and thrusting spears, but the median TCSA value is now more towards the mean for stabbing spears compared to being at the upper margin of this weapon-tip range during MIS 7 (Figure 3). Of all the MIS 6 assemblages Pyre, however, still has the highest frequency of artefacts with TCSA values conforming to the possible use of tipping bimanual thrusting spears. The other two sites associated with Neanderthal and probable Neanderthal occupation show a similar pattern. However, for Bouheben in France we used the descriptive statistics published in Villa and Lenoir [84] so that outliers and how they affect the mean/median values are invisible, and the Tabun sample (n = 12) is probably too small to draw any robust conclusions from.

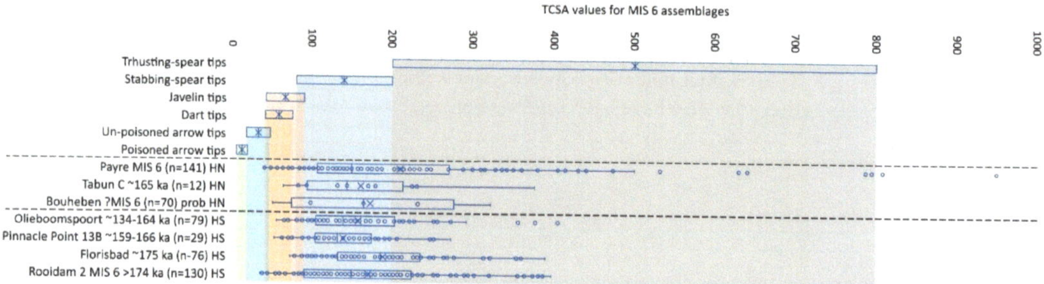

Figure 3. Comparative analysis of TCSA values for assemblages dating to MIS 6. Key: Same as for Figure 2, but HS = assemblages made by *Homo sapiens*, because these are the only known humans on the African landscape at the time. The key to boxplot information is the same as for Figure 2. Data sources: Bouheben [84]; Tabun [40,41]; South African assemblages [36].

For Africa, we currently have directly comparable data for four assemblages from South Africa only. The TCSA values for all of these indicate hunting with stabbing spears, but a simultaneous employment of bimanual thrusting spears, should the points have been used to tip hunting weapons. The use of thrusting spears seems especially frequent at the pan-scape sites of Rooidam at 36.2% and Florisbad at 42.1% [36,85,86], and less so at the river-ravine site of Olieboomspoort at 26.6% [36,87]. By this time, *H. sapiens* is accepted to be the only humans living in southern Africa, so that if TCSA values are indeed able to reveal something about weapon-assisted hunting behaviours, it would seem that *H. neanderthalensis* in the Levant and France and *H. sapiens* populations in southern Africa followed similar hunting strategies in terms of their choice and variation in hunting weapons. Interestingly, all the assemblages now have convergent pieces that would have been suited ballistically for javelin tips. Especially assemblages with >15% of its TCSA values falling in this category, such as Payre in France (17%) and Rooidam in South Africa (22.3%) may represent a new trend, wherein populations started to experiment with lighter spears that can be thrown at greater velocity over longer distances.

5.3. Marine Isotope Stage 5

MIS 5 is a climatically complex phase with oscillating interglacial (a, c, e) and glacial (b and d) substages, and the only assemblage from this phase directly associated with *H. neanderthalensis* remains is that of Tabun in Israel [40]. Its mean TCSA value is still within the stabbing spear range, but it is lower compared to the MIS 6 assemblage from the same site (Figure 4). The mean TCSA value for the small assemblage from Abrigo de la Quebrada in Spain [88,89] is the lowest of the assemblages probably associated with Neanderthal groups, whilst that of Saint-Marcel stands out as the one with the largest proportion falling in the thrusting spear-tip category. Interpretation of all three of these assemblages can only be thought of as preliminary, based on the small number of pointed artefacts recorded (Tabun n = 9, Abrigo de la Quebrada n = 6, Saint-Marcel n = 8). The Abri du Maras assemblage has a wide distribution, centred on making tips suitable for stabbing spears, but also includes pieces that would work well for tipping heavier thrusting weapons or lighter javelins (Figure 4).

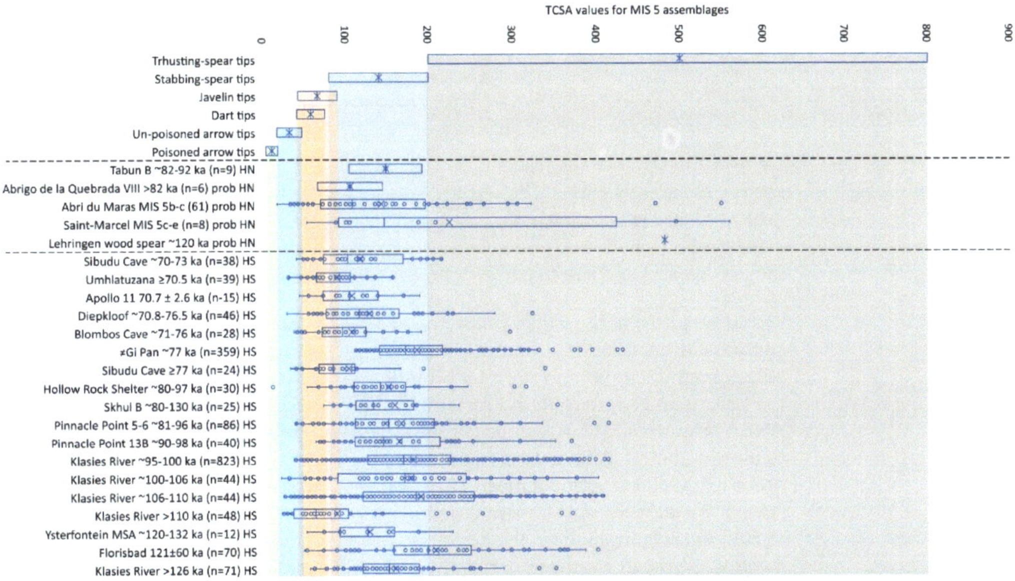

Figure 4. Comparative analysis of TCSA values for assemblages dating to MIS 5. Keyfor hominin abbreviations is the same as for Figure 3. The key to boxplot information is the same as for Figure 2. Data sources: Tabun [40]; Abrigo de la Quebrada [89]; Lehringen [41]; South African assemblages [36]; ≠Gi Pan [43].

The assemblage from Skhul B in Israel is directly associated with *H. sapiens* remains but has a roughly similar age estimate and TCSA value distribution as the Tabun assemblage, which is associated with a Neanderthal population at the time. All the other assemblages with available TCSA data come from southern Africa, showing complex variability in the use of possible stone-tipped weapon choices across the region. This variability and its possible implications have been discussed previously [36,41,43], and we do not repeat those details here.

5.4. Marine Isotope Stage 4

The Grotte du Figuier chamber two assemblage spans MIS 3–4. Here we first compare it to other assemblages with MIS 3–4 age estimates and those from MIS 4 contexts, before doing the same with MIS 3 below. For the first time, we see an assemblage in south eastern France assumed to have a Neanderthal context with no pointed artefacts suited hypothetically for hunting with bimanual thrusting spears (Figure 5). This is in contrast with the *H. neanderthalensis* associated assemblage from Kebara in Israel that has several artefacts suited for such hunting [40,41]. At Rosh Ein Mor, for which the human population remains unknown, there is a proportion of points most suitable for thrusting spears. In all three of these assemblages some of the points would also be well-suited for use as javelin tips based on their TCSA values (Figure 5).

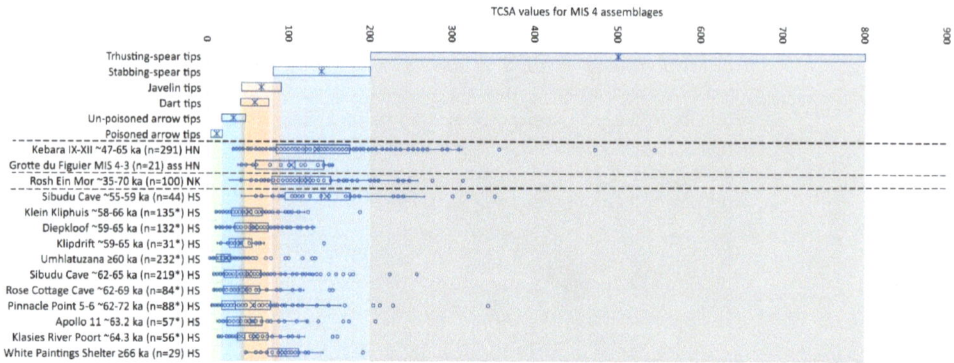

Figure 5. Comparative analysis of TCSA values for assemblages dating to MIS 4. Hominin abbreviation key: Same as for Figure 3, but ass HN = assemblages with no human fossil material but assumed to be associated with *H. neanderthalensis* based on technology or dating only, even though *H. sapiens* may have been on the landscape. The key to boxplot information is the same as for Figure 2. Data sources: Kebara [40,41]; Rosh Ein Mor [40]; Southern African assemblages [36].

It was during the MIS 4 phase that, based on the current assemblages with directly comparable data, the TCSA values indicate a pronounced difference between point assemblages manufactured and used by Neanderthal or assumed Neanderthal populations in the Levant and south eastern France, and those made and used by *H. sapiens* in southern Africa (Figure 5). TCSA values for the oldest southern African assemblage from White Paintings Shelter in Botswana has a median TCSA value that falls just within the upper limits for javelin tips, and a mean TCSA value at the lower limit for hunting with stabbing spears. Most of the subsequent assemblages distributed across the region have median TCSA values falling within the arrow range with some pieces also well-suited for tipping javelins. MIS 4 in southern Africa therefore represents the earliest known phase during which *H. sapiens* groups may have been hunting habitually with bows and arrows—an inference supported by multiple strands of evidence from independent studies and approaches [90–94]. If the artefacts measured were used to tip hunting weapons, amongst other things, these groups did not only hunt with bows and arrows, but also with javelins as suggested by the spread of the TCSA values and some outliers.

By the transition from the glacial MIS 4 to the interglacial MIS 3, however, the TCSA values for the youngest assemblage from southern Africa (Sibudu Cave dating to ~55–59 ka) indicate a return to hunting mostly with stabbing spears (Figure 5). Noteworthy here is also a return to discoidal and Levallois flaking at long-sequence Middle Stone Age sites such as Sibudu Cave during this phase [95]. A trend that already started by the end of the previous phase (Sibudu assemblage dated to ~62–65 ka) that was strongly blade based with backed implements [96]. At Sibudu Cave faunal data too indicate a gradual shift from hunting smaller forest/woodland prey to large, herding plains game such as zebra [97].

5.5. Marine Isotope Stage 3

We include here again the assemblages that span MIS 3–4. Similar to the previous inter-glacial (MIS 5), the TCSA values for the MIS 3 assemblages show variability around the use of spear hunting (Figure 6). The *bona fide* Neanderthal artefacts from Le Moustier in central-western France stand out as the only MIS 3 assemblage with TCSA values almost exclusively associated with bimanual thrusting spears. This assemblage is curated at the Harvard Peabody Museum, and we do not know how it was selected, so that future analyses on a more comprehensive sample may provide different results. The other two assemblages directly associated with *H. neanderthalensis* remains, namely Bagratashen 1 and

Kebara, both show TCSA values most consistent with stabbing-spear hunting in tandem with the possible use of javelins (Figure 6).

Most assemblages assumed to have been made and used by Neanderthals have sizable TCSA frequencies falling within the hypothetical range of javelin tips. For example, assemblages from the Spanish sites of Amalda and Abrigo de la Quebrada [89,98,99] have mean and median TCSA values that fall within this category, whilst that from Boker Tachtit in Israel [100] is just within the range of stabbing spears with a large proportion of tips also suitable for hunting with javelins. The assemblages from Mirak in Iran [101] and Tor Faraj in Jordan [102] both have mean TCSA values most consistent with hunting with one-handed stabbing spears. The assemblage from Abri du Maras stands out amongst the assumed Neanderthal assemblages with its TCSA range also spanning that of arrows and darts, and those from Rosh Ein Mor and Mirak roughly follow the Neanderthal and assumed Neanderthal trend (Figure 6).

Amongst the assemblages associated or assumed to be associated with *H. sapiens*, that of Üçagizli in Turkey and Ksar Akil in Lebanon [40] stand out with median TCSA values within the arrow range—similar to Abri du Maras (Figure 6). The French Chatelperonian assemblage that consists of points from the sites of Trou de La Chevre, La Ferrassie and Grotte du Portel points [40] and Rose Cottage Cave in South Africa [36], both have median and mean TCSA values consistent with a preference for javelin hunting. For the remaining South African assemblages, namely that of Umbeli Belli, Umhlatuzana and Sibudu Cave, the marked differences in TCSA ranges between *H. sapiens* and Neanderthal assemblages observed for MIS 4 are no longer visible in the MIS 3 record. It seems that if stone tips were used to tip hunting weapons during this phase, people in southern Africa preferred to hunt with stabbing spears, often tipped with points produced with Levallois-like knapping methods similar to some of the assumed Neanderthal groups.

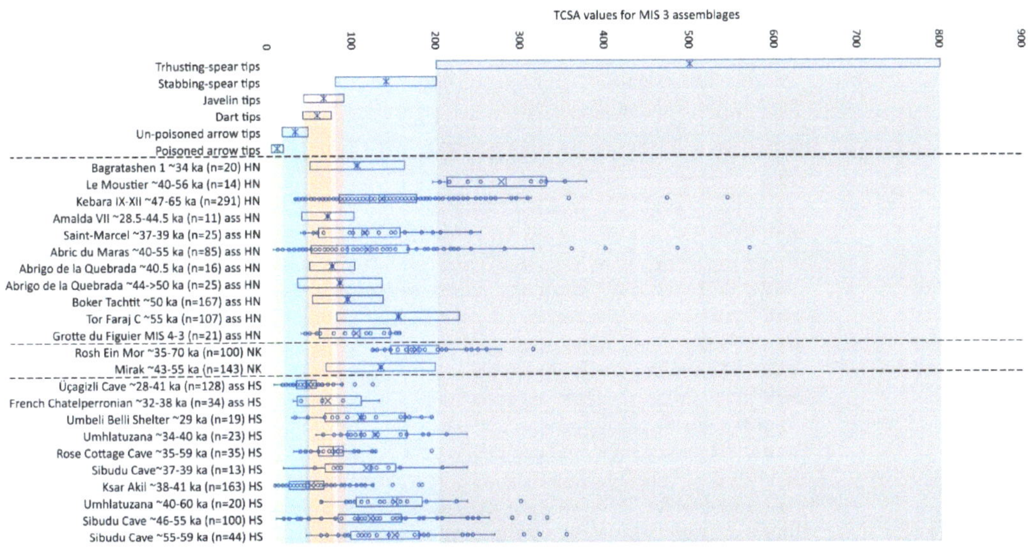

Figure 6. Comparative analysis of TCSA values for assemblages dating to MIS 3. The hominin abbreviation key is the same as for Figure 5. The key to boxplot information is the same as for Figure 2. Data sources: Bagratashen [103]; Kebara [40,41]; Amalda [98,99]; Abrigo de la Quebrada [89]; Boker Tachtit [100]; Tor Faraj [102]; Mirak [101]; Üçagizli, Chatelperonian, Ksar Akil [40]; South African assemblages [36].

6. Discussion

In our introduction we listed five assumptions about Neanderthal weapon-assisted hunting that the TCSA approach may help re-assess. Below we discuss each of these topics based on the results of our study presented above. Other approaches and additional lines of evidence may provide different outcomes or support our inferences.

6.1. Assumption 1: Neanderthal Stone-Tipped Weaponry Was Mostly Limited to Double-Handed Thrusting Spears

Berger and Trinkaus [25] (p. 841) suggested that if stone points were used as weapon tips in Neanderthal contexts, "the resultant spears would have been thick and heavy, usable only as thrusting spears and not as effective throwing projectiles", and that the Neanderthal wooden spears were best suited for "close quarter predation on ungulates". They link the use of thrusting spears with morphological evidence for infrequent throwing behaviour in Neanderthal populations despite the strength of their upper limbs, and with high frequencies of head and neck trauma suffered by Neanderthals [25]. Churchill [26] and Schmitt et al. [104] also see the humeral morphology of Neanderthals and early *H. sapiens* as consistent with the habitual and forceful use of bimanual thrusting spears.

Our TCSA work on bimanual thrusting spears of known use [15], and the TCSA values of the Middle Palaeolithic wooden spears (Figure 2) corroborate their inferences, showing that some pointed stone artefacts from Neanderthal assemblages and the wooden spears are best-suited ballistically for use as bimanual thrusting spears. In the assemblages from south eastern France, such tips are relatively frequent in the ≥MIS 7–5e assemblages, where they constitute the second-most represented weapon-tip type after one-handed stabbing spears (Table 2). They become less frequent in the younger Neanderthal or assumed Neanderthal assemblages. The current TCSA data therefore supports the notion that early Neanderthal groups may have hunted habitually with bimanual thrusting spears. However, it does not support a scenario wherein this was their exclusive way of hunting, especially after MIS 5e. Instead, our TCSA data indicate that strategies around stone-tipped weapon use amongst Neanderthal or assumed Neanderthal groups changed through time to also include other approaches to spear-hunting.

6.2. Assumption 2: Neanderthals Used Stone-Tipped Spears for Both Thrusting and Short-Distance Throwing

Hardy et al. [27] examined artifacts from two sites in the Crimea, namely Starosele dating to ~80–40 ka and Buran Kaya III with an age estimate of 37–32 ka, assuming that they were made and used by Neanderthals based on the 'cultural material' and contemporaneous *H. neanderthalensis* remains from elsewhere. They suggested the use of both thrusting and throwing in terms of spear use. In the context of a functional analysis of Mousterian and Micoquian assemblages from Sesselfelsgrotte in Germany dating to ~48–36 ka assumed to be of Neanderthal origin, Rots [28] also found proxy evidence for both thrusting and throwing spear use on pointed artefacts. She argues that the same hunting weapon may have been used for both purposes but highlights that the weight of Neanderthal spears would limit their effective throwing distance to well below 10 m [28]. In both their discussions no distinction is made between bimanual thrusting and single-handed stabbing spears. The mechanics and biomechanics of one-handed stabbing are fundamentally different from bimanual thrusting [105], and the two hunting strategies may be associated with different evolutionary trajectories in hunting behaviours [104], and/or with adaptive advantages [37,41]. It is therefore important to consider both weapon-assisted hunting strategies in the Neanderthal context.

Our TCSA analyses support the use of bimanual thrusting weapons and single-handed stabbing spears that can be thrown effectively over short (< 10 m) distances for Neanderthal groups. For example, the TCSA results of the point assemblages from south eastern France demonstrate that in all seven instances there were more artefacts suited ballistically for tipping single-handed stabbing spears than any other weapon-tip category (Table 2). In the three older assemblages (Payre MIS 7, Payer MIS 6, Saint-Marcel MIS 5e) this category was followed by spear tips suited for bimanual thrusting. The four younger assemblages (Abri du Maras MIS 5 a-b, Grotte du Figuier MIS 3–4, Saint-Marcel MIS 3, Abri du Maras MIS 3), however, follow a different pattern wherein tips best-suited for javelin hunting becomes the second-most represented category (Table 2). We suggest that this indicates a development in hunting strategies on the south eastern margins of the Massif Central Mountains in France from contact or close-quarter hunting to hunting from a relatively safer distance. This trend also becomes evident in other Neanderthal or assumed Neanderthal assemblages dating to MIS 3 (Figure 6), so that we may accept that the trend observed for south eastern France also played out elsewhere in Europe and the Levant.

It is necessary at this stage to reflect on the adaptive advantages of hunting with single-handed stabbing spears or assegais. Before the recent introduction of firearms to Africa, various types of assegais represented the bulk of hunting and warfare weaponry across the continent because of the effectiveness and flexible nature of this weapon [15]. For example, in hunting scenarios single-handed stabbing spears or assegais:

(a) Provide better balance and movement to weapon-yielding hunters compared to large and heavy bimanual thrusting spears;
(b) Provide enough shaft and tip strength to stab powerfully and effectively at dangerous antagonists without hunters losing their weapons;
(c) Can be flung from a short distance providing some flexibility in risk management;
(d) Free up one hand to carry a backup or defence weapon such as a club or another spear or spear type.

Because each person can carry no more than one bimanual thrusting spear and no more than two or three stabbing spears, these weapons are unsuitable for hunting as a single person or in a pair. Instead, all spear-hunting strategies require multi-person cooperation [41]. The communal success of such social weapon-use strategies may explain why the stabbing-spear category became the preferred weapon for Neanderthal groups at least since MIS 7. Greater flexibility and development in Neanderthal or assumed Neanderthal stone-tipped weapon use may be indicated by the TCSA outcomes that suggest that stabbing spears may have been used in tandem with bimanual thrusting spears early on, and later together with lighter javelins that could be thrown over longer distances (Table 2). *Bona fide* TCSA indicators for habitual javelin hunting, however, only appear later in the Neanderthal record, and most often at sites without corroborating *H. neanderthalensis* fossil remains—a point we return to below (Figures 5 and 6). This interpretation is supported by work that shows a shift in Middle Palaeolithic technical behaviour between the earlier (MIS 9–6) and the later (MIS 4–3) phases [54,55].

6.3. *Assumption 3: Neanderthals Did Not Make or Use Mechanically Projected Technologies Such as the Spear-Thrower and Dart or the Bow and Arrow for Killing at a Distance*

Based on neurological differences between *H. neanderthalensis* and *H. sapiens*, Coolidge et al. [29] argue that Neanderthals did not produce or use complex weaponry, such as harpoons launched with spear-throwers or bow and arrow sets [106,107]. According to them it suggests that Neanderthals may have been limited by an inability to aim long-distance weapons accurately, because of constraints involving hand/movement control, vision, and spatial awareness and judgment [108]. One of us have reached similar conclusions in work relating to a co-evolutionary approach to human cognition [30], wherein we built from Bruner et al. [109]. They explain the role of the precuneus in a process called 'visuospatial integration', which is integral to bimanual manipulation and operating over time and distance—as required for bow hunting—and which varies between *H. neanderthalensis* and *H. sapiens* [110–113].

Our TCSA analyses reveals that for the assemblages from south eastern France dating to MIS 5e and older, the frequencies of tips with values falling in the dart and arrow categories are all below the 15% benchmark of probable use. For the later assemblages the possibility of dart use with spear-throwers increases to 21–36% (Table 2). Yet, because of the overlap between the TCSA ranges for dart tips and javelin tips (Table 1), and because there is little to no definitive corroborating evidence for spear-thrower use in the Neanderthal record, we must follow a parsimonious interpretation wherein the data is most consistent with javelin hunting. The MIS 3 assemblage from Abri du Maras is the only one with artefacts that have TCSA values in the arrow range for >15%. Again, it is only during MIS 3 that other Neanderthal or assumed Neanderthal assemblages also show similarly low TCSA values, especially in the case of Kebara in Israel and Abrigo de la Quebrada in Spain (Figure 6). Here it is noteworthy that the Kebara assemblage is directly associated with *H. neanderthalensis* remains, but those from Abri du Maras and Abrigo de la Quebrada thus far lacks fossil hominin evidence. Thus, based on current evidence, it cannot be claimed that Neanderthals hunted with long-distance weapon systems such as the spear-thrower and dart or the bow and arrow—instead, the possibility of javelin hunting requires more exploration.

6.4. Assumption 4: Neanderthals Used String-Thrown Darts

Above we indicated that there is little to no corroborating evidence for spear-thrower use in the Neanderthal record. We have also presented Hardy et al.'s [31] interpretation that twisted fibres observed on a few small, pointed artefacts may be seen as proxy evidence for Neanderthals hunting with string-thrown darts at Abri du Maras during MIS 3. The Hardy study did not consider hunting with lightweight javelins, which would result in the production of relatively small weapon tips, similar in dimension to dart tips (Table 1). Thus, the string could have simply been used to tighten javelin tips in their shafts.

Our TCSA results show that even if darts were used in the manner described by Hardy et al. [31], the preferred stone-tipped hunting weapon used at the site during this phase was the trustworthy stabbing spear, possibly supported by javelin hunting (Table 2). Whilst there is currently no supporting evidence for the use of darts or arrows at Abri du Maras, or any of the other Neanderthal or assumed Neanderthal sites, seen together with the presence of small game such as lagomorphs, the Abri du Maras assemblage can be considered a 'smoking gun' for the use of complex hunting technologies. However, the timing and the fact that there are no directly associated *H. neanderthalensis* remains, require a more nuanced consideration than simply attributing it to Neanderthals based on age or technology.

6.5. Assumption 5: Neanderthal Weapons Were Similar to Those Used by Broadly Contemporaneous Homo sapiens Hunters in Africa

Villa and Soriano [32] set out to show that "Neanderthals were not inferior hunters" and that they used hunting weapons similar to those used by their concomitant *H. sapiens* counterparts in South Africa. They argue that both Middle Palaeolithic (Neanderthal) and Middle Stone Age (*H. sapiens*) assemblages indicate the use of hand-delivered spears only—including backed pieces that dominate southern African MIS 4 assemblages. In contrast to their findings, work conducted by Africa-based researchers at the same time and after their study provides multi-stranded proxy evidence suggesting bow-hunting behaviour in southern Africa from perhaps as early as ~80 ka (see evidence listed in [15,90]; but see [32] for different view). Our comparative TCSA analyses support an inference of broadly similar weapon-assisted hunting behaviours for *H. neanderthalensis* in Europe and the Levant and probable *H. heidelbergensis* and early *H. sapiens* groups in Africa before and until roughly MIS 6 (Figures 2 and 3). By MIS 6, both groups seem to start experimenting with lighter javelins, whilst still depending heavily on the adaptive advantages of the stabbing spear.

For MIS 5—the period between ~130 ka and 71 ka—the TCSA data do not support the assumption that Neanderthal weapon use in Eurasia was the same as that of *H. sapiens* in Africa. During this phase, African *H. sapiens* weapon-assisted hunting strategies became more variable (Figure 4), possibly in response to climatic oscillations between the interglacial and glacial substages. The variation may also reflect increasingly flexible trends in socio-technical behaviour amongst the *H. sapiens* groups [30,114]. The comparative

TCSA results for the subsequent glacial MIS 4 demonstrate clear differences between most *H. sapiens* assemblages from southern Africa and the contemporaneous Neanderthal or assumed Neanderthal assemblages (Figure 5). During this phase it seems that the bow and arrow became the weapon of choice in southern Africa, supplemented with javelins and some stabbing spears [36].

By MIS 3 the TCSA results indicate that hunting with stone-tipped weapons became largely variable in both Neanderthal and *H. sapiens* populations (Figure 6). The evolutionary narrative of Middle Palaeolithic Neanderthal weapon use in Eurasia compared to Middle Stone Age *H. sapiens* weapon use in Africa can therefore not be reduced to being 'similar' across the board. In the TCSA record we see trends of similarity before MIS 5, divergence in hunting with stone-tipped weapons during MIS 5–4, seemingly converging again during MIS 3 when *H. sapiens* appear and stay on the European landscape.

7. Conclusions

The TCSA method cannot determine the function of each artefact measured. However, if we accept that some convergent stone tools were used to tip hunting weapons, it becomes a powerful, standardised tool for generating directly comparable data about variability in weapon-assisted hunting strategies through time and across space. Because the ballistic characteristics of weapon tips (represented by their TCSA values) articulate directly with the contextual effectiveness of weapon-delivery systems, they can reveal aspects of adaptive success. Depending on the level of analysis and interpretation, comparative TCSA data can be used to generate hypotheses about intra- and inter-site weapon use on a regional or a global scale. With this contribution we reassessed some interpretations and assumptions about Neanderthal hunting with stone-tipped weaponry. We showed that in south eastern France, between MIS 7 starting by ~243 ka and MIS 3 ending by ~29 ka, the TCSA values of stone point assemblages indicate interesting developments that require future testing. For example:

- By using TCSA standards that distinguish between three spear-hunting strategies, we can argue that early Neanderthals probably hunted with both bimanual thrusting spears and single-handed stabbing spears.
- Collectively, the TCSA and faunal records of Payre suggest that its early Neanderthal population probably adapted their spear-use strategies (relying more or less on bimanual thrusting and single-handed stabbing) according to prey type and hunting season.
- The adaptive advantages of using smaller stabbing spears, and the fact that they could be thrown more effectively over short distances, seems to have encouraged experimentation with lighter javelin-type weapons, perhaps starting as early as MIS 6, but becoming increasingly pronounced in younger populations.

Especially during MIS 3 (~57–29 ka) the TCSA record for assemblages assumed to have been made and used by Neanderthals show a tendency towards hunting with lightweight javelins or even darts and/or arrows in combination with single-handed stabbing spears. Most of these assemblages are not directly associated with *H. neanderthalensis* fossil remains. One of us [41], has argued that for humans to throw lightweight javelins over longer distances with the necessary force and accuracy to become effective hunting weapons, several factors needed to converge. For example, a powerful thumb-tip to finger-tip prehension, the ability to throw objects from a distance at high speed and with great accuracy, and neurocognitive underpinnings associated with the ability to mentally map force transmission through time and across space [112]. The latter requires effective visuo-spatial integration associated with brain areas such as the precuneus that expanded to its full capacity only in *H. sapiens* by ~100 ka or during MIS 5 [115], when we see TCSA data showing an increase in possible javelin use in southern Africa associated with a *H. sapiens* population only.

Successful hand-eye coordination, spatial awareness and distance judgment probably developed incrementally throughout our genus, and the evolution of increasingly efficient visuo-spatial feedback was necessary for developing and choosing between the use of different weapon-delivery systems. Our TCSA data suggest that both *H. neanderthalensis* and *H. sapiens* started to independently develop a capacity for hunting at a distance. Yet, whereas javelin and bow hunting (supported by the use of stabbing spears) may have become habitual for *H. sapiens* during MIS 5–4 in the southern African record, these strategies seem not to have evolved to the same extent in the concurrent Neanderthal populations. It is therefore possible that the variation in weapon choice between the two populations reflect differences in neurocognitive development between ~130 ka and 57 ka. TCSA data for MIS 3, however, indicate a convergence in weapon choice between Neanderthal or assumed Neanderthal groups and *H. sapiens* groups.

The African TCSA record indicates that by MIS 3 *H. sapiens* hunters returned to the preferred use of stabbing spears in tandem with javelin hunting. This may reflect the opening of the landscape after the MIS 4 glacial and hunting larger animals in the context of group hunting, instead of focussing on smaller prey hunted alone, or with only a few hunters participating. African *H. sapiens* MIS 3 weapons were often tipped with stone points made with Levallois-like knapping techniques—similar to the weapon tips of contemporaneous Neanderthal populations further north.

Genetic research highlights multiple admixture events between *H. neanderthalensis* and *H. sapiens* populations, from as early as ~120 ka in the Near East [116], most likely in western Asia from ~60–50 ka [117], and in present-day Romania at ~42–37 ka [118,119]. Even if interaction or co-existence was limited to 1400–2900 years during the Châtelperronian (~43–40 ka) in France and northern Spain [120], it still spans 56–116 human generations at 25 years per generation. We do not know yet whether some of the *H. sapiens* groups who entered Eurasia from Africa also used Levallois knapping techniques. Picin et al. [121] unpack dilemmas and possibilities associated with so-called 'transitional industries' of the Near East and Europe containing Levallois elements, suggesting that it is perhaps not feasible to ascribe such assemblages to either Neanderthals or *H. sapiens* only. Socio-economical overlap and exchange, behavioural flexibility, as well as genetic mixing between the populations must be considered as an explanation for some of the aspects associated with the MIS 3 archaeological record in Eurasia. This is especially necessary when behavioural trends—such as hunting with long-distance weapons—are observed without the presence of decisive human remains.

Author Contributions: Conceptualization, M.L.; methodology, M.L.; validation, M-H.M; formal analysis, M.L.; investigation, M.L. and M.-H.M.; resources, M.L. and M.-H.M.; data curation M.-H.M.; writing—original draft preparation, M.L.; writing—review and editing, M.-H.M.; visualization, M.L.; project administration, M.-H.M.; funding acquisition, M.-H.M. All authors have read and agreed to the published version of the manuscript.

Funding: Fieldwork on the Middle Palaeolithic sites of the southeast margins of the Massif Central were awarded to M-H. Moncel by the "Service Régional de l'Archéologie, Auvergne-Rhône-Alpes", French Ministry of Culture.

Data Availability Statement: Data for the French assemblages can be requested from M-H. Moncel. Comparative raw data for weapon tips of known use can be downloaded from https://link.springer.com/article/10.1007/s10816-022-09595-1 (accessed on 19 February 2023).

Acknowledgments: We thank Ran Barkai for inviting us to participate in this special volume, John Shea for sharing data with M.L., and four anonymous reviewers for their time and input that helped to improve our contribution. Any mistakes or omissions, however, remain our own.

Conflicts of Interest: The authors declare no conflict of interest.

References

1. Hardy, B.L. Climatic variability and plant food distribution in Pleistocene Europe: Implications for Neanderthal diet and subsistence. *Quat. Sci. Rev.* **2010**, *29*, 662–679. [CrossRef]
2. Hardy, B.L.; Moncel, M.H. Neanderthal use of fish, mammals, birds, starchy plants and wood 125–250,000 years ago. *PLoS ONE* **2011**, *6*, e23768. [CrossRef] [PubMed]

3. Henry, A.G.; Brooks, A.S.; Piperno, D.R. Plant foods and the dietary ecology of Neanderthals and early modern humans. *J. Hum. Evol.* **2014**, *69*, 44–54. [CrossRef] [PubMed]
4. Nabais, M.; Dupont, C.; Zilhão, J. The exploitation of crabs by Last Interglacial Iberian Neanderthals: The evidence from Gruta da Figueira Brava (Portugal). *Front. Environ. Archaeol.* **2023**, *2*, 1097815. [CrossRef]
5. Thieme, H.; Veil, S.; Meyer, W.; Moller, J.; Plisson, H. Neue Untersuchungen zum eemzeitlichen Elefanten-Jagdplatz Lehringen, Ldkr. Verden. *Die Kunde* **1985**, *36*, 11–58.
6. Boëda, E.; Geneste, J.M.; Griggo, C.; Mercier, N.; Muhesen, S.; Reyss, J.L.; Taha, A.; Valladas, H. A Levallois point embedded in the vertebra of a wild ass (Equus africanus): Hafting, projectiles and Mousterian hunting weapons. *Antiquity* **1999**, *73*, 394–402. [CrossRef]
7. Gaudzinski-Windheuser, S.; Kindler, L.; MacDonald, K.; Roebroeks, W. Hunting and processing of straight-tusked elephants 125,000 years ago: Implications for Neanderthal behavior. *Sci. Adv.* **2023**, *9*, eadd8186. [CrossRef] [PubMed]
8. Lazuén, T. Please do not shoot the pianist. Criteria for recognizing ancient lithic weapon use. *J. Archaeol. Sci.* **2014**, *46*, 1–5. [CrossRef]
9. Hovers, E.; Belfer-Cohen, A. Are lithics and fauna a match made in (prehistoric) heaven? *J. Paleolit. Archaeol.* **2020**, *3*, 108–125. [CrossRef]
10. Klein, R.G. Archeology and the evolution of human behavior. *Evol. Anthropol.* **2000**, *9*, 17–36. [CrossRef]
11. Stiner, M.C. *Honor among Thievesa Zooarchaeological Study of Neandertal Ecology*; Princeton University Press: Princeton, NJ, USA, 1994.
12. Shea, J.J. Spear points from the Middle Paleolithic of the Levant. *J. Field Archaeol.* **1988**, *15*, 441–450.
13. Shea, J.J.; Brown, K.S.; Davis, Z.J. Controlled experiments with Middle Palaeolithic spear points: Levallois points. In *Experimental Archaeology: Replicating Past Objects, Behaviors, and Processes*; Mathieu, J.R., Ed.; BAR International Series: Oxford, UK, 2002; Volume 1035, pp. 55–72.
14. Dusseldorp, G.L. *A View to a Kill: Investigating Middle Palaeolithic Subsistence Using an Optimal Foraging Perspective*; Sidestone Press: Leiden, The Netherlands, 2009.
15. Lombard, M.; Lotter, M.G.; Caruana, M.V. The tip cross-sectional area (TCSA) method strengthened and constrained with ethno-historical material from sub-Saharan Africa. *J. Archaeol. Method Theory* **2022**. [CrossRef]
16. Lazuén, T. European Neanderthal stone hunting weapons reveal complex behaviour long before the appearance of modern humans. *J. Archaeol. Sci.* **2012**, *39*, 2304–2311. [CrossRef]
17. Rots, V. Projectiles and hafting technology. In *Multidisciplinary Approaches to the Study of Stone Age Weaponry*; Iovita, R., Sano, K., Eds.; Springer: Berlin/Heidelberg, Germany, 2016; pp. 167–185.
18. Gaudzinski-Windheuser, S.; Noack, E.S.; Pop, E.; Herbst, C.; Pfleging, J.; Buchli, J.; Jacob, A.; Enzmann, F.; Kindler, L.; Iovita, R.; et al. Evidence for close-range hunting by last interglacial Neanderthals. *Nat. Ecol. Evol.* **2018**, *2*, 1087–1092. [CrossRef]
19. Britton, K.; Grimes, V.; Niven, L.; Steele, T.E.; McPherron, S.; Soressi, M.; Kelly, T.E.; Jaubert, J.; Hublin, J.J.; Richards, M.P. Strontium isotope evidence for migration in late Pleistocene Rangifer: Implications for Neanderthal hunting strategies at the Middle Palaeolithic site of Jonzac, France. *J. Hum. Evol.* **2011**, *61*, 176–185. [CrossRef]
20. Julien, M.A.; Rivals, F.; Serangeli, J.; Bocherens, H.; Conard, N.J. A new approach for deciphering between single and multiple accumulation events using intra-tooth isotopic variations: Application to the Middle Pleistocene bone bed of Schöningen 13 II-4. *J. Hum. Evol.* **2015**, *89*, 114–128. [CrossRef] [PubMed]
21. Kuitems, M.; van der Plicht, J.; Drucker, D.G.; Van Kolfschoten, T.; Palstra, S.W.; Bocherens, H. Carbon and nitrogen stable isotopes of well-preserved Middle Pleistocene bone collagen from Schöningen (Germany) and their paleoecological implications. *J. Hum. Evol.* **2015**, *89*, 105–113. [CrossRef] [PubMed]
22. Rivals, F.; Julien, M.A.; Kuitems, M.; Van Kolfschoten, T.; Serangeli, J.; Drucker, D.G.; Bocherens, H.; Conard, N.J. Investigation of equid paleodiet from Schöningen 13 II-4 through dental wear and isotopic analyses: Archaeological implications. *J. Hum. Evol.* **2015**, *89*, 129–137. [CrossRef] [PubMed]
23. White, M.; Pettitt, P.; Schreve, D. Shoot first, ask questions later: Interpretative narratives of Neanderthal hunting. *Quat. Sci. Rev.* **2016**, *140*, 1–20. [CrossRef]
24. Mithen, S. The hunter-gatherer prehistory of human-animal interactions. *Anthrozoös* **1999**, *12*, 195–204. [CrossRef]
25. Berger, T.D.; Trinkaus, E. Patterns of trauma among the Neandertals. *J. Archaeol. Sci.* **1995**, *22*, 841–852. [CrossRef]
26. Churchill, S.E. Of assegais and bayonets: Reconstructing prehistoric spear use. *Evol. Anthropol.* **2002**, *11*, 185–186. [CrossRef]
27. Hardy, B.L.; Kay, M.; Marks, A.E.; Monigal, K. Stone tool function at the paleolithic sites of Starosele and Buran Kaya III, Crimea: Behavioral implications. *Proc. Natl. Acad. Sci. USA* **2001**, *98*, 10972–10977. [CrossRef] [PubMed]
28. Rots, V. The functional analysis of the Mousterian and Micoquian assemblages of Sesselfelsgrotte, Germany: Aspects of tool use and hafting in the European Late Middle Palaeolithic. *Quartär–Int. Jahrb. Zur Erforsch. Des Eiszeitalt. Und Der Steinzeit* **2009**, *56*, 37–66.
29. Coolidge, F.L.; Wynn, T.; Overmann, K.A.; Hicks, J.M. Cognitive archaeology and the cognitive sciences. In *Human Paleoneurology*; Bruner, E., Ed.; Springer: Berlin/Heidelberg, Germany, 2015; pp. 177–208.
30. Lombard, M.; Högberg, A. Four-field co-evolutionary model for human cognition: Variation in the Middle Stone Age/Middle Palaeolithic. *J. Archaeol. Method Theory* **2021**, *28*, 142–177. [CrossRef]

31. Hardy, B.L.; Moncel, M.H.; Daujeard, C.; Fernandes, P.; Béarez, P.; Desclaux, E.; Navarro, M.G.C.; Puaud, S.; Gallotti, R. Impossible Neanderthals? Making string, throwing projectiles and catching small game during Marine Isotope Stage 4 (Abri du Maras, France). *Quat. Sci. Rev.* **2013**, *82*, 23–40. [CrossRef]
32. Villa, P.; Soriano, S. Hunting weapons of Neanderthals and early modern humans in South Africa: Similarities and differences. *J. Anthropol. Res.* **2010**, *66*, 5–38. [CrossRef]
33. Beck, C. Projectile point types as valid chronological units. In *Unit Issues in Archaeology: Measuring Time, Space, and Material*; Ramenofsky, A.F., Steffen, A., Eds.; University of Utah Press: Salt Lake City, UT, USA, 1998; pp. 21–40.
34. Knecht, H. *Projectile Technology*; Plenum Press: New York, NY, USA, 1997.
35. Yu, P.L. From atlatl to bow and arrow: Implicating projectile technology in changing systems of hunter-gatherer mobility. In *Archaeology and Ethnoarchaeology of Mobility*; Sellet, F., Greaves, R., Yu, P.L., Eds.; University Press Florida: Gainsville, FL, USA, 2006; pp. 201–220.
36. Lombard, M. Variation in hunting weaponry for more than 300,000 years: A tip cross-sectional area study of Middle Stone Age points from southern Africa. *Quat. Sci. Rev.* **2021**, *264*, 107021. [CrossRef]
37. Lombard, M. Re-considering the origins of Old World spearthrower-and-dart hunting. *Quat. Sci. Rev.* **2022**, *293*, 107677. [CrossRef]
38. Hughes, S.S. Getting to the point: Evolutionary change in prehistoric weaponry. *J. Archaeol. Method Theory* **1998**, *5*, 345–408. [CrossRef]
39. Sitton, J.; Story, B.; Buchanan, B.; Eren, M.I. Tip cross-sectional geometry predicts the penetration depth of stone-tipped projectiles. *Sci. Rep.* **2020**, *10*, 13289. [CrossRef]
40. Shea, J.J. The origins of lithic projectile point technology: Evidence from Africa, the Levant, and Europe. *J. Archaeol. Sci.* **2006**, *33*, 823–846. [CrossRef]
41. Lombard, M. A standardized approach to the origins of lightweight-javelin hunting. *Lithic Technol.* **2022**, 1–11, *in press*. [CrossRef]
42. Cattelain, P. Hunting during the Upper Paleolithic: Bow, spearthrower, or both? In *Projectile Technology*; Knecht, H., Ed.; Plenum Press: New York, NY, USA, 1997; pp. 213–240.
43. Lombard, M.; Churchill, S. Revisiting Middle Stone Age hunting at≠GI, Botswana: A tip cross-sectional area study. *S. Afr. Field Archaeol.* **2022**, *17*, 11381. [CrossRef]
44. Thomas, D.H. Arrowheads and atlatl darts: How the stones got the shaft. *Am. Antiq.* **1978**, *43*, 461–472. [CrossRef]
45. Shott, M.J. Stones and shafts redux: The metric discrimination of chipped-stone dart and arrow points. *Am. Antiq.* **1997**, *62*, 86–101. [CrossRef]
46. Combier, J. *Le Paléolithique de l'Ardèche dans son Cadre Paléoclimatique*, 4th ed.; l'Institut de Préhistoire de l'Université de Bordeaux: Bordeax, France, 1967.
47. Moncel, M.-H. *Des Occupations Humaines de la Moyenne Vallée du Rhône de la Fin du Pléistocène Moyen et du Début du Pléistocène Supérieur*; Mémoire de la Société Préhistorique Française: Paris, France, 2008.
48. Moncel, M.H.; Condemi, S. Des restes humains dans le site Paléolithique moyen ancien de Payre (Ardèche): Dents et pariétal. *Bull. De La Société Préhistorique Française* **1997**, *94*, 168–171. [CrossRef]
49. Smith, T.M.; Austin, C.; Green, D.R.; Joannes-Boyau, R.; Bailey, S.; Dumitriu, D.; Fallon, S.; Grün, R.; James, H.F.; Moncel, M.-H.; et al. Wintertime stress, nursing, and lead exposure in Neanderthal children. *Sci. Adv.* **2018**, *4*, eaau9483. [CrossRef]
50. Verna, C.; Détroit, F.; Kupczik, K.; Arnaud, J.; Balzeau, A.; Grimaud-Hervé, D.; Bertrand, S.; Riou, B.; Moncel, M.-H. The Middle Pleistocene hominin mandible from Payre (Ardèche, France). *J. Hum. Evol.* **2020**, *144*, 102775. [CrossRef]
51. Masaoudi, H. *Application des Méthodes du Déséquilibre des Familles de l'uranium (230Th/234U) et de la Résonance de Spin Éléctronique (ESR) à la Datation des Sites d'Orgnac 3, de Payre et de l'abri des Pêcheurs (Ardèche)*; Muséum National d'Histoire Naturelle: Paris, France, 1995.
52. Valladas, H.; Mercier, N.; Ayliffe, L.K.; Falguères, C.; Bahain, J.J.; Dolo, J.M.; Froget, L.; Joron, J.L.; Masaoudi, H.; Reyss, J.L.; et al. Radiometric dates for the Middle Palaeolithic sequence of Payre (Ardèche, France). *Quat. Geochron.* **2008**, *3*, 377–389. [CrossRef]
53. Richard, M.; Pons-Branchu, E.; Genuite, K.; Jaillet, S.; Joannes-Boyau, R.; Wang, N.; Genty, D.; Cheng, H.; Price, G.J.; Pierre, M.; et al. Timing of Neanderthal occupations in the southeastern margins of the Massif Central (France): A multi-method approach. *Quat. Sci. Rev.* **2021**, *273*, 107241. [CrossRef]
54. Baena, J.; Moncel, M.-H.; Cuartero, F.; Chacón Navarro, M.G.; Rubio, D. Late Middle Pleistocene genesis of Neanderthal technology in Western Europe: The case of Payre site (south-east France). *Quat. Int.* **2017**, *436*, 212–238. [CrossRef]
55. Carmignani, L.; Moncel, M.-H.; Fernandes, P.; Wilson, L. Technological variability during the Early Middle Palaeolithic in Western Europe: Reduction systems and predetermined products at the Bau de l'Aubesier and Payre (South-East France). *PLoS ONE* **2017**, *12*, e0178550. [CrossRef]
56. Daffara, S.; Borel, A.; Moncel, M.-H. Conditioning of the raw materials on discoid exploitation strategies during the Early Middle Palaeolithic: The example of Payre level D (South-East France). *Archaeol. Anthropol. Sci.* **2019**, *11*, 4681–4695. [CrossRef]
57. Debard, E. *Le Quaternaire du Bas-Vivarais: Dynamique Sédimentaire, Paléoclimatologie et Chronologie d'après l'étude Sédimentologique des Remplissages d'avens, de Porches de Grottes et d'abris sous Roche: Comparaisons Avec le Velay Oriental*; Université Claude Bernard: Lyon, France, 1988.
58. Szmidt, C.C.; Moncel, M.-H.; Daujeard, C. New data on the Late Mousterian in Mediterranean France: First radiocarbon (AMS) dates at Saint-Marcel Cave (Ardèche). *Comptes Rendus Palevol* **2010**, *9*, 185–199. [CrossRef]

59. Crégut-Bonnoure, E.; Boulbes, N.; Daujeard, C.; Fernandez, P.; Valensi, P. Nouvelles données sur la grande faune de l'Eémien dans le Sud-Est de la France. *Quaternaire. Rev. De L'association Française Pour L'étude Du Quat.* **2010**, *21*, 227–248. [CrossRef]
60. Moncel, M.-H.; Michel, V. Première datation radiométrique par U-Th d'un niveau moustérien de l'Abri du Maras (Ardèche, France). *Bull. De La Société Préhistorique Française* **2000**, *97*, 371–375. [CrossRef]
61. Richard, M.; Falguères, C.; Pons-Branchu, E.; Bahain, J.J.; Voinchet, P.; Lebon, M.; Valladas, H.; Dolo, J.M.; Puaud, S.; Rué, M.; et al. Contribution of ESR/U-series dating to the chronology of late Middle Palaeolithic sites in the middle Rhône valley, southeastern France. *Quat. Geochron.* **2015**, *30*, 529–534. [CrossRef]
62. Moncel, M.-H.; Puaud, S.; Daujeard, C.; Lartigot-Campin, A.-S.; Millet, J.-J.; Theodoropoulou, A.; Cregut-Bonnoure, É.; Gely, B.; Vercoutère, C.; Desclaux, E.; et al. La Grotte du Figuier (Saint-Martin-d'Ardèche): Bilan des travaux récents sur un site du Paléolithique moyen et supérieur de la moyenne vallée du Rhône (Sud-Est de la France). *Bull. De La Société Préhistorique Française* **2012**, *109*, 35–67.
63. Moncel, M.-H.; Daujeard, C.; Cregut-Bonnoure, É.; Boulbes, N.; Puaud, S.; Debard, É.; Bailon, S.; Desclaux, E.; Escude, É.; Roger, T.; et al. Nouvelles données sur les occupations humaines du début du Pléistocène supérieur de la moyenne vallée du Rhône (France). Les sites de l'Abri des Pêcheurs, de la Baume Flandin, de l'Abri du Maras et de la Grotte du Figuier (Ardèche). *Quaternaire* **2010**, *21*, 385–411.
64. Moncel, M.-H.; Chacón, M.G.; Coudenneau, A.; Fernandes, P. Points and convergent tools in the European Early Middle Palaeolithic site of Payre (SE, France). *J. Archaeol. Sci.* **2009**, *36*, 1892–1909. [CrossRef]
65. Rivals, F.; Moncel, M.-H.; Patou-Mathis, M. Seasonality and intra-site variation of Neanderthal occupations in the Middle Palaeolithic locality of Payre (Ardèche, France) using dental wear analyses. *J. Archaeol. Sci.* **2009**, *36*, 1070–1078. [CrossRef]
66. Marín, J.; Daujeard, C.; Saladié, P.; Rodríguez-Hidalgo, A.; Vettese, D.; Rivals, F.; Boulbes, N.; Crégut-Bonnoure, E.; Lateur, N.; Gallotti, R.; et al. Neanderthal faunal exploitation and settlement dynamics at the Abri du Maras, level 5 (south-eastern France). *Quat. Sci. Rev.* **2020**, *243*, 106472. [CrossRef]
67. Geist, V. Of reindeer and man, modern and Neanderthal: A creation story founded on a historic perspective on how to conserve wildlife, woodland caribou in particular. *Rangifer* **2003**, *23*, 57–63. [CrossRef]
68. Daujeard, C.; Moncel, M.-H. On Neanderthal subsistence strategies and land use: A regional focus on the Rhone Valley area in southeastern France. *J. Anthropol. Archaeol.* **2010**, *29*, 368–391. [CrossRef]
69. Moncel, M.-H.; Daujeard, C. The variability of the Middle Palaeolithic on the right bank of the Middle Rhône Valley (southeast France): Technical traditions or functional choices? *Quat. Int.* **2012**, *247*, 103–124. [CrossRef]
70. Moncel, M.-H.; Daujeard, C.; Crégut-Bonnoure, E.; Fernandez, P.; Faure, M.; Guérin, C. L'occupation de la grotte de Saint-Marcel (Ardèche, France) au Paléolithique moyen: Stratégie d'exploitation de l'environnement et type d'occupation de la grotte. L'exemple des couches i, j et j'. *Bull. De La Société Préhistorique Française* **2004**, *101*, 257–304. [CrossRef]
71. Moncel, M.-H.; Chacón, M.G.; Vettese, D.; Courty, M.-A.; Daujeard, C.; Eixea, A.; Fernandes, P.; Allué, E.; Hardy, B.; Rivals, F.; et al. Late Neanderthals short-term and specialized occupations at the Abri du Maras (South-East France, level 4.1, MIS 3). *Anthropol. Archaeol. Sci.* **2021**, *13*, 45. [CrossRef]
72. Rots, V. Insights into early Middle Palaeolithic tool use and hafting in Western Europe. The functional analysis of level IIa of the early Middle Palaeolithic site of Biache-Saint-Vaast (France). *J. Archaeol. Sci.* **2013**, *40*, 497–506. [CrossRef]
73. Voormolen, B. Ancient Hunters, Modern Butchers: Schöningen 13II-4, a Kill-Butchery Site Dating from the Northwest European Lower Palaeolithic. Ph.D. Thesis, Leiden University, Leiden, The Netherlands, 2008.
74. Gaudzinski, S.; Roebroeks, W. Adults only: Reindeer hunting at the Middle Palaeolithic site Salzgitter Lebenstedt, northern Germany. *J. Hum. Evol.* **2000**, *38*, 497–521. [CrossRef]
75. Sahle, Y.; Hutchings, W.K.; Braun, D.R.; Sealy, J.C.; Morgan, L.E.; Negash, A.; Atnafu, B. Earliest stone-tipped projectiles from the Ethiopian Rift date to> 279,000 years ago. *PLoS ONE* **2013**, *8*, e78092. [CrossRef] [PubMed]
76. Porat, N.; Chazan, M.; Grün, R.; Aubert, M.; Eisenmann, V.; Horwitz, L.K. New radiometric ages for the Fauresmith industry from Kathu Pan, southern Africa: Implications for the Earlier to Middle Stone Age transition. *J. Archaeol. Sci.* **2010**, *37*, 269–283. [CrossRef]
77. Wilkins, J.; Schoville, B.J.; Brown, K.S.; Chazan, M. Evidence for early hafted hunting technology. *Science* **2012**, *338*, 942–946. [CrossRef] [PubMed]
78. Dusseldorp, G.; Lombard, M.; Wurz, S. Pleistocene Homo and the updated stone age sequence of South Africa. *S. Afr. J. Sci.* **2013**, *109*, 46–52. [CrossRef]
79. Lotter, M.G.; Gibbon, R.J.; Kuman, K.; Leader, G.M.; Forssman, T.; Granger, D.E. A geoarchaeological study of the middle and upper Pleistocene levels at Canteen Kopje, Northern Cape Province, South Africa. *Geoarchaeology* **2016**, *31*, 304–323. [CrossRef]
80. Grün, R.; Brink, J.S.; Spooner, N.A.; Taylor, L.; Stringer, C.B.; Franciscus, R.G.; Murray, A.S. Direct dating of Florisbad hominid. *Nature* **1996**, *382*, 500–501. [CrossRef]
81. Bruner, E.; Lombard, M. The skull from Florisbad: A paleoneurological report. *J. Anthropol. Sci.* **2020**, *98*, 89–97.
82. Dirks, P.H.; Roberts, E.M.; Hilbert-Wolf, H.; Kramers, J.D.; Hawks, J.; Dosseto, A.; Duval, M.; Elliott, M.; Evans, M.; Grün, R.; et al. The age of Homo naledi and associated sediments in the Rising Star Cave, South Africa. *Elife* **2017**, *6*, e24231. [CrossRef]
83. Dusseldorp, G.L.; Lombard, M. Constraining the likely technological niches of late Middle Pleistocene hominins with Homo naledi as case study. *J. Archaeol. Method Theory* **2021**, *28*, 11–52. [CrossRef]

84. Villa, P.; Lenoir, M. Hunting weapons of the Middle Stone Age and the Middle Palaeolithic: Spear points from Sibudu, Rose Cottage and Bouheben. *S. Afr. Humanit.* **2006**, *18*, 89–122.
85. Kuman, K.A. Florisbad and ≠Gi: The Contribution of Open-Air Sites to Study of the Middle Stone Age in Southern Africa. Ph.D. Thesis, University of Pennsylvania, Philadelphia, PA, USA, 1989.
86. Eltzholtz, A.K. Investigating Temporal Change in Fauresmith Technology: Insights from Rooidam 2, Northern Cape Province, South Africa. Master's Thesis, University of Cape Town, Cape Town, South Africa, 2020.
87. Val, A.; De la Peña, P.; Duval, M.; Bansal, S.; Colino, F.; Culey, J.; Hodgskiss, T.; Morrissey, P.; Murray, A.; Murungi, M.; et al. The place beyond the trees: Renewed excavations of the Middle Stone Age deposits at Olieboomspoort in the Waterberg Mountains of the South African Savanna Biome. *Archaeol. Anthropol. Sci.* **2021**, *13*, 116. [CrossRef]
88. Real, C.; Eixea, A.; Sanchis, A.; Morales, J.V.; Klasen, N.; Zilhão, J.; Villaverde, V. Abrigo de la Quebrada level IV (Valencia, Spain): Interpreting a Middle Palaeolithic palimpsest from a zooarchaeological and lithic perspective. *J. Paleolit. Archaeol.* **2020**, *3*, 187–224. [CrossRef]
89. Eixea, A.; Giner, B.; Jardón, P.; Zilhão, J.; Villaverde, V. Elementos líticos apuntados en el yacimiento de Paleolítico Medio del Abrigo de la Quebrada (Chelva, Valencia): Caracterización tecno-tipológica y análisis de las macrofracturas. Espacio, tiempo y forma. *Prehist. Arqueol.* **2015**, *8*, 78–109.
90. Lombard, M.; Shea, J.J. Did Pleistocene Africans use the spearthrower-and-dart? *Evol. Anthropol.* **2021**, *30*, 307–315. [CrossRef] [PubMed]
91. Lombard, M. Quartz-tipped arrows older than 60 ka: Further use-trace evidence from Sibudu, KwaZulu-Natal, South Africa. *J. Archaeol. Sci.* **2011**, *38*, 1918–1930. [CrossRef]
92. Backwell, L.; Bradfield, J.; Carlson, K.J.; Jashashvili, T.; Wadley, L.; d'Errico, F. The antiquity of bow-and-arrow technology: Evidence from Middle Stone Age layers at Sibudu Cave. *Antiquity* **2018**, *92*, 289–303. [CrossRef]
93. de la Peña, P.; Taipale, N.; Wadley, L.; Rots, V. A techno-functional perspective on quartz micro-notches in Sibudu's Howiesons Poort indicates the use of barbs in hunting technology. *J. Archaeol. Sci.* **2018**, *93*, 166–195. [CrossRef]
94. Bradfield, J.; Lombard, M.; Reynard, J.; Wurz, S. Further evidence for bow hunting and its implications more than 60,000 years ago: Results of a use-trace analysis of the bone point from Klasies River Main site, South Africa. *Quat. Sci. Rev.* **2020**, *236*, 106295. [CrossRef]
95. Cochrane, G.W. The transition from Howieson's Poort to post-Howieson's Poort industries in southern Africa. *S. Afr. Archaeol. Soc. Goodwin Ser.* **2008**, *10*, 157–167.
96. de la Peña, P.; Wadley, L. Technological variability at Sibudu Cave: The end of Howiesons Poort and reduced mobility strategies after 62,000 years ago. *PLoS ONE* **2017**, *12*, e0185845. [CrossRef]
97. Clark, J.L.; Plug, I. Animal exploitation strategies during the south African middle stone Age: Howiesons Poort and post-Howiesons Poort fauna from sibudu cave. *J. Hum. Evol.* **2008**, *54*, 886–898. [CrossRef] [PubMed]
98. Rios-Garaizar, J. Economic organization among Neanderthals: Amalda (Basque Country, Spain) level VII Mousterian. *Zephyrus* **2010**, *65*, 15–37.
99. Sánchez-Romero, L.; Benito-Calvo, A.; Marín-Arroyo, A.B.; Agudo-Pérez, L.; Karampaglidis, T.; Rios-Garaizar, J. New insights for understanding spatial patterning and formation processes of the Neanderthal occupation in the Amalda I cave (Gipuzkoa, Spain). *Sci. Rep.* **2020**, *10*, 8733. [CrossRef] [PubMed]
100. Yaroshevich, A.; Kaufman, D.; Marks, A. Weapons in transition: Reappraisal of the origin of complex projectiles in the Levant based on the Boker Tachtit stratigraphic sequence. *J. Archaeol. Sci.* **2021**, *131*, 105381. [CrossRef]
101. Feizi, N.; Nasab, H.V. Symmetry and Middle Paleolithic Points: A Case Study of Iran's Middle Paleolithic Points. Research Square 2022. Available online: https://assets.researchsquare.com/files/rs-257471/v3_covered.pdf?c=1647444337 (accessed on 26 December 2022).
102. Groucutt, H.S. Middle Palaeolithic point technology, with a focus on the site of Tor Faraj (Jordan, MIS 3). *Quat. Int.* **2014**, *350*, 205–226. [CrossRef]
103. Egeland, C.P.; Gasparian, B.; Fadem, C.M.; Nahapetyan, S.; Arakelyan, D.; Nicholson, C.M. Bagratashen 1, a stratified open-air Middle Paleolithic site in the Debed river valley of northeastern Armenia: A preliminary report. *Archaeol Res. Asia* **2016**, *8*, 1–20. [CrossRef]
104. Schmitt, D.; Churchill, S.E.; Hylander, W.L. Experimental evidence concerning spear use in Neandertals and early modern humans. *J. Archaeol. Sci.* **2003**, *30*, 103–114. [CrossRef]
105. Milks, A.; Champion, S.; Cowper, E.; Pope, M.; Carr, D. Early spears as thrusting weapons: Isolating force and impact velocities in human performance trials. *J. Archaeol. Sci. Rep.* **2016**, *10*, 191–203. [CrossRef]
106. Wynn, T.; Coolidge, F. Why not cognition? *Curr. Anthropol.* **2008**, *49*, 895–897. [CrossRef]
107. Wynn, T.; Coolidge, F.L. A Stone-Age Meeting of Minds: Neandertals became extinct while Homo sapiens prospered. A marked contrast in mental capacities may account for these different fates. *Am. Sci.* **2008**, *96*, 44–51. [CrossRef]
108. Coolidge, F.L.; Overmann, K.A. Numerosity, abstraction, and the emergence of symbolic thinking. *Curr Anthropol.* **2012**, *53*, 204–225. [CrossRef]
109. Bruner, E.; Spinapolice, E.; Burke, A.; Overmann, K.A. Visuospatial integration: Paleoanthropological and archaeological perspectives. In *Evolution of Primate Social Cognition*; Di Paolo, L., Di Vincenzo, F., De Petrillo, F., Eds.; Springer: Berlin/Heidelberg, Germany, 2018; pp. 299–326.

110. Bruner, E.; Lozano Ruiz, M. Extended mind and visuo-spatial integration: Three hands for the Neandertal lineage. *J. Anthropol. Sci.* **2014**, *92*, 273–280. [PubMed]
111. Bruner, E.; Lozano Ruiz, M. Three hands: One year later. *J. Anthropol. Sci.* **2015**, *93*, 191–195.
112. Gärdenfors, P.; Lombard, M. Causal cognition, force dynamics and early hunting technologies. *Front. Psychol.* **2018**, *9*, 87. [CrossRef]
113. Lombard, M. On the minds of bow hunters. In *Squeezing Minds from Stones*; Coolidge, F.L., Overmann, K.A., Wynn, T., Eds.; University of Oxford Press: Oxford, UK, 2019; pp. 473–496.
114. Högberg, A.; Lombard, M. 'I can do it' becomes 'we do it': Kimberley (Australia) and Still Bay (South Africa) points through a socio-technical framework lens. *J. Paleolit. Archaeol.* **2020**, *3*, 633–663. [CrossRef]
115. Bruner, E.; Preuss, T.M.; Chen, X.; Rilling, J.K. Evidence for expansion of the precuneus in human evolution. *Brain Struct. Funct.* **2017**, *222*, 1053–1060. [CrossRef]
116. Kuhlwilm, M.; Gronau, I.; Hubisz, M.J.; De Filippo, C.; Prado-Martinez, J.; Kircher, M.; Fu, Q.; Burbano, H.A.; Lalueza-Fox, C.; de La Rasilla, M.; et al. Ancient gene flow from early modern humans into Eastern Neanderthals. *Nature* **2016**, *530*, 429–433. [CrossRef]
117. Nielsen, M.; Subiaul, F.; Galef, B.; Zentall, T.; Whiten, A. Social learning in humans and nonhuman animals: Theoretical and empirical dissections. *J. Comp. Psychol.* **2012**, *126*, 109–113. [CrossRef]
118. Fu, Q.; Hajdinjak, M.; Moldovan, O.; Constantin, S.; Mallick, S.; Skoglund, P.; Patterson, N.; Rohland, N.; Lazaridis, I.; Nickel, B.; et al. An early modern human from Romania with a recent Neanderthal ancestor. *Nature* **2015**, *524*, 216–219. [CrossRef]
119. Gokcumen, O. Archaic hominin introgression into modern human genomes. *Am. J. Phys. Anthropol.* **2019**, *171*, 60–73. [CrossRef]
120. Djakovic, I.; Soressi, M.; Key, A. Optimal linear estimation models predict 1400-2800 years of co-existence between Neandertals and Homo sapiens in western Europe. *bioRxiv* **2022**. [CrossRef]
121. Picin, A.; Moroni, A.; Benazzi, S. The arrival of Homo sapiens in the Near East and Europe. In *Updating Neanderthals*; Romagnoli, F., Rivals, F., Benazzi, S., Eds.; Academic Press: Cambridge, MA, USA; Elsevier: Amsterdam, The Netherlands, 2022; pp. 321–347.

quaternary

MDPI

Article

Ecological Constraints and Drivers for Human Dispersals and Adaptations in the Late Pleistocene and Early Holocene Environments of the East Siberian Arctic

Vladimir V. Pitulko [1,2,*] and Elena Y. Pavlova [3,*]

1 Institute for the History of Material Culture, Russian Academy of Sciences, 18 Dvortsovaya Nab., 191186 St. Petersburg, Russia
2 Peter the Great Museum of Anthropology and Ethnography (Kunstkamera), Russian Academy of Sciences, 3, Universitetskaya Nab., 199034 St. Petersburg, Russia
3 Arctic & Antarctic Research Institute, 35 Bering St., 199397 St. Petersburg, Russia
* Correspondence: pitulko.vladimir@gmail.com (V.V.P.); pavloval@rambler.ru (E.Y.P.)

Abstract: Starting roughly 50,000 years ago, the Arctic region of East Siberia remained continuously populated by groups of anatomically modern humans including the most uncomfortable episodes in the development of the late Quaternary environment; for some of them, human presence in the area became ephemeral. At present, archaeological fossil records allow for distinguishing three main stages in human occupation of the area: Early (~50 to ~29 ka, MIS 3), middle (~29 to ~11.7 ka, MIS 2), and late (from 11.7 to ~8 ka). For most of the time, they the populated open landscapes of the Mammoth Steppe, which declined at the onset of the Holocene. Human settlement of the Arctic was driven by various abiotic and biotic factors and thus archaeologically visible cardinal cultural and technological changes correspond to the most important paleoclimatic and habitat changes in the late Pleistocene and early Holocene. Successful peopling of the Arctic was largely facilitated by the adoption of critically important innovations such as sewing technology based on the use of the eyed bone needle and the manufacture of long shafts and pointed implements made of mammoth tusks. Mammoth exploitation is seen in mass accumulations of mammoths formed by hunting. An obvious connection between archaeological materials and such accumulations is observed in the archaeological record. In the lithic technology, the early stage is presented by archaic-looking flake industries. Starting the LGM, the wedge-core based-microblade technology known as the Beringian microblade tradition spread widely following the shrinkage of the mammoth range. At the late stage, starting at the Holocene boundary, microprismatic blade technology occurs. In all stages, the complex social behavior of the ancient Arctic settlers is revealed. The long-distance transport of products, knowledge, and genes occurs due to the introduction of the land transportation system. Initial human settlement of this region is associated with carriers of the West Eurasian genome who became replaced by the population with East Asian ancestry constantly moving North under the pressure of climate change.

Keywords: arctic Siberia; late Pleistocene; early Holocene; Stone Age; Upper Palaeolithic; human dispersal; adaptations; mammoth; critical technology; complex human behavior

Citation: Pitulko, V.V.; Pavlova, E.Y. Ecological Constraints and Drivers for Human Dispersals and Adaptations in the Late Pleistocene and Early Holocene Environments of the East Siberian Arctic. *Quaternary* **2023**, *6*, 56. https://doi.org/10.3390/quat6040056

Academic Editor: Ran Barkai

Received: 18 July 2023
Revised: 28 August 2023
Accepted: 28 September 2023
Published: 6 November 2023

1. Introduction

The Arctic territory is an area occupied by many variations of the tundra biome from the sparse Arctic, common on the coasts of the Arctic Ocean and some islands, to relatively richer variants of shrub tundra and forest tundra in the southern areas of the region [1]. Since the beginning of the scientific study of this area of the planet approximately 400 years ago, European and Russian travelers and researchers have learned that these territories, despite the severity of the climate and limited natural resources, were actually inhabited by humans throughout their entire length. At the same time, the assumption occurred that people had lived there for a long time—in a certain sense, always—and a natural desire developed to find out when people first came to the Arctic, and who these people were.

The volumes of archaeological evidence on the basis of which it is possible to judge the processes of human settlement in the Arctic as a whole (from the initial stage to historical modernity) are quite large, but they are unevenly distributed [2]. In addition, sometimes there is a significant chronological gap between the evidence of the initial human occupation of the area and the continuous archaeological chronicle documenting their permanent stay in certain regions of the Arctic. In some cases, such gaps span tens of thousands of years, which is typical, first, for the western edge of the East Siberian Arctic—the Taimyr Peninsula [3].

The chronicle of human settlement in the Arctic spans approximately 50,000 years [4–7]; hereafter, the calendar ages are used. Geographically, archaeological finds are associated with Arctic territories to the west of the Beringian Land Bridge, which once connected the Euro-Asian and North American continents—a paleo-geographical, zoogeographic, and floral phenomenon of the Late Pleistocene of the Holarctic [8–16]. These data characterize the final stage of the global process of settlement of anatomically modern humans. Thus, the Arctic regions of Northern Eurasia were settled, human migration to the New World took place [17,18], and finally, the Arctic territories of America and Greenland became inhabited. As a result, the Eastern Hemisphere was completely mastered by people; the Western Hemisphere was also discovered and populated [19,20], although the participants of that process did not suspect this. Thus, the Arctic territories turned out to be the scene of the most important events in the history of mankind. On the scale of geological time, this process covers a significant part of MIS 3 and MIS 2 (the second half of the late Pleistocene) and ends in geological modernity, in the Holocene (MIS 1) (Figure 1). At the same time, in the second half of the Holocene, the population was mainly redistributed in previously populated areas, often associated with the influx of new human groups moving in a northerly direction. The main goal of the article is to determine the most important events in the history of the human settlement of East Arctic Siberia based on the ages of known archaeological evidence and the geographic position of archaeological sites, discuss these in relation to the past environment and climate change, and reconstruct the population history of the study area.

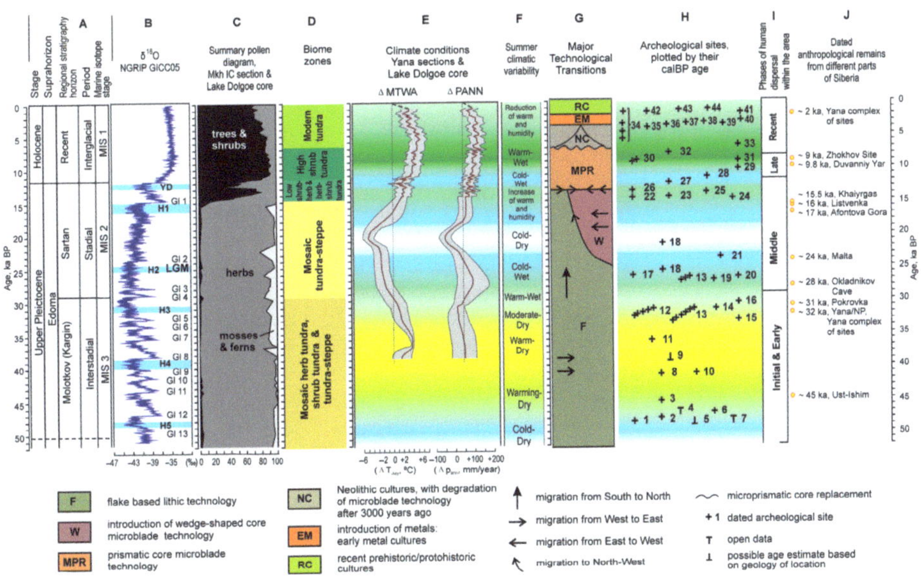

Figure 1. Climatic, paleoenvironmental, and archaeological record of the East Siberian Arctic. (**A**) Regional stratigraphic scheme of the Yano-Kolyma lowland and its mountain framing, after Stratigraphic

Code... [21]; (**B**) NorthGRIP δ18O scale and the sequence of Greenland interstadials (GI1–GI13), according to Svensson et al. [22], blue bands mark the approximate position of the Younger Dryas cooling (YD) and Heinrich events (H1–H5) according to Tierney et al. [23]; (**C**) general composition of palynospectra retrieved from Quaternary sediments of Mkh IC section, according to Sher et al. [24] and bottom sediment core from Dolgoe Lake, after Pisaric et al. [25], Klemm et al. [26]; (**D**–**G**), after Pavlova and Pitulko [16]: (**D**) biomes; (**E**) paleoclimatic reconstructions based on paleofloristic analysis: ΔMTWA—deviation of air temperatures of the warmest month (°C), ΔPANN—average annual precipitation (mm/year); (**F**) total climate change; (**G**) archaeological record of the East Siberian Arctic; (**H**) archaeological objects according their calendar age; (**I**) phases of human dispersal within the area; (**J**) dated anthropological remains from various regions of Siberia. For (**H**), dated archaeological sites are indicated: 1—Sopkarga mammoth, 2—the locality of Bunge-Toll-1885, 3—Kyuchus, 4—Upstream Point, Yana complex of sites (YCS), 5—Zyryanka, 6—Irelyakh-Siene, 7—Bolshoy Anyui, 8—New Siberia/West, 9—New Siberia/East, 10—AL044-2005 site, 11—Omoloy, 12—Yana complex of sites (YCS): Northern Point, Yana B area and Yana mammoth "graveyard"/YMAM, 13—Diring-Ayan, 14—Buor-Khaya/Ortho-Stan, 15—Kastykhtakh mammoth, 16—Tabayuryakh mammoth, 17—Lagernyi Point/YCS, 18—Yana A area/YCS, 19—Ilin-Syalakh 034, 20—Wrangel island, 21—Zyryanka 1, 22—Urez-22, 23—Ilin-Syalakh, 24—Achchaghyi-Allaikha, 25—Berelekh geoarchaeological complex; 26—Nikita Lake, 27—Cape Kamennyi, 28—Tytylvaam IV, 29—Naivan, 30—Zhokhov site, 31—Chelkun IV, 32—Tuguttakh, 33—Tagenar VI, 34—Siktyakh I, 35—Rodinka burial, 36—Chertov Ovrag, 37—Burulgino, 38—Rauchuagytgyn I, 39—Pegtymel, 40—Aachim-base, 41—Aachim-lighthouse, 42—Cape Baranov, 43—Pegtymel cave, 44—Shalaurova Izba, after Pavlova and Pitulko ([16]: Table 1) (with modifications based on Pitulko et al., 2015 [27]), Pitulko and Pavlova 2016 [28], Novgorodov et al. [29], Cheprasov et al. [30,31], Pavlov and Suzuki 2020 [32], Chlachula et al. [33], Kirillova at al. [34]; Pitulko et al. [35,36], Dikov [37], K [38], Khlobystin [39], Gusev [40]. Geographic locations of the sites are indicated on the maps organized by time slices.

2. The Early Stage of Human Settlement in the East Siberian Arctic (~50–29 ka)

The specified interval (termed Molotkovskii (Karginskii) interstadial in Russian chronostratigraphic scheme of NE Siberia) was warmer and wetter than the beginning of MIS 3 in terms of temperature and humidity; in Arctic West Beringia, an ecosystem of open spaces was formed, dominated by herbaceous vegetation, [41], which was favorable for the existence of megafauna. Throughout its entire stretch from the northern to the inner southern regions at that time, against the background of an arid climate, dry grass-sedge tundra-steppe prevailed under developing warming conditions (Figure 1A–F), where woolly mammoth (*Mammuthus primigenius* Blumenbach), Pleistocene bison (*Bison priscus* Bojanus), and Pleistocene horse (*Equus caballus* L.) lived [24,42–46].

For local mammoth and bison populations, based on the frequency of dating, a close increasing trend of changes in relative abundance was established, quickly reaching maximum values [47]. At this time, the relative abundance of the mammoth population was rapidly increasing, and approximately 45,000 years ago reached a historical maximum in its history of existence [48]. A biome of the mammoth steppe formed, which played an important role in human settlement in Northern Eurasia. Its development was determined by the alternation of relatively warmer and relatively colder periods, during which the features of similarity in vegetation and landscapes were preserved throughout the territory it occupied, but with simultaneous diversity of local conditions [49].

The early phase of the MIS 3 interstadial (57–45.7 ka) includes the archaeological materials listed here from west to east, obtained from the following locations: Sopochnaya Karga (~48.3 ka), Bunge-Toll-1885 (~48 ka), Kyuchus (~45.2 ka), the Upstream Point of the Yana complex of sites (>46.6 ka), Zyryanka (~49 ka), Irelyakh-Siene (~47 ka), and Bolshoy Anyui (~48.4 ka) (Figure 1H—1–7, accordingly). The most reliable among them are Sopochnaya Karga and Bunge-Toll'-1885 locations, the Upstream Point of the Yana complex of sites, and finds from Zyryanka and Irelyakh-Siene sites. An age estimate for the first three sites is supported by direct dates on the artefacts or bone remains associated with human activity, while the geological age of the archaeological materials from Zyryanka and Irelyakh-Siene is controlled by the stratigraphic context of the locations. All of them indicate for this interval the presence of man in the area from the present mouth of the Yenisei River to Western Chukotka (Figure 2A). This evidence was dispersed over ~3000 km and appeared approximately simultaneously, within one to three thousand years. The pace of human settlement in the territory, therefore, was very fast.

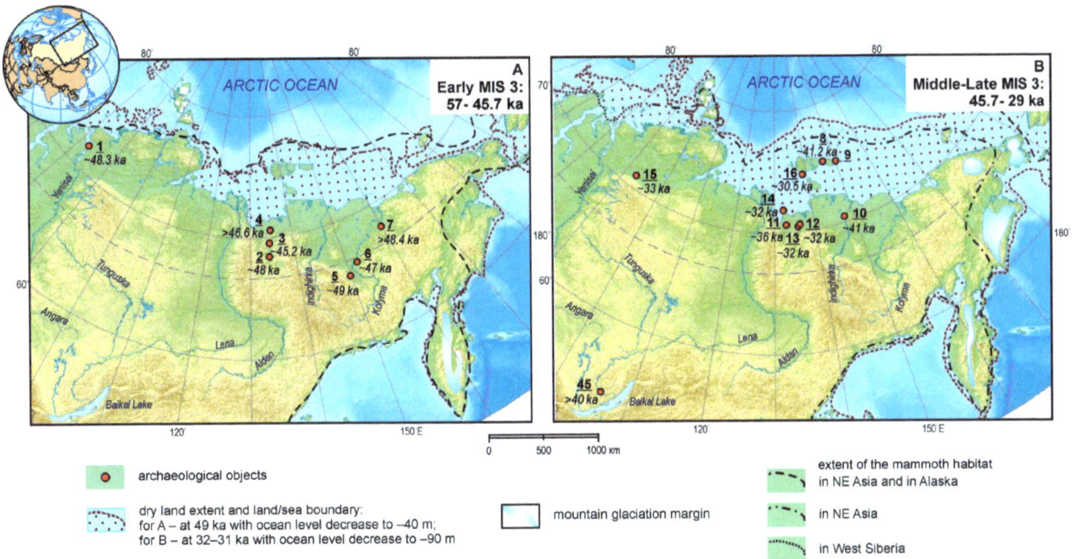

Figure 2. Early stage of the human settlement of the arctic East Siberia and MIS 3 paleoenvironments of the area. (**A**) Locations that yielded evidence of human presence during the early MIS 3 interstadial (57–45.7 ka). Paleoenvironments: dry land extent and land/sea boundary at 49 ka with ocean level decrease to −40 m is based on Pico et al. [50]; mountain glaciation margin, after Barr and Clark [51], Galanin [52]; extent of the mammoth habitat: in NE Asia and in Alaska, after McDonald et al. [53]. Dated archaeological sites are indicated: 1—Sopohnaya Karga mammoth, 2—the locality of Bunge-Toll-1885, 3—Kyuchus, 4—Upstream Point, Yana complex of sites (YCS), 5—Zyryanka, 6—Irelyakh-Siene, 7—Bolshoy Anyui. (**B**) Locations that yielded evidence of human presence during middle and late MIS 3 interstadial (45.7–29 ka). Paleoenvironments: dry land extent and land/sea boundary at 32—31 ka with ocean level decrease to −90 m is based on Pico et al. [50]; mountain glaciation margin, after Barr and Clark [51], Galanin [52]; extent of the mammoth habitat: in NE Asia and in Alaska, after McDonald et al. [53], in NE Asia, after Pitulko, Nikolskiy [54], in West Siberia, after Kahlke [55]. Dated archaeological sites are indicated: 8—New Siberia/West, 9—New Siberia/East, 10—AL044-2005 site, 11—Omoloy, 12—Yana complex of sites (YCS): Northern Point, Yana B area and Yana mammoth "graveyard"/YMAM, 13—Diring-Ayan, 14—Buor-Khaya/Ortho-Stan, 15—Kastykhtakh mammoth, 16—Tabauyriakh mammoth. Archaeological sites of that time in South Siberia: 45—Makarovo IV. Location of the sites and their age estimate is based on Pitulko et al., 2016b [3], Pitulko [4], Pavlova and Pitulko [16], Cheprasov et al., 2015 [30], Cheprasov et al., 2018 [31], Pavlov, Suzuki 2020 [32], Chlachula et al., 2021 [33], Kirillova et al., 2012 [34], Novgorodov et al., 2014 [29], Pitulko et al., 2015 [27], Pitulko et al., 2015 [56], Pitulko et al., 2017 [57], Goebel and Aksenov 1995 [58], Derevianko 1998 [59], Goebel et al., 2000 [60], Rybin and Khatsenovich [61]. Note numbers provided for archaeological sites are the same as shown in Figure 1H.

This is possible when occupying an ecological niche that was not occupied by anyone at the time of the arrival of the first settlers. Since there is no reliable evidence of the settlement of Neanderthals or other representatives of archaic forms of *Homo* in the Arctic, that is, no anthropological remains or archaeological finds north of 55° N exist [62–64], the assumption that there is a colossal free ecological niche in the Arctic region of the continent is at least acceptable. In such a situation, the rapid movement of population groups occurs in search of the best free land according to the scenario described for the historical Eskimo migrations [65], and it is possible only in a familiar landscape that does not require adaptation to an environment different from already known conditions.

As an example, it is necessary to cite Arctic Sea mammal hunters, who were very picky in the practice of using their ecological niche. The choice of a place for their settlement has always been conditioned by a number of necessary requirements (surface height, the scope of sea-level run-up fluctuations, the quality of the beach, freshwater, availability of basic resources, availability of alternative and/or complementary resources, availability of wood, etc.). Respectively, in the first place, locations were occupied whose physical characteristics corresponded to the list of requirements as much as possible. For this reason, such sites have existed for millennia; then as others, less convenient, began to be used forcibly, when the demographic capacity of the territory of long-term settlements was exhausted. In this regard, Krupnik [65] notes that the distance between settlements with long continuous habitation is, as a rule, very large.

This scenario may well be extrapolated on the model of settlement of the Early Upper Palaeolithic people, who began to explore the Arctic zone of Eastern Siberia at the beginning of the MIS 3 interstadial, with an almost simultaneous manifestation of the archaeological signal within the boundaries of the large region from Western Taimyr to Western Chukotka is connected (Figures 1H and 2A). The conditions for rapid human settlement were the relative uniformity of the landscape and faunal groupings living there, among which the most important for humans were large herbivores and the lack of competition for resources with other hominins. As a result, it turned out to be quickly dominated by people with an extremely low demographic density, which remained the norm throughout all subsequent periods and is true to the present-day Arctic.

Thus, approximately 2.3 million people currently live in the Arctic zone of Russia, with an average density of 0.63 people per square kilometer. Based on the historical records, for the territories of the whole of Siberia at the time of contact with the Russians, the local population is estimated at 250,000 people [66], and for the Siberian Arctic, this figure is an order of magnitude lower. It is impossible to reliably estimate the number of Arctic first settlers of the Upper Paleolithic; the only reference point in this regard is the result of a genome-wide study performed by Sikora and colleagues [67]. It is shown that the number of the entire human group, to which the residents of the Yana site belonged, did not exceed 500 people; at the same time, they knew how to circumvent the problem of inbreeding [67]. This knowledge was not unique since the same ability was established for the Upper Paleolithic inhabitants of the Russian Plain [68], i.e., its possession is one of the fundamental behavioral traits of anatomically modern humans.

The small amount of archaeological evidence, among which there are practically no settlement contexts, does not allow us to fully assess the features of the material culture of the population of the East Siberian Arctic at the initial stage of human settlement within its borders. Based on the findings from the Upstream Point of the Yana site and the location of Kyuchus [4,57], it can be said that, at that time, there were flake-based lithic industries of archaic appearance based on the splitting of pebbles, which were accompanied by bone-processing techniques and the technology of producing long shafts (rods, long points, and full-size spears) from mammoth tusks. Finds from sites at the early stage suggest that the basis of the life of these people was the hunting of various large herbivores—bison, horse, red deer (*Cervus elaphus* L.), reindeer (*Rangifer tarandus* L.), woolly rhinoceros (*Coelodonta antiquitatis* Blumenbach), and woolly mammoth, but not the hunting of small animals such a hare (*Lepus tanaiticus* Gureev), i.e., in general, an economic model that fully corresponded to the one that existed subsequently for tens of thousands of years with changes corresponding to the fluctuation in the number of members of the faunal complex of the biome [69,70].

The most favorable natural climatic conditions of MIS 3 (MIS 3 optimum, that is, the interval covering 45.7–37.8 ka) are associated with the highest sea-level rise for the time of MIS 3 transgression [50]. The onset of the most favorable conditions and the duration of the MIS 3 optimum differed in time in different parts of the East Siberian Arctic, which was likely due to the degree of remoteness of these areas from the transgression front, the increase in the area occupied by the sea, and the effects of the deglaciation of the territories occupied by the Scandinavian/Barents Sea glacier [71,72].

During this period, sedge-grass and sedge-grass-forb tundra-steppe with a noticeable participation of shrubby willow were widespread [24,43–45,73,74]. The MIS 3 optimum in the northern part of the Arctic Western Beringia (Bolshoi Lyakhovskii Island) and its second half after 40.6 ka in the inner regions (within the present-day Bykovskii Peninsula and Yana-Indighirka lowland) were distinguished by the greatest number and diversity of mammoth fauna, represented by woolly mammoth, the Pleistocene horse, the Pleistocene bison, and reindeer [24,42–45]. In the range of 40–37 ka, a maximum number of woolly mammoths is noted everywhere [48,75].

Favorable conditions of the climatic optimum of MIS 3 apparently gave an additional impetus to the process of development of these territories by man. The population movement within the East Siberian Arctic at this time—new migrations into the area and possibly displacement of the population in already inhabited areas (Figure 2B)—can be judged by several facts, based on the dating obtained for the locations AL044–2005, Omoloi, New Siberia West, and New Siberia East (Figure 1H—10, 11, 8, 9, respectively), found in the space from the Yana site to the Indighirka River and on the New Siberian Islands [4,16,27,56,57]. Based on these few data, it can be concluded that the inhabited territory expanded to the north and embraced the area of the modern New Siberian Islands, for which there is correspondence in the above observations on the spatial dynamics of the habitats of the fauna contemporary with these people and the growth of the relative number of populations of the main hunted species.

In terms of the economic behavior of these people, it can be assumed that it has not undergone significant changes, but the data for this conclusion are minimal. The lithic industry is still flake-based, evident from finds at the Omoloi locality; rock-crystal is used (based on the finds from locality AL044-2005 in the Indighirka basin). In addition, on the island of New Siberia (localities West and East), cores from mammoth tusks were found 670 km northeast of the Yana complex of sites, from which long rods (points) were obtained, executed in classical Yana technology [56]. The age of one of these cores has been determined at ~41.2 ka.

The dominant process of transformation of the natural environment in the late phase of the interstage MIS 3 (37.8–29 ka) is the renewed decline in sea level. After 38 ka, it took on a growing exponential character, and at the end of MIS 3 by 29 ka, the decline reached marks 109–110 m relative to the modern sea level [50,76–78], in whose connection there was an increase in land area continued along the northern edges of the continent with a simultaneous increase in aridity in its interior (Figure 2B). Differences in the degree of moistening and, as a consequence, differences in the nature of vegetation cover, were determined by the position of individual territories in relation to the sea basin. During the late MIS 3 interstadial, sharply continental dry conditions existed in most of the territory, and in some time periods, wet conditions alternated with warm summer temperatures and severe winters with little snow [43,45,79]. This contributed to the widespread development of mosaic tundra landscapes at this time (Figure 1A–F). Plant associations were represented by forb-grass-sedge, wormwood-grass-sedge, and grass-wormwood-forbs communities, xeropetrophytic communities, local tundra communities in waterlogged areas, and sparse thickets of shrubby willow [16,80,81].

The rich herbage of the graminoid-dominated tundra steppe served as an excellent food base supporting the fauna of the mammoth complex, widespread throughout the territory of Western Beringia during the MIS 3 interstadial [24,42–44,48,57,71,82–85]. The most common species were the woolly mammoth, Pleistocene horse, reindeer, and Pleistocene bison. Woolly rhinoceros and musk ox were also encountered; among the predators were the wolf (*Canis lupus* L.), the polar fox (*Alopex lagopus* L.), the brown bear (*Ursus arctos* L.), the Pleistocene lion [*Panthera spelaea* (Goldfuss)], and a wolverine (*Gulo gulo* L.) [86,87]. During the late phase of MIS 3, there is a distinct dynamic change in the number and diversity of mammoth fauna in space and time [57]. In the inner regions of the Arctic of West Beringia (the Bykovskii Peninsula and Yana-Indighirka lowland), initially high numbers and diversity of mammoth fauna animals, represented by woolly mammoth, Pleistocene horse, bison, reindeer, gray wolf, and brown bear, were shrinking. After 31.3 ka, only woolly mammoth, Pleistocene horse, and reindeer remained in communities. The northern regions (the Bolshoy Lyakhovskii Island) are also characterized by a decrease in the diversity and abundance of mammoth fauna. Thus, if the woolly mammoth, Pleistocene bison, Pleistocene horse, musk ox (*Ovibos moschatus* Zimmerman), and woolly rhinoceros were represented between 37.8 and 33.2 ka, then only the woolly mammoth and Pleistocene horse were present by 29 ka [45,57]. The woolly mammoth population was characterized by the greatest constancy in spatial distribution and variations in abundance in the late MIS 3 thermochron, depending on climatic fluctuations [48,57,75].

Cultural remains of this time are represented in only two districts of the northwest Yana-Indighirka lowland. This is the Yana complex of sites [4,56,87–89] and the location of the Buor-Khaya/Orto-Stan on the west side of the Buor-Khaya Peninsula [35]; their age fits into the interval 33–31 ka (Figure 1H—12, 14, respectively; Figure 2B). The discovery of the Katystakh mammoth on the Taimyr belongs to the same time [34]; hunting lesions on its skeletal elements possibly indicate the presence of a human there on the eve of the last glacial maximum (Figure 1H—15).

The findings from the Buor-Khaya allow us to talk about the anthropogenic contribution to the formation of the bone-bed, which yielded modified mammoth bones, bones with traces of hunting impact, and skeletal elements with traces of butchering, as well as bison and horse bones. This is a rather expressive trace of human activity, replicating the behavior of the Yana site dwellers and indicating their widespread movement within the territory. The possibility of people moving great distances is documented by the finds of pendants made of exotic stone raw materials: Anthraxolite and amber. The nearest source of the first is located 300 km to the northwest in the area of the northeastern spurs of the Kharaulakh Range, and the source of amber is available only on the island of New Siberia, 670 km northeast of the Yana site [90]. The presence of people in the territory of modern New Siberian Islands on the eve of LGM is documented by the discovery of the Tabayuryakh mammoth (Figure 1H—16; Figure 2B) with a fragment of a throwing point embedded in its scapula [32]; in Taimyr, the most likely evidence of this kind is the injuries on the bones of the Katystakh mammoth [34].

The Yana complex of sites has a complicated spatial structure and contains several archaeological objects/localities, including seasonal residential zones—spring–summer, the Northern Point locality [91], and winter, the site of Yana "B" [88], as well as the mass accumulation of mammoth bone remains, YMAM. This object was formed as a result of human activity, as an area for storing and preliminary preparation for processing the results of mammoth hunting, that is, different body parts including complete heads with tusks [56,92,93].

The material cultural remains of the Yana complex, according to observations made at the Yana "mammoth graveyard" (YMAM) and during excavations at the Northern Point and Yana "B" sites, include four main technological contexts: (1) The production of multifunctional tools (scrapers) used for processing hunted prey and various materials, of which the most remarkable are backed forms [90,94]; (2) the production of micro tools for processing diverse osseous materials such as mammoth bones and tusks (ivory), and reindeer antler, as well as for creating elements of hunting equipment [89]; (3) the production of artifacts from mammoth tusk, bone, and antler [56,91,95]; and (4) the production of red "ochre" [90]. There are no visible dwelling structures; however, there are hearths and evidence related to them, or of a "fire use context".

The stone industry, based on the splitting of pebbles, has a pronounced flaking character and an archaic appearance—discoid, pyramidal, and orthogonal cores made of pebbles readily available on the beach next to the site area comprising a variety of local rocks. It is complemented by the limited use of rock crystal (available in the area approximately 50 km southward), mainly for processing tusks, as well as the production of ochre. The processing of the tusk and bone is represented by a massive series of functional (points, needles, and awls) and non-utilitarian objects (beads, pendants, 'diadems', bracelets, and figurative art), forming a phenomenal quality context of symbolic behavioral manifestations. On the basis of the materials from the Yana site, critical technologies that provided the possibility for the stable existence of these people in the conditions of the Late Pleistocene of the East Siberian Arctic have been reconstructed [56,91,96].

The basis of the economic activity of this population was hunting for Pleistocene bison, horses, reindeer, woolly rhinoceros, and woolly mammoth. Hares were caught in a variety of ways, but only as raw materials for sewing production; in the cultural sediments of the site, there are many skeletal deposits of hares from which the skin was removed [90]. In the cultural practice of the Arctic peoples, it is a valuable raw material for tailoring undergarments [97].

Mammoth hunting was carried out for its tusks, which were an important raw material resource as a substitute for wood [56], completely absent in the open landscapes of the mammoth steppe but extremely necessary for the creation of objects of the hunting equipment complex. In the series of dating mammoth bone remains from the YMAM, there are no indications of mass one-time hunting. In contrast, they indicate a gradual accumulation of bone remains in that special area of the Yana site complex [56,93]. Despite the fact that the mammoth was obviously eaten, it was not the main food resource. As such, bison, horses, and reindeer have appeared at different times [90–92].

3. The Middle Stage of Human Settlement in the East Siberian Arctic (~29–11.7 ka)

The MIS 2 spanning 29–11.7 ka (for East Siberia, it is also termed Sartan cryochron in Russian nomenclature) is characterized by the development of environmental conditions in the regime of progressive global climate cooling [98,99] against the background of ongoing deepening of the sea level, which by 24 ka had decreased to its minimum and reached a position of 129–130 m relative to the modern level [50,76–78]. In the inner part of the Arctic Western Beringia in the area of the Yana complex of sites, the MIS 2 stadial was characterized by the spread of mosaic landscapes of low-yielding grass-sedge tundra steppes (Figure 1A–F), where Arctic, aquatic, coastal, meadow, and steppe taxa were combined. In the conditions of a dry cold continental climate, there were widespread cryophytic grass-wormwood and wormwood-grass communities with the inclusion of forbs, dryad tundra, and xeropetrophytic and nival communities, which together had a tundra-steppe appearance [80,81,100]. The coldest and driest climatic conditions of the LGM in this area appeared after 24 ka and persisted until 18.2 ka.

During MIS 2, local climates of the East Siberian Arctic developed under conditions of increasing sharp continentality and at the initial stage of 29–24 ka was cold and wet. The average July temperatures were 1–4 °C lower than current temperatures and the annual precipitation exceeded the current values by 100–175 mm [28,100–103].

The time of the glacial maximum was characterized by a decrease in the number and diversity of mammoth fauna. However, woolly mammoths were widespread everywhere in the Western Arctic of Beringia; the presence of woolly rhinoceros and Pleistocene bison was noted in some periods in its interior. In the north, Pleistocene horses were often found, and occasionally woolly rhinoceros were found [43,45,57]. Based on the analysis of the mass results of radiocarbon dating of mammoth bone remains, changes in the relative number of populations of these animals and the dynamics of their areas of settlement have been reconstructed. Thus, an increase in the number of mammoths is noted at the beginning of MIS 2 in the extreme north of the Arctic Western Beringia (New Siberia Islands); the relative population size reaches peak values at the same time with the onset of extreme climatic conditions of the LGM. During the last glacial maximum, the number of mammoths decreased, and approximately 18.2 ka, they became invisible in the paleontological record. After the end of extremely harsh conditions, the number of mammoths began to recover [48]. In the inland continental areas (the Yana-Indighirka lowland), there was a progressive decrease in the number of mammoths from the beginning of the MIS 2 stadial to the end of the LGM cooling maximum around 18.2 ka [48,57].

The East Siberian Arctic includes a rather large number of locations and sites represented by two groups (Figure 3A,B). The first characterizes the presence of people at an early stage and specifically at the maximum of cooling; these are the Diring-Ayan, Yana "A" area, and Lagernyi localities in the Yana cluster of sites, Ilin-Syalakh 034 [27,57], the upper horizon of the location of Zyryanka-1 [33], and the discovery of a punctured scapula of a mammoth on Wrangel Island [84] (Figure 1H—13, 17, 18, 19, 21, 20, respectively; Figure 3A). They belong to the era of the harsher environmental conditions, the Last Glacial Maximum, and their age is in the range of 27–23 ka. The listed localities serve as evidence of the presence of people, but it is gradually decreasing; among these locations, there is not one whose materials would allow for judging the culture of these people in detail. Nevertheless, it is obvious that they steadily carried out hunts for the mammoth (the scapula from Wrangel Island and evidence from the Ilin-Syalakh 034 site); they had a technology of production of long points and full-size spears of mammoth tusk, which can be argued on the basis of the discovery of a fragment of the mammoth tusk core at the Lagernyi locality of the Yana site. Furthermore, at least in some cases, they used the technology for the production of small blades) known from the Diring-Ayan location near the Yana complex of sites; their prey, in addition to the mammoth, became any available animals, primarily reindeer [57].

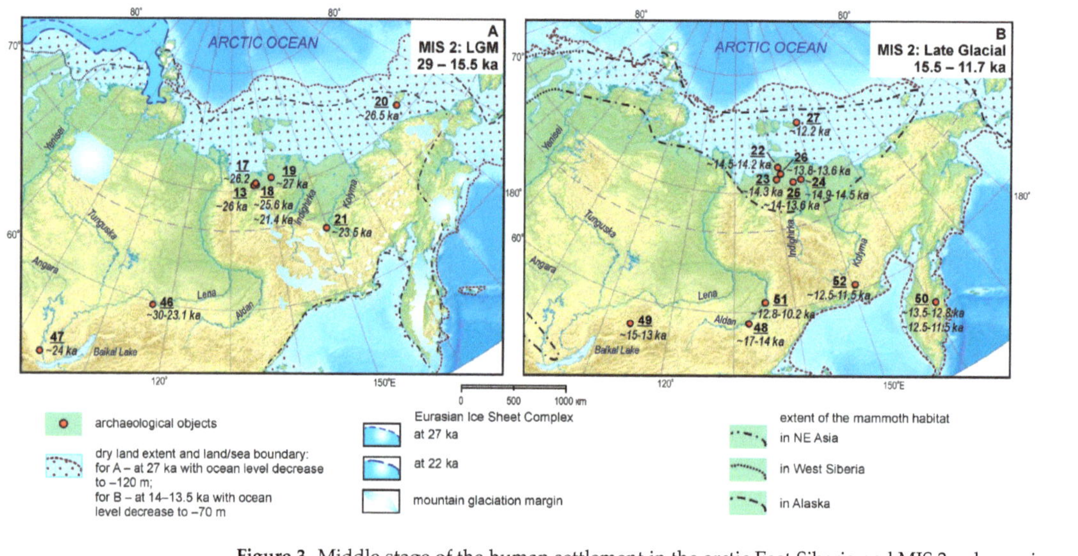

Figure 3. Middle stage of the human settlement in the arctic East Siberia and MIS 2 paleoenvironments. (**A**) Archaeological localities of the extreme cold phase of MIS 2 stadial (MIS 2 pessimum, 29–15.5 ka). Paleoenvironments: dry land extent and land/sea boundary at 27 ka with ocean level decrease to −120 m is based on Pico et al. [50]; Eurasian Ice Sheet Complex at 27 ka and 22 ka, after Hughes et al. [67] mountain glaciation margin, after Barr and Clark [51], Galanin [52], Glushkova [104]; extent of the mammoth habitat: in NE Asia, after Pitulko, Nikolskiy [54], in West Siberia, after Kahlke [55], in Alaska, after McDonald et al. [53]. Dated archaeological sites of the East Siberian Arctic are indicated: 13—Diring-Ayan, 17—Lagernyi Point/YCS, 18—Yana A area/YCS, 19—Ilin-Syalakh 034, 20—Wrangel island, 21—Zyryanka 1, after Pavlova and Pitulko [16], Cheprasov et al., 2018 [31]; Pitulko et al., 2016a [36]; Pitulko et al., 2015 [27]; Pitulko et al., 2017 [57]. Archaeological site of that time in South Siberia: 46—Khaiyrgas Cave, 47—Malta [102]. Site location

and age estimate provided based on Kuzmin et al., 2017 [105], Derevianko et al., 2003 [106], Raghavan et al., 2014 [107], Sitlivy et al., 1997 [108]. (**B**) Archaeological locations belonging to the late glacial stage of MIS 2 (15.5–11.7 ka). Paleoenvironments: dry land extent and land/sea boundary at 14–13.5 ka with ocean level decrease to −70 m is based on Pico et al. [50]; extent of the mammoth habitat: in NE Asia, after Pitulko, Nikolskiy [54], in West Siberia, after Kahlke [55], in Alaska, after McDonald et al. [53]. Dated archaeological sites are indicated: 22—Urez-22, 23—Ilin-Syalakh, 24—Achchaghyi-Allaikha, 25—Berelekh geoarchaeological complex; 26—Nikita Lake, 27—Cape Kamennyi, after Pavlova and Pitulko [16], Pitulko et al., 2016a [36], Pitulko et al., 2017 [57]. Archaeological sites in South Siberia confined to the same time span: 48—Dyuktai Cave, 49—Bolshoy Iakor, 50—Ushki site; 51—Ezhantsy; 52—Kheta. Site location and age estimate for them is based on Mochanov 1977 [109]; Ineshin and Tetenkin 2017 [110], Dikov 1979 [111], Slobodin 1999 [112] Note geographic pattern in which northern group of sites with a single microblade site (Urez-22) are located within the range of the local population of woolly mammoths, while in the archaeological contexts of the southern group outside the range of mammoths, narrow-front wedge-shaped core technology is confidently presented.

It should be emphasized that there is a chronological gap of approximately 10,000 years between the first and second groups of archaeological sites belonging to the MIS 2 (Figures 1H,I and 3A,B). However, there is no need to return to the discussion about the possible human depopulation of the region—the presence of people in it becomes ephemeral; thus, most of the evidence relates to the beginning of the extreme cold time slice of MIS 2 and there are only two in its middle. One of them comes from the Yana "A" locality of the Yana complex of sites, where the evidence for human presence dates to 25.6 and 21.4 ka [57] (Figure 1H—14; Figure 3A). Another one, documented by direct dating of a core made of a mammoth tusk, was found at Zyryanka 1 (Figure 1H—21; Figure 3A); it is ~23,500 years old [31]. In fact, this is a complete analogy to the gap observed in the paleontological chronicle of the local population of woolly mammoths, which, approximately 18,200 years ago, fell out of the biome but reappeared after the end of the period of extremely harsh conditions, but clearly, they were in the area all the time [48].

The late MIS 2 (15.5–11.7 ka) in Arctic Western Beringia is characterized by dynamic changes in the natural environment in accordance with the general global climate warming trend [98,99]. This is in relation to the progress and rate of deglaciation of the glacial cover of the Northern Hemisphere [71,113–116], the dynamics and amplitude of the rise of the sea level in general [50,78], and the development of the transgression of the Arctic Ocean [77,117–122].

The beginning of a fundamental restructuring of the natural environment in Western Beringia corresponds to the warming of the Bølling-Allerød Interstadial in Europe during the interval of 14.8–12.8 ka. In the Late Glacial period, tundra-steppe communities still retained their positions. However, the areas of moist meadow-shrub tundra with dwarf birch, tundra, and waterlogged meadows expanded [81]. Late glacial warming in the range of 13.7–12.8 ka, corresponding to the Allerød in Europe, was quite clearly manifested in the West Beringian Arctic.

The vegetation cover of the warm period of 13.7–12.8 ka was distinguished by a variety of tundra associations. In the north of the Western Beringian Arctic, dwarf shrubs sedge-grass tundra communities were developed with the inclusion of dwarf birch and shrub willow on the Bunge Land [16,123]; on Bolshoy Lyakhovskii Island, there were widespread grass-sedge and sedge-grass communities with the participation of shrubby willow and dwarf birch [41,45]. Vegetation in the western part of the Western Beringian Arctic was represented by birch shrub tundra in the lower reaches of the Lena River [25,26]; grass-sedge communities with the addition of wormwood, and shrub species of birch and willow in the area of the Oiyagos Yar [124]. Summer temperatures were above current temperatures by ~1–4 °C, and precipitation was 50–100 mm/year more than now [26,101].

Against this natural and climatic background was a second group of archaeological sites of the East Siberian Arctic, belonging to the final stage of MIS 2 (Figures 1A–I and 3B). It is composed of geoarchaeological objects of the northwest and north of the Yana-Indighirka lowland, whose geological age is in the range of 16.6–12.8 ka, while the episodes of human presence are chronologically even more compact and fit into the interval of 14.8–12.8 ka. These are the localities of Achchaghyi-Allaikha [125], Berelekh complex of geoarchaeological objects [126], locations of Nikita Lake, Urez-22, and Ilin-Syalakh [36,57] (Figure 1H—24, 25, 26, 22, 23 respectively). At the turn of the Holocene, the presence of people was noted in the New Siberian Islands, on the island of New Siberia, at Cape Kamenny locality, ~12.2 ka [16] (Figure 1H—27; Figure 3B). In the same direction, the southern border of the area of the local mammoth population shifted, and on the island of New Siberia, the presence of the last mammoths is noted, which finally fell out of the biome approximately 10,000 years ago [75].

A characteristic feature of the archaeological sites of the Late Glacial period in Arctic Eastern Siberia is their association with mass concentrations of mammoth bone remains (Figure 3B). For each of them, with the exception of Berelekh, a conclusion is made about their anthropogenic origin [36,92,125] as a result of human exploitation of local populations of these animals, i.e., hunting, which, as before that, was carried out for the sake of their tusks, which were used as raw materials for the production of long points. This conclusion is supported by various signs, including the undoubted sorting of bones [125], evidence of the production of long points from mammoth tusks, and direct evidence of hunting these animals [36].

The increase in the number of objects indirectly indicates that the human population has grown. Thus, people have become more active in this hunting, in which there are no signs of food exploitation of woolly mammoths. At the same time, each of the items contains the remains of clearly food species, sometimes in noticeable quantities: These are mainly reindeer and horses. At Berelekh, a huge quantity of bone remains is represented by hare, which, as at the Yana site, was most likely taken for the sake of the pelt for sewing raw materials. The material culture of this stage remains largely unknown. However, it can be noted that the technology of production of long rods from tusk is still relevant. In addition, for this time at the location of Urez-22, the presence of a developed form of the microblade industry [36] was noted for the first time in the East Siberian Arctic, likely in the form of lithic technology based on the wedge-shaped core idea. Interestingly, this is one of the earliest sites in the group; its age is ~14,500–14,200 years.

Sites in this group are characterized by the presence of specific small points, whose closest analogies are found outside the East Siberian Arctic (Western Beringia). Such examples are found in the American Northwest (eastern Beringif) at the sites of the Nenana complex (13.5–13 ka). They are quite numerous (for example, Dry Creek, Owl Ridge, Moose Creek, and Walker Road), and their number likely includes the sites of Broken Mammoth, Little John, and some others. The main feature of the stone industry in sites of the Nenana complex is the absence or small number of microblades and burins and the presence of teardrop-shaped and triangular points made of flakes, that is, incomplete bifaces known as Chindadn points [127–133].

In Western Beringia, they are represented at the sites of Berelekh, Nikita Lake, and Achchagyi-Allaikha. Thus, the time of existence of these well-recognized varieties is only approximately 1000 years, despite the fact that they are found from the Yana-Indighirka interfluve in Eurasia to the rivers of the Yukon basin in the northwest American continent. Interestingly, such expressive, short-lived cultural phenomena are not exclusive, so no less expressive Clovis sites have an even shorter history, in the interval of only ~13–12.7 ka, but very wide spatial distribution across mid-latitude North America [134,135]. Apparently, in both cases, the appearance of diagnostic artifacts is associated with changes in adaptations rather than with the appearance of new population groups. Regarding East Siberian sites with Chindadn points, it can be said that in Western Beringia, they appear in its Arctic region on the eve of the disappearance of the local mammoth population [36,93].

4. Late Stage of Human Settlement in the East Siberian Arctic (~11.7–8 ka)

In the Holocene, the development of the natural environment of Arctic Western Beringia occurred dynamically under the active influence of the oceanic post-glacial transgression, whose course and speed largely determined the appearance of the landscapes and the climate of this territory. At the beginning of the Holocene, approximately 11.7 ka, the sea level reached 52–53 m relative to the modern one [50]. Flooding and disintegration of the Arctic margin of Western Beringia have significantly accelerated [136].

A distinctive feature of the natural and climatic changes in the Holocene of the West Beringian Arctic is the early onset of the optimum in the interval of 11.5–10.2 ka on the New Siberian islands, first identified for Bolshoy Lyakhovskii Island [137] and Kotelny Island [138,139]. At this time, due to the significant warming and moisterization of the climate (Figure 1A–F), there was a significant movement of large shrubs (alder) and even tree-like birch to the north up to 75–76° N. July temperatures were 5 to 6 °C higher than at present, and landscapes were represented by subarctic grass-sedge-cereal-moss tundra and shrub tundra, where shrubby willow, birch (*Betula exilis*), and alder grew.

After 12.5 ka on the Yana-Indighirka lowland and the New Siberian Islands, according to all data available, tundra vegetation elements, represented by heather, appear in the landscapes, and the contribution of shrubby birch, willow, and alder sharply increases. Actually, this time was the boundary, after which the increase in temperature and humidity led to the final formation of tundra landscapes, the intensive development of thermoerosion, the discharge of thermokarst lakes, the formation of thaw-lakes in the territory, and the activation of peat accumulation [140–142]. The development of these processes led to the degradation of forage lands and provoked a reduction in populations of large herbivores in the biomes of the mammoth steppe—important for humans—and caused a shift in their habitats to the north [54,93].

Further changes in the landscapes of Northeast Siberia at the turn and at the beginning of the Holocene (Figures 1A–D and 4) were marked by the northward movement of the forest boundary, which, for a straight-stemmed birch, is approximately the same as the current position of the coastline [53,143]. In some areas, forest vegetation moved northward, occupying some modern Arctic islands [144].

At this time, the existence of populations of many species of Pleistocene herbivore fauna ended, which had served as a resource base for the population of the East Siberian Arctic for tens of thousands of years [53,69,70,75]. Actually, very soon, the reindeer turned out to be the most accessible species. In the dynamics of the ranges of large herbivore species, which subsequently fell out of the mammoth steppe biome [69], and the degradation of the biome itself [70], the northern trend became clearly visible. Thus, the last 'patches' of the mammoth steppe existed in the north of the New Siberian islands still approximately 12,000 years old. Simultaneously, the existence of the local population of woolly mammoths ended [16,75].

These events certainly caused stress for the population of the East Siberian Arctic. Human groups likely moved in the same direction, staying in the zone of well-known landscapes of open spaces, where animals familiar to them were still preserved. From the south, both those and others were squeezed out by woody vegetation, whose northern boundary of distribution at ~10.2 ka reached the position of the modern coastline.

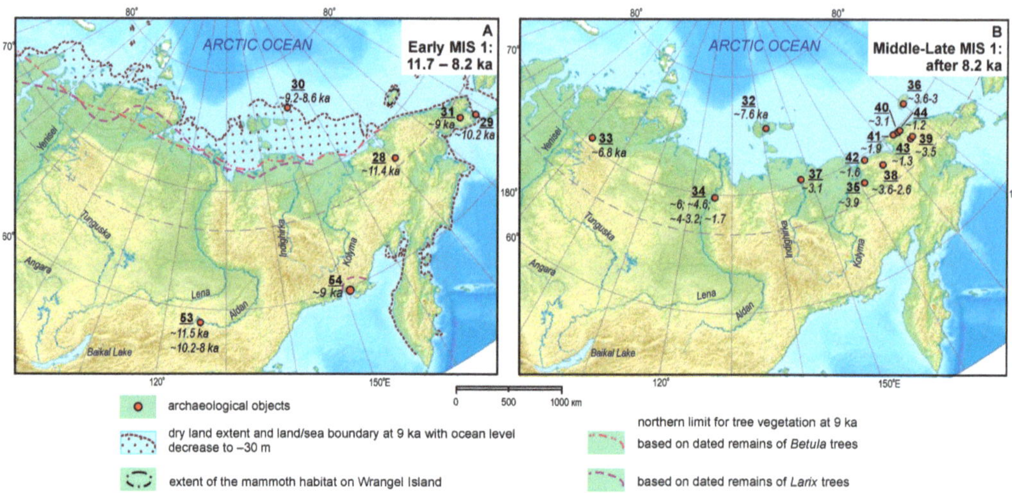

Figure 4. Late stage of the human settlement in the arctic East Siberia and paleoenvironments of MIS 1: (**A**) The Early Holocene archaeological sites (11.7–8.2 ka). Paleoenvironments: dry land extent and land/sea boundary at 9 ka with ocean level decrease to −30 m is based on Pico et al. [50], Bauch et al. [119]; extent of the mammoth habitat on Wrangel Island, after Vartanyan et al. [145]; northern limit for tree vegetation based on dated remains of Betula trees and Larix trees at 9 ka, after Kremenetski, et al. [146], Binney, et al. [147,148]. Dated archaeological sites of the East Siberian Arctic are indicated: 28—Tytylvaam IV, 29—Naivan, 30—Zhokhov site, 31—Chelkun IV, after Pavlova and Pitulko [16], Kiryak et al., 2003 [38], Dikov 1993 [37], Gusev 2002 [40], Pitulko 1999 [149]; Pitulko, Pavlova 2016 [28]. Archaeological site of that time in South Siberia: 53—Ust-Timpton; 54—Uptar I. Site location and their age estimate are provided based on Mochanov 1977 [109], Slobodin 1999 [112]. Objects are shown to estimate the time of the occurrence of microprismatic blade technology in the region. (**B**) Archaeological objects of Middle-Late MIS 1 (after 8.2 ka), dated archaeological sites are indicated: 32—Tuguttakh, 33—Tagenar VI, 34—Siktyakh I, 35—Rodinka burial, 36—Chertov Ovrag, 37—Burulgino, 38—Rauchuagytgyn I, 39—Pegtymel, 40—Aachim-base, 41—Aachim-lighthouse, 42—Cape Baranov, 43—Pegtymel cave, 44—Shalaurova Izba, after Pitulko and Pavlova 2016 [28]; Pavlova and Pitulko 2020 [16], Khlobystin 1998 [39], Mochanov 1977 [109].

Climate changes after ~12.5 ka in the arctic Western Beringia were largely due to the development of the marine Holocene transgression. It proceeded more dynamically at the initial stages up to ~8.3 ka, which led to rapid large-scale flooding of the shelf to the modern −20 m isobath [150]. The highest rates of sea level rise are calculated between 9.8 and 8.9 ka [119]. The development of transgression dynamically changed the outlines of the coastline and actively influenced the increase in the humidity of the climate. Rapid movement of the marine transgression front by ~8.3 ka not only caused an increase in humidity but also had a cooling effect on the environment enforcing an 8.2K event. As a result, the dry continental climate was replaced by an excessively humid marine one, which caused significant cooling and the termination of the development of thermokarst. After ~8.3 ka, the rate of sea level rise slowed down, and the sea level reached its current position at ~5 ka [119].

Archaeological materials that relate to this stage of human settlement in the East Siberian Arctic are few in number (Figure 4). These are, first, the Zhokhov site (Zhokhov Island, New Siberian Islands) (Figure 1H—30), whose excavations brought unique evidence of human adaptation to the conditions of the high-latitude Arctic [149,151]; this site was visited by people repeatedly in the interval of 9.2–8.6 ka [152]. At the time of human presence on the island, it represented the outskirts of the residual area of the former Great

Arctic Plain [20]. It was abandoned after 8.3 ka, when this portion of dry land completely lost its connection to the mainland [117].

The Zhokhov site represents the extreme northern point of distribution of sites, whose contexts contain massive evidence of the use of microprismatic technology for microblade production. The signs of its early presence are known from Taimyr at the Tagenar VI site to Eastern Chukotka seen in the contexts of Chel'kun IV, Naivan, and some others. These sites appear in the region almost simultaneously within the range of ~9 to ~7 ka (Figure 4A) [28,37,39,40,153]. The Zhokhov site settlers hunted reindeer, and in winter, they hunted polar bears (*Ursus maritimus* Phipps) in dens [154]. This is a unique case of mass procurement of polar bears in the world. Excavations of the site also yielded serial finds of morphologically true domestic dog (*Canis familiaris* L.) remains, the earliest in Siberia. A striking feature of this culture is the possession of ground transport technologies—sled dogs and sleds [155]—with whose help they made long journeys. This ability is confirmed by their participation in the system of long-range obsidian exchange from the Anadyr Valley, carried out at a distance of more than 1000 km [156]. The availability of such vehicles ensured high mobility of the population and the rapid spread of technological knowledge.

In addition to these sites, it is necessary to name the dated subsurface contexts of the sites of Podgornaya and Tytyl' IV in the Tytyl' sites group [38] located near Lake Tytyl' in the valley of the Tytyl'vaam River, the right tributary of the Malyi Anyui in its upper course, Western Chukotka (Figure 1H—28; Figure 4A). In the materials of these sites, wedge-shaped core-based splitting occurs, and this is, at present, the oldest dated reliable context of this kind in the Arctic zone at ~11.2 ka.

5. Paleodemography of the East Siberian Arctic in the Late Pleistocene and Early Holocene

Sharp changes in the material culture of the ancient population of the Arctic of Eastern Siberia are associated with both natural and climatic changes as well as the influx of new populations to these territories (Figures 1 and 5). Anthropological remains from the Pleistocene age in Siberia are very rare [157], and in the Arctic, they are isolated (Figure 1J). Materials from the Eurasian Arctic and the Sub-Arctic indicate that the population, which dominated the northern edge of the continent approximately 50,000–45,000 years ago, was physically represented by anatomically modern humans. Thus, the age of a human femur found near Ust-Ishim, at 57° N, was approximately 45,000 years ago [158]. In the genetic sense, the find belongs to a representative of the ancient population of Northern Eurasia, who occupied this territory on the eve of the major split between Western (European) and Eastern (Asian) lineages of modern people [158], with a slight admixture of a Neanderthal genome absorbed by them much earlier than 45,000 years ago and observed at the same level in the genomes of today's Europeans.

In addition to the Ust-Ishim man, only the findings of human milk teeth from the Yana site are known in the north of the continent [67,159]. The preservation of DNA in two of them made it possible to sequence the genome of the inhabitants of the Yana site with high resolution [67]. It has been established that the population to which the Yana people belonged replaced the undifferentiated Ust-Ishim population in northern Siberia (Figure 5).

This group, called the "Ancient North Siberians" (ANS), separated from the European line approximately 38,000 years ago, shortly after the divergence of the major split that happened approximately 43,000 years ago. ANS likely separated from the West Eurasian lineage; however, their genome kept, at the same time, a significant (up to 30%) East Asian admixture that is associated both with common ancestors and close geographical proximity, as well as the proximity of events in time.

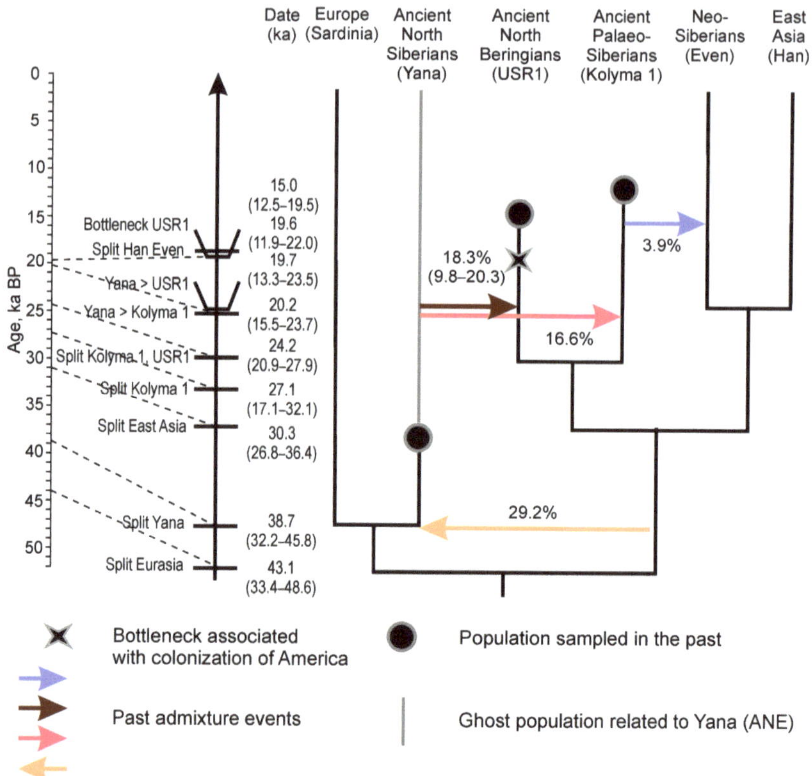

Figure 5. Demographic modeling for Siberian and North American ancient native population: Identified patterns, the time of event (in thousand years ago), and assessment of connection of various genetic lineages (in percent) (after Sikora et al., 2019 [62]).

During MIS 3 and MIS 2, the population of the East Siberian Arctic was most likely represented by the ANS population, to which the inhabitants of the Yana site also belonged [67]. These people made a significant contribution to the formation of the ancient Beringian population that colonized the New World but did not participate in it directly. It was formed as a result of the unification of the ancient local genetic lineage of the North Siberian population and a powerful genetic impulse of East Asian origin, which moved northward following the deterioration of natural and climatic conditions in the interior of the continent during the LGM [160,161]. The genetic characteristics of migrants who displaced or assimilated the Ancient North Siberians are known, thanks to the discovery of an incomplete human skull on the lower reaches of the Kolyma River in the well-known Duvannyi Yar exposure [159]. The find is ~10,000 years old and marks the complete replacement of the former inhabitants of the territory by the bearers of genomes of the East Asian lineage [67].

6. Adaptations of the Arctic East Siberian Population in the Late Pleistocene and Early Holocene

Based on the listed materials, it can be seen that people who entered the East Siberian Arctic around 50,000 years ago have never left it, although, at certain intervals, they experienced demographic stress seen as the bottle-necks in the genetic record, during the LGM and near the Pleistocene/Holocene boundary. Since the early stage of human penetration in the area, the territory has been widely inhabited [162]. A significant contribution to the historical and cultural development of the territory was made by migrations in a south-to-north direction associated with the movement of carriers of the genomes of the East Asian lineage (Figure 5). The reason for these movements of the human population is the frontier of natural and climatic changes of the Late Pleistocene–Early Holocene (Figure 1).

One of the most important cultural and economic characteristics of the epoch is 'human–mammoth' relationships. Throughout the entire period of the joint existence of humans and mammoths, these animals remained the object of the hunt, as is attested by direct evidence observed on the bone remains of woolly mammoths, for example, found in a number of sites, such as Sopochnaya Karga, YMAM locality in the Yana complex of sites, Tabayuryakh, Wrangel Island, Ilin–Syalakh 034, and Nikita Lake. They are present in all time slices, which is supported by the evidence of the existence of the production technology of long points from mammoth tusks, contemporary to them.

Many of the locations of the early and middle stages of settlement (at least 11 of them) are associated with massive accumulations of mammoth bone remains. On the basis of materials in the Upper Paleolithic of Northeast Europe, there was previously the opinion [163] that the people of this epoch exploited the "graveyards" of mammoths, that is, natural mass concentrations of their bone remains, in connection with which there was a specific localization of the sites of ancient man, confined to such clusters. Materials from the East Siberian Arctic show that this is not the case. Such concentrations have an anthropogenic nature and arise as a result of human hunting of these animals, which leads to the formation of "warehouses" (stocks of biogenic raw materials, primarily osseous materials).

The nutritional value of this animal, most likely, was limited since these people extracted animals of other species of fauna from the Late Pleistocene in multitude: Pleistocene bison, horse, reindeer, hare, and predators: Bear, wolf, arctic fox, and wolverine. The use of these animals is typical for the entire Late Pleistocene, but at the turn of the Holocene, the resource base had been reduced to reindeer and elk (*Alces alces* L.), and possibly musk ox, and therefore exotic resources had also been used, for example, polar bear in the areas of maternity dens [154]. There are no early Holocene traces of marine adaptations (hunting of sea mammals and fishing) in the culture of the population of the East Siberian Arctic for MIS 3 and MIS 2 since there was no environment in which it was possible: The formation of the marine ecosystem of the shelf zone of the Eastern Arctic began after 9000 years ago following the flooding of the coastal plains and the opening of Bering Strait, and continued until the middle of the Holocene.

In terms of the development of the lithic industry, there are fundamental differences between the sites of various stages, coinciding in time with the main climatic boundaries (Figure 1A–I). So, for the early stage, the presence of a flake industry of archaic appearance is characteristic, based on the splitting of pebbles and preserving some Middle Paleolithic elements, even the idea of splitting based on a simplified (opportunistic) scheme; these features are present in the Yana site industry in the final segment of the early stage of the human settlement of the arctic east Siberia. However, there is no reason to think that it was preceded by some advanced variant of technology with the production of blades. The materials of the LGM locations are characterized only by the actual evidence of human presence, which at that moment was weak. The only object that contained stone artifacts of this time enables one to think that in the East Siberian Arctic at this time, the industry of small blades was spreading, for which cores with a circular removal system were used. At the end of this stage, the technology of end-wedge-shaped core splitting appears in the region, present in rare sites of this stage in Western Chukotka and, possibly, in the northern Yana-Indighirka lowland.

The successful mammoth population of Eastern Siberia, which was widespread within its borders, experienced a noticeable decline in the number during the end segment of MIS 3, and during the LGM, it decreased even more. The described features of its spatial dynamics [54] suggest that it was shrinking in a northerly direction. It is noteworthy that a new technology of stone processing (wedge-shaped core-based microblade technology) was spreading in the same direction (to the north and east) from the regions of Northern China and Mongolia adjacent to Western Beringia.

The spread of this lithic technology is most likely due to the migration of the population from the inland region, faced with a sharp cooling and aridization of the climate due to the manifestation of the global climatic trend during the LGM (Figure 1B), with a decrease in the average global temperature by 8 to 9 °C [164]. This process was reinforced by the increase in the dry land surface area in North Eurasia due to the decrease in ocean level, thanks to which the coastline in the LGM shifted more than 1000 km north of its current position [136].

The consequence of these events was the desertification of the intracontinental area [160,161], migration in the northern direction of the boundaries of vegetation zones, changing landscapes, and spatial redistribution of populations of fauna that served as a resource base for the population of these areas. In fact, the very appearance of microblade production technology based on wedge-shaped cores is most likely a consequence of environmental and climate change and is associated with the fall-out of mammoth from the biome of the southern part of the tundra-steppe of Northern Eurasia since tusks served as an important raw material for the production of hunting equipment—long points and/or full-size spears of mammoth tusks [54].

In the terminal Pleistocene, complexes with Chindadn points (small incomplete bifaces of a predominantly teardrop shape) appear in the East Siberian Arctic, which are widespread in Northwest North America (Eastern Beringia); this form represents the only archaeologically visible connection between these territories. At the turn of the Holocene, these differences disappeared, and the lithic technology changed again. Microprismatic blade technology is rapidly spreading across the entire region, from Taimyr to Chukotka and from its southern regions to high latitudes (Figures 1G–I and 5), associated with the settlement of carriers of East Asian genetic lineages in the territory of Eastern Siberia [20,67]. This process was likely accelerated and simplified by the presence of the land transportation system among populations of the early Holocene arctic Siberia in the form of sled dogs and sleds [155]. Their appearance is associated with the completion of the dog/wolf domestication process in the terminal Pleistocene, and thus this land transportation system is the most important innovation in the human culture of the Arctic at the turn of the Holocene [6].

It was thanks to a set of Upper Paleolithic innovations that the Upper Paleolithic people were capable of mastering the boundless open spaces of the mammoth steppe. More precisely, these innovations reflect the technological complexity of the ancient man's culture [165,166]. Each of the innovations from the 'Mellars list' [167] hides a set of hierarchically organized particular technologies, which is easy to imagine using the example of the technology of manufacturing clothing and other products from hides and skins [91]. The main technologies listed by Mellars [167] do not have a hierarchy, but it is obvious that the most important were the actions directly related to the technology of survival of human groups in the open spaces of the tundra-steppe belt of Northern Eurasia. Evidence of such technologies is fully present in the materials of the Yana site.

Initially, three of them are the most significant: Hunting for food, making clothes, and building dwelling structures. The extraction of animals by hunting in all chronological references supplied raw materials for producing products from hides and leather. It also satisfied the need for raw materials for the production of bone products; the most important were hunting weapons and sewing tools.

The garment industry was obviously one of the two critical technologies necessary for human development of the coldest regions of the planet. Its development is associated with the appearance of the eyed needle, the most ancient of which was found in Siberia [168,169]. In Siberia, there are known archaeological sites where mass production and use of such implements in MIS 3 are documented, especially in the Yana complex of sites [91]. The introduction of eyed needles made it possible to produce multi-layered clothing and adjust it to size, as well as to create a full range of sewn products—shoes, sleeping bags, soft containers, and bags—as well as dwellings. The latter, judging by the evidence from excavations of the Yana site, were light ground structures with hearths [89], for which bones of large animals, including mammoths, were used as fuel in winter [88].

The presence of a developed sewing culture is an indispensable condition for successful human activity in the subarctic and Arctic zones, where the current average annual temperatures vary from 0 to −16 °C, and any slight decrease in the global average temperature will significantly worsen environmental conditions. Such changes occurred at different times; the most famous are the LGM and the sharp Younger Dryas cooling. However, sudden short-term cold spells, defined in the climatic record as Greenland stadials, were much more numerous [170]. They undoubtedly contributed to the spatial distribution of human groupings and caused certain changes in the material culture of this population as a cultural response to changes in the habitat.

During MIS 3 and MIS 2, the studied territories of Northern Eurasia were occupied by a tundra-steppe biome (see, for example: [49,148,171]), which differed by significant variability in space and time. The main feature of the natural environment of these landscapes was the complete or almost complete absence of woody vegetation [41,147]. This circumstance was extremely important for ancient man in the Arctic since wood served not only and not so much as fuel but also as a construction material used for the frames of dwellings. Finally, it served as a material for the manufacture of hunting weapons, namely, the shafts of spears and thrown spears.

If in the first two cases, it can be replaced, i.e., to use shrub branches, peat, bones, and animal fat as fuel, and use large shoots of shrubby plants as a building material for dwelling frames, as such shoots can easily reach 3 to 4 m in length in habitats favorable for growth (for example, willow shoots in river valleys), then a much higher quality material is needed to make a spear shaft. As in the case of mastering sewing technologies, which look like everyday ordinary work, with little visibility in the culture, incomparable with the true meaning and role of this technology in adapting to the harsh environment, the use of wood in any situation looks simple and natural [172]. It is possible to fully assess the degree of importance of this resource only if it is completely absent. This is exactly the situation that arose at the time of the initial penetration of people into the open spaces of the tundra-steppe belt of the Eurasian Subarctic and Arctic.

To overcome the lack of wood, an original solution was found. It consisted of the development of a special technology for the longitudinal splitting of mammoth tusks, which helped make it possible to obtain long strong points up to 2 m long, reaching the length of a full-sized spear. Traces of this technology are best documented in the materials of the Yana site [56] and at Berelekh [173], although there are relatively many in the area [56]. In any case, with its help, it was possible to obtain serial long massive points necessary for hunting large Pleistocene herbivores, including mammoths.

Mammoth hunting was an important activity. However, it was carried out on a limited basis with a very low volume (1 to 2 animals per year). It was undoubtedly connected to the need to create a stock of important bone raw material [56,93] and possibly fat, which, along with bones, could have been used as fuel in winter. Meat was used for food, but it did not play a significant role in the diet of the ancient hunters of the Siberian Arctic, who extracted Pleistocene bison, horses, and reindeer in large numbers [93]. There is no evidence of a mass one-time mammoth hunt [3,92,126].

The interaction of "man–mammoth" is a fundamental characteristic of the Late Pleistocene epoch, in connection to issues related to its disappearance as a result of exterminating hunting of it by ancient man often discussed. This issue was studied using the example of the historical dynamics of the population of North Siberian woolly mammoths [75]. The relative size of the mammoth population changed over time, but no anthropogenic contribution is visible in these changes. As was shown earlier [75], such fluctuations in mammoth often exactly repeated changes in the proportion of xerophytic beetles in the entomofauna, which, in turn, are an indicator of the presence of steppe vegetation in the phytocenoses of the late Pleistocene Western Beringia [24,174,175].

After the end of the last glacial maximum, the population of Siberia gradually increased. However, it affected the fate of mammoths to a very small extent. The number of dates for mammoths attributable to the interval from 15,000 years ago to the turn of the Holocene increases rapidly and only then decreases abruptly. The growth of the mammoth population occurs against the background of a simultaneous increase in the human population and its density. Thus, the presence of humans had no noticeable effect on the stable mammoth population.

Changes in the natural environment for a long time affected the mammoth population more than coexistence with man. Their action at the very end of the Pleistocene and Holocene led to the collapse of the mammoth population and their concentration in a narrow northern refugium, which for this species, is a normal strategy of surviving unfavorable climatic epochs. Along with the biological mechanisms of adaptation (the appearance of small forms), such a strategy could have worked this time but did not. The final extinction of mammoths, in fact, occurred due to their interaction with humans in a limited space occupied by a small population, and this means not only the direct extermination of mammoths by humans. It is possible that for the disappearance of the remnants of an oppressed isolated mammoth population, it was enough just to have a joint presence in the same territory with people. In a certain sense, their disappearance is a consequence of the successful adaptation of the ancient inhabitants of the East Siberian Arctic to the conditions of the environment.

7. Conclusions

The existing archaeological chronicle of the Stone Age of the East Siberian Arctic covers approximately 50,000 years, from the early phase of MIS 3 (late Pleistocene) to the Early Holocene (beginning of MIS 1). These data are not numerous; however, three chronological groupings can be confidently distinguished: (1) Early (~50 to ~29 ka, MIS 3); (2) Middle (~29 to ~11.7 ka, MIS 2), and (3) Late stage (Early Holocene, from 11.7 to ~8 ka). In the Early Stage, flake-based lithic technology of archaic appearance using a variety of local rocks was widespread; in the Middle Stage, lithic technology based on wedge-shaped core splitting (the Beringian microblad tradition) spread into the Arctic zone, and its appearance in the Arctic regions follows the reduction in the range of woolly mammoths of Eastern Siberia during the LGM. During the Early and Middle stages, the flake-based lithic industries were accompanied by the technology of manufacturing long points from a mammoth tusk. With the reduction in the number of mammoths and the gradual loss of the source of that biogenic raw material, this technology was replaced by the production of grooved tools, initially realized through the production of microblades in the framework of end-wedge-shaped core technology, with a subsequent transition to microprismatic splitting. In the late stage, from the turn of the Holocene to its middle, in Arctic Beringia, variations in lithic industries with microprismatic core-based splitting spread. The connection of the archaeological material with massive accumulations of mammoth bone remains, which are man-made objects formed as a result of human exploitation of the local woolly mammoth population at that time, is clearly manifested. Archaeological materials allow us to reconstruct the human dispersal and adaptation to the Late Pleistocene and Early Holocene to the conditions of the natural environment.

The Arctic region of Eastern Siberia, starting from the moment of initial human exploration shortly after 50,000 years ago, has continuously remained inhabited by human populations, including the least favorable periods that are difficult from the point of view of natural and climatic conditions. Cardinal changes in archaeological cultures and technologies correspond to the most important paleoclimatic boundaries of the Late Pleistocene and Early Holocene. In all cases, they are associated with migration to the Arctic region of the southern population belonging to the East Asian lineages with the exception of the initial stage of development of the territory associated with the settlement of carriers of the Western Eurasian gene pool. These changes should be considered a form of feedback connection—the adaptation of Stone Age hunters to the conditions of their habitat.

During the Late Pleistocene, the population of the East Siberian Arctic carried out economic activities within the framework of the model of continental hunters who exploited all available resources in the form of local populations of Pleistocene fauna, among which Pleistocene bison and horses, as well as reindeer, were the most important as food resources, while the mammoth served primarily as a source for raw material (tusks, bones, and fat). At the turn of the Holocene, species diversity was reduced to a state close to modern. In the Arctic zone, reindeer became the main source of human existence, but exotic resources were also used, for example, the polar bear in the early Holocene on Zhokhov Island.

The condition for the initial development of Arctic territories and the successful life of people of the Stone Age of the Late Pleistocene in a changing natural environment was the introduction into economic practice of important technological innovations—comprehensive/complex technologies. In the adaptations of the Late Pleistocene, an important role was played by technologies for the production of hunting equipment (long points and full-sized spears) from a mammoth tusk and advanced sewing technologies based on using the eyed needles, the use of which was crucially important. The most important innovation at the turn of the Holocene was the completion of dog/wolf domestication, the formation of a breed of sled dogs that served as the genetic source of all currently existing sled breeds, and the creation of land-based transportation resources (sled dog teams). This achievement ensured the mobility of the population at the turn of the Holocene and contributed to the rapid dissemination of cultural knowledge and population exchange. This innovation ensured the emergence of large socio-cultural systems, whose spatial expression reaches millions of square kilometers.

Settlement of the Eastern Siberian Arctic of the Late Pleistocene is associated with the initial settlement in the region of people whose genome is dominated by the Western Eurasian lineage. A significant contribution to the historical and cultural development of the territory was made by human migrations in a south–north direction associated with the relocation of carriers of the genomes of the East Asian lineage. The reasons for these relocations of the human population were the milestones of the natural and climatic changes of the Late Pleistocene–Early Holocene.

Author Contributions: Conceptualization, V.V.P. and E.Y.P.; methodology, V.V.P.; validation, V.V.P. and E.Y.P.; formal analysis, V.V.P. and E.Y.P.; investigation, V.V.P. and E.Y.P.; resources, V.V.P.; data curation, E.Y.P.; writing—original draft preparation, V.V.P.; writing—review and editing, V.V.P.; visualization, E.Y.P..; supervision, V.V.P.; project administration, V.V.P.; funding acquisition, V.V.P. All authors have read and agreed to the published version of the manuscript.

Funding: The present study was funded by the Russian Science Foundation project no. 21-18-00457.

Data Availability Statement: The data presented in this study are openly available in cited published sources.

Acknowledgments: The authors are more than grateful to all participants of the long-term investigations in New Siberian Islands and northern Yana-Indighirka lowlands from 2000 through 2022 whose hard work and effort made this research possible. Special thanks to Richard L. Bland of the Oregon State University who kindly agreed to help with the translation of the initial manuscript. The authors are more than grateful to the Reviewers who took the trouble to review the article for their valuable comments, which helped to improve the article.

Conflicts of Interest: The authors declare no conflict of interest.

References

1. Blinnikov, M.S.; Gaglioti, B.V.; Walker, D.A.; Wooller, M.J.; Zazula, G.D. Pleistocene graminoid-dominated ecosystems in the Arctic. *Quat. Sci. Rev.* **2011**, *30*, 2906–2929. [CrossRef]
2. Hollesen, J.; Callanan, M.; Dawson, T.; Fenger-Nielsen, R.; Friesen, T.M.; Jensen, A.M.; Markham, A.; Martens, V.V.; Pitulko, V.V.; Rockman, M. Climate change and the deterioration of the Arctic's archaeological and environmental archives. *Antiquity* **2018**, *92*, 574–587. [CrossRef]
3. Pitulko, V.V.; Tikhonov, A.N.; Pavlova, E.Y.; Nikolskiy, P.A.; Kuper, K.E.; Polozov, R.N. Early human presence in the Arctic: Evidence from 45,000-year-old mammoth remains. *Science* **2016**, *351*, 260–263. [CrossRef]

4. Pitulko, V.V. A race against time: Looking for the initial stage of the human occupation in the arctic Siberia. In *The Past of Humankind as Seen by the Petersburg Archaeologists at the Dawn of the Millenium (to the Centennial of the Russian Academic Archaeology)*; Vinogradov, Y.A., Vasiliev, S.A., Stepanova, K.V., Eds.; St. Petersburg Centre for Oriental Studies Publishers: St. Petersburg, Russia, 2019; pp. 103–136. (In Russian)
5. Pitulko, V.V.; Pavlova, E.Y. Human settlement of the Arctic in the Late Pleistocene and Early Holocene: The most important events and processes. In *Archaeology of the Arctic. Issue 6*; Tupakhin, D.S., Fedorova, N.V., Eds.; YaNAO International Dept., Scientific Center for the Arctic Sudies: Salekhard, Russia, 2019; pp. 22–50. (In Russian)
6. Pitulko, V.V.; Pavlova, E.Y. Colonization of the Eurasian Arctic. In *Encyclopedia of the World's Biomes*, 1st ed.; Goldstein, M.I., DellaSalla, D.A., Eds.; Volume 2 Deserts—Life in the Extremes Ice Sheets and Polar Deserts—Ice of Life; Elsevier: Amsterdam, The Netherlands, 2020; pp. 374–391.
7. Pitulko, V.V.; Pavlova, E.Y. Colonization of the Arctic in the New World. In *Encyclopedia of the World's Biomes*, 1st ed.; Goldstein, M.I., DellaSalla, D.A., Eds.; Volume 2 Deserts—Life in the Extremes Ice Sheets and Polar Deserts—Ice of Life; Elsevier: Amsterdam, The Netherlands, 2020; pp. 392–408.
8. Slobodin, S.B. On the question of the toponym "Beringia" and the role of Petr petrovich Sushkin in disseminating this term. *Vestn. Far East. Branch Russ. Acad. Sci.* **2016**, *1*, 90–98. (In Russian)
9. Sher, A.V. *Mammals and Stratigraphy of the Pleistocene of the Extreme North-East of USSR and North America*; Nauka Publ.: Moscow, Russia, 1971. (In Russian)
10. Yurtsev, B.A. *Problems of Botanical Geography of North-East Asia*; Nauka Publ.: Leningrad, Russia, 1974. (In Russian)
11. Yurtsev, B.A. The Pleistocene "Tundra–Steppe" and the productivity paradox: The landscape approach. *Quat. Sci. Rev.* **2001**, *20*, 165–174. [CrossRef]
12. Guthrie, R.D. Origin and causes of the mammoth steppe: A story of cloud cover, woolly mammal tooth pits, buckles, and inside-out Beringia. *Quat. Sci. Rev.* **2001**, *20*, 549–574. [CrossRef]
13. Hoffecker, J.F. *A Prehistory of the North: Human Settlement of the Higher Latitudes*; Rutgers University Press: New Brunswick, NJ, USA, 2005.
14. Hopkins, D.M. *The Bering Land Bridge*; Stanford University Press: Stanford, CA, USA, 1967.
15. Hultén, E. *Outline of the History of Arctic and Boreal Biota during the Quaternary Period*; Lehre, J., Ed.; Cramer: New York, NY, USA, 1937.
16. Pavlova, E.Y.; Pitulko, V.V. Late Pleistocene and Early Holocene climate changes and human habitation in the arctic Western Beringia based on revision of palaeobotanical data. *Quat. Int.* **2020**, *549*, 5–25. [CrossRef]
17. Hoffecker, J.F.; Powers, W.R.; Goebel, T. The Colonization of Beringia and the Peopling of the New World. *Science* **1993**, *259*, 46–53. [CrossRef]
18. Hoffecker, J.F.; Elias, S.A. *The Human Ecology of Beringia*; Columbia University Press: New York, NY, USA, 2007.
19. Hoffecker, J.F.; Pitulko, V.V.; Pavlova, E.Y. Climate, technology and glaciers: The settlement of the Western Hemisphere. *Vestn. St. Petersburg Univ. Hist.* **2019**, *64*, 327–355. [CrossRef]
20. Hoffecker, J.F.; Pitulko, V.V.; Pavlova, E.Y. Beringia and the Settlement of the Western Hemisphere. *Vestn. St. Petersburg Univ. Hist.* **2022**, *67*, 882–909. [CrossRef]
21. Girshgorn, L.S.; Zhamoida, A.I.; Kovalevskii, O.P.; Oleinikov, A.N.; Prozorovskaya, E.L.; Khramov, A.N.; Skatova, V.K. *Stratigraphy Code of Russia*, 3rd ed.; VSEGEI: St. Petersburg, Russia, 2019. (In Russian)
22. Svensson, A.; Andersen, K.K.; Bigler, M.; Clausen, H.B.; Dahl-Jensen, D.; Davies, S.M.; Johnsen, S.J.; Muscheler, R.; Parrenin, F.; Rasmussen, S.O.; et al. A 60,000 year Greenland stratigraphic ice core chronology. *Clim. Past* **2008**, *4*, 47–57. [CrossRef]
23. Tierney, J.E.; Russell, J.M.; Huang, Y.; Damsté, J.S.S.; Hopmans, E.C.; Cohen, A.S. Northern hemisphere controls on tropical southeast African climate during the past 60,000 years. *Science* **2008**, *322*, 252–255. [CrossRef] [PubMed]
24. Sher, A.V.; Kuzmina, S.A.; Kuznetsova, T.V.; Sulerzhitsky, L.D. New insights into the Weichselian environment and climate of the East Siberian Arctic, derived from fossil insects, plants, and mammals. *Quat. Sci. Rev.* **2005**, *24*, 533–569. [CrossRef]
25. Pisaric, M.F.J.; MacDonald, G.M.; Velichko, A.A.; Cwynar, L.C. The Lateglacial and Postglacial vegetation history of the northwestern limits of Beringia, based on pollen, stomate and tree stump evidence. *Quat. Sci. Rev.* **2001**, *20*, 235–245. [CrossRef]
26. Klemm, J.; Herzschuh, U.; Pisaric, M.F.J.; Telford, R.J.; Heim, B.; Pestryakova, L.A. A pollen-climate transfer function from the tundra and taiga vegetation in Arctic Siberia and its applicability to a Holocene record. *Palaeogeogr. Palaeoclim. Palaeoecol.* **2013**, *386*, 702–713. [CrossRef]
27. Pitulko, V.V.; Pavlova, E.Y.; Nikolskiy, P.A.; Ivanova, V.V.; Basilyan, A.E.; Anisimov, M.A.; Remizov, S.O. Human colonization of Siberian Arctic in the late Neo-Pleistocene and Holocene: New archaeological map materials. In *IV Northern Archaeological Congress. Papers*; Chairkina, N.M., Ed.; Institute of History and Archaeology of Ural Branch of RAS: Ekaterinburg/Khanty-Mansiisk, Russia, 2015; pp. 153–177.
28. Pitulko, V.V.; Pavlova, E.Y. *Geoarchaeology and Radiocarbon Chronology of Stone Age Northeast Asia*; Texas A&M University Press: College Station, TX, USA, 2016.
29. Novgorodov, G.P.; Gtigoriev, S.E.; Cheprasov, M.Y.; Savvinov, G.N. Overview of mammoth fauna sites in the Yana Basin. *Arctic XXI Century Nat. Sci.* **2014**, *1*, 66–73. (In Russian)
30. Cheprasov, M.Y.; Obada, T.F.; Grigoriev, S.E.; Novgorodov, G.P.; Marareskul, V.A. New locations of the mammoth fauna and Palaeolithic sites in the basin of the Kolyma River. *Vestn. North-East. Fed. Univ. Yakutsk* **2015**, *6*, 53–68. (In Russian)

31. Cheprasov, M.Y.; Chlachula, J.; Obada, T.F.; Grigoriev, S.E.; Novgorodov, G.P. New records on the Palaeolithic of the Central Kolyma Basin, Yakutia. In *Man and the North–Anthropology, Archaeology and Ecology, Proceedings of the Materials of All-Russian Scientific Conference, Tyumen, Russia, 2–6 April 2018*; FIC TyumNC SO RAN: Tyumen, Russia, 2018; Volume 4, pp. 267–273. (In Russian)
32. Pavlov, I.; Suzuki, N. Tabayuriakhsky mammoth (Mammuthis primigenius Blum., 1799), from the island of Kotelny, Novosibirskiy archipelago. *Arct. Subarct. Nat. Res.* **2020**, *25*, 56–66. (In Russian)
33. Chlachula, J.; Cheprasov, M.Y.; Novgorodov, G.P.; Obada, T.F.; Little, E. The MIS 3–2 environments of the middle Kolyma Basin: Implications for the Ice Age peopling of northeast Arctic Siberia. *Boreas* **2021**, *50*, 556–581. [CrossRef]
34. Kirillova, I.K.; Shidlovskiy, F.K.; Titov, V.V. Kastykhtakh mammoth from Taimyr (Russia). *Quat. Int.* **2012**, *276–277*, 269–277. [CrossRef]
35. Pitulko, V.; Yakshina, I.; Strauss, J.; Schirrmeister, L.; Kuznetsova, T.; Nikolskiy, P.; Pavlova, E. A MIS 3 Kill-Butchery Mammoth Site on Buor-Khaya Peninsula, Eastern Laptev Sea, Russian Arctic. *Sci. Ann. Sch. Geol. Aristotle Univ. Thessalon.* **2014**, *102*, 158–159.
36. Pitulko, V.V.; Pavlova, E.Y.; Basilyan, A.E. Mass accumulations of mammoth (mammoth 'graveyards') with indications of past human activity in the northern Yana-Indighirka lowland, Arctic Siberia. *Quat. Int.* **2016**, *406*, 202–217. [CrossRef]
37. Dikov, N.N. *At the Ancient Crossroad of Asia and America: Stone Age of the Chukchi Peninsula*; Nauka Publ.: Saint Petersburg, Russia, 1993. (In Russian)
38. Kiryak, M.A.; Glushkova, O.Y.; Brown, T.A. Upper Paleolithic Complexes of the Tytylvaam River Valley (Polar Chukotka). *Archaeol. Ethnogr. Anthropol. Eurasia* **2003**, *3*, 2–15.
39. Khlobystin, L.P. *Taimyr. The Archaeology of Northernmost Eurasia*; Fitzhugh, W.W., Pitulko, V.V., Eds.; Contributions to Circumpolar Anthropology 5; National Museum of Natural History, Smithsonian Institution: Washington, DC, USA, 2005.
40. Gusev, S.V. The Early Holocene site of Naivan: The earliest dated site in Chukotka. *Univ. Or. Anthropol. Pap.* **2002**, *59*, 111–126.
41. Andreev, A.A.; Schirrmeister, L.; Tarasov, P.E.; Ganopolski, A.; Brovkin, V.; Siegert, C.; Wetterich, S.; Hubberten, H.-W. Vegetation and climate history in the Laptev Sea region (Arctic Siberia) during Late Quaternary inferred from pollen records. *Quat. Sci. Rev.* **2011**, *30*, 2182–2199. [CrossRef]
42. Kuznetsova, T.V.; Wetterich, S.; Matthes, H.; Tumskoy, V.E.; Schirrmeister, L. Mammoth Fauna Remains From Late Pleistocene Deposits of the Dmitry Laptev Strait South Coast (Northern Yakutia, Russia). *Front. Earth Sci.* **2022**, *10*, 757629. [CrossRef]
43. Schirrmeister, L.; Siegert, C.; Kuznetsova, T.; Kuzmina, S.; Andreev, A.A.; Kienast, F.; Meyer, H.; Bobrov, A.A. Paleoenvironmental and paleoclimatic records from permafrost deposits in the Arctic region of Northern Siberia. *Quat. Int.* **2002**, *89*, 97–118. [CrossRef]
44. Schirrmeister, L.; Grosse, G.; Kunitsky, V.; Magens, D.; Meyer, H.; Dereviagin, A.; Kuznetsova, T.; Andreev, A.; Babiy, O.; Kienast, F.; et al. Periglacial landscape evolution and environmental changes of Arctic lowland areas for the last 60,000 years (Western Laptev Sea coast, Cape Mamontov Klyk). *Polar Res.* **2008**, *27*, 249–272. [CrossRef]
45. Andreev, A.; Grosse, G.; Schirrmeister, L.; Kuznetsova, T.V.; Kuzmina, S.A.; Bobrov, A.A.; Tarasov, P.E.; Novenko, E.Y.; Meyer, H.; Derevyagin, A.Y.; et al. Weichselian and Holocene palaeoenvironmental history of the Bolshoy Lyakhovsky Island, New Siberian Archipelago, Arctic Siberia. *Boreas* **2009**, *38*, 72–110. [CrossRef]
46. Sher, A.V. On the history of the mammal fauna of Beringida. *Quärtarpaleontologie* **1986**, *8*, 185–193.
47. Nikolskiy, P.A.; Pitulko, V.V. West Beringian bison population dynamics during the Late Pleistocene and Early Holocene. In *XI International Conference on Permafrost*; Günther, F., Morgenstern, A., Eds.; Book of Abstracts, 20–24 June 2016; Alfred Wegener Institute Helmholtz Centre for Polar and Marine Research: Potsdam, Germany, 2016; pp. 512–513.
48. Nikolskiy, P.A.; Pitulko, V.V. Correlation between climate fluctuations and relative number of mammoth population in the light of their extinction (according to the results of mass ^{14}C dating of mammoth bone remains collected in Arctic Siberia. *Strat. Plus* **2013**, *1*, 133–165.
49. Hubberten, H.-W.; Andreev, A.; Astakhov, V.; Demidov, I.; Dowdeswell, J.; Henriksen, M.; Hjort, C.; Houmark-Nielsen, M.; Jakobsson, M.; Kuzmina, S.; et al. The periglacial climate and environment in northern Eurasia during the Last Glaciation. *Quat. Sci. Rev.* **2004**, *23*, 1333–1357. [CrossRef]
50. Pico, T.; Creveling, J.R.; Mitrovica, J.X. Sea-level records from the U.S. mid-Atlantic constrain Laurentide Ice Sheet extent during Marine Isotope Stage 3. *Nat. Commun.* **2017**, *8*, 15612. [CrossRef] [PubMed]
51. Barr, I.D.; Clark, C.D. Late Quaternary glaciations in Far NE Russia; combining moraines, topography and chronology to assess regional and global glaciation synchrony. *Quat. Sci. Rev.* **2012**, *53*, 72–87. [CrossRef]
52. Galanin, A.A. Age of the Last Glacial Maximum on Asian North-East. *Earth's Cryosphere* **2012**, *XVI*, 39–52.
53. MacDonald, G.M.; Beilman, D.W.; Kuzmin, Y.V.; Orlova, L.A.; Kremenetski, K.V.; Shapiro, B.; Wayne, R.K.; van Valkenburgh, B. Pattern of extinction of the woolly mammoth in Beringia. *Nat. Commun.* **2012**, *3*, 893. [CrossRef]
54. Pitulko, V.V.; Nikolskiy, P.A. Extinction of wooly mammoth in Northeastern Asia and the archaeological record. *World Archaeol.* **2012**, *44*, 21–42. [CrossRef]
55. Kahlke, R.-D. The origin of Eurasian Mammoth Faunas (Mammuthus—Coelodonta Faunal Complex). *Quat. Sci. Rev.* **2014**, *96*, 32–49. [CrossRef]
56. Pitulko, V.V.; Pavlova, E.Y.; Nikolskiy, P.A. Mammoth Ivory Technologies in the Upper Palaeolithic Arctic Siberia: A Case Study based on the materials from Yana RHS site. *World Archaeol.* **2015**, *47*, 333–389. [CrossRef]
57. Pitulko, V.; Pavlova, E.; Nikolskiy, P. Revising the archaeological record of the Upper Pleistocene Arctic Siberia: Human dispersal and adaptations in MIS 3 and 2. *Quat. Sci. Rev.* **2017**, *165*, 127–148. [CrossRef]

58. Goebel, T.; Aksenov, M. Accelerator radiocarbon dating of the initial Upper Palaeolithic in southeast Siberia. *Antiquity* **1995**, *69*, 349–357. [CrossRef]
59. Derevianko, A.P. *The Paleolithic of Siberia: New Discoveries and Interpretations*; Univ. of Illinois Press: Urbana, IL, USA, 1998.
60. Goebel, T.; Waters, M.R.; Mescherin, M.N. Masterov Kliuch and the Early Upper Palaeolithic of the Transbaikal, Siberia. *Asian Perspect.* **2000**, *39*, 47–70. [CrossRef]
61. Rybin, E.P.; Khatsenovich, A.M. The Mystery of Makarovo Stratum: The Earliest Upper Palaeolithic in Eurasia or a Cis-Baikalian Type of the Initial Upper Palaeolithic Technocomplex of MIS-3 Time? *Strat. Plus* **2020**, *1*, 279–304. (In Russian)
62. Vishnyatsky, L.B.; Pitulko, V.V. Susiluola, Byzovaya, and the question of the northern limit of the Neanderthal area. *IHMC RAS Trans.* **2012**, *7*, 3–15.
63. Hoffecker, J.F. *Modern Humans: Their African Origin and Global Dispersal*; Columbia University Press: New York, NY, USA, 2017.
64. Zwyns, N.; Roebroeks, W.; McPherron, S.P.; Jagich, A.; Hublin, J.-J. Comment on "Late Mousterian Persistence near the Arctic Circle". *Science* **2012**, *335*, 167. [CrossRef]
65. Krupnik, I.I. *Arctic Ethnoecology*; Nauka Publ.: Moscow, Russia, 1989. (In Russian)
66. Dolgikh, B.O. *The Kinship and Tribal Composition of the Indigenous Population of Siberia in the XVIIth Century*; Series: Proceedings of the Institute for Ethnography, Academy of Sciences of the USSR, New Series, 55; Academy of Sciences of the USSR: Moscow, Russia, 1960. (In Russian)
67. Sikora, M.; Pitulko, V.V.; Sousa, V.C.; Allentoft, M.E.; Vinner, L.; Rasmussen, S.; Margaryan, A.; de Barros Damgaard, P.; de la Fuente, C.; Renaud, G.; et al. The population history of northeastern Siberia since the Pleistocene. *Nature* **2019**, *570*, 182–188. [CrossRef]
68. Sikora, M.; Seguin-Orlando, A.; Sousa, V.C.; Albrechtsen, A.; Korneliussen, T.; Ko, A.; Rasmussen, S.; Dupanloup, I.; Nigst, P.R.; Bosch, M.D.; et al. Ancient genomes show social and reproductive behavior of early Upper Paleolithic foragers. *Science* **2017**, *358*, 659–662. [CrossRef]
69. Lorenzen, E.D.; Nogues-Bravo, D.; Orlando, L.; Weinstock, J.; Binladen, J.; Marske, K.A.; Ugan, A.; Borregaard, M.K.; Gilbert, M.T.P.; Nielsen, R.; et al. Species-specific responses of Late Quaternary megafauna to climate and humans. *Nature* **2011**, *479*, 359–365. [CrossRef]
70. Wang, Y.; Pedersen, M.W.; Alsos, I.G.; De Sanctis, B.; Racimo, F.; Prohaska, A.; Coissac, E.; Owens, H.L.; Merkel, M.K.F.; Fernandez–Guerra, A.; et al. Late Quaternary dynamics of Arctic biota from ancient environmental genomics. *Nature* **2021**, *600*, 86–92. [CrossRef]
71. Hughes, A.L.C.; Gyllencreutz, R.; Lohne, Ø.S.; Mangerud, J.; Svendsen, J.I. The last Eurasian ice sheets—A chronological database and time-slice reconstruction, DATED-1. *Boreas* **2015**, *45*, 1–45. [CrossRef]
72. Svendsen, J.I.; Alexanderson, H.; Astakhov, V.I.; Demidov, I.; Dowdeswell, J.A.; Funder, S.; Gataullin, V.; Henriksen, M.; Hjort, C.; Houmark-Nielsen, M.; et al. Late Quaternary ice sheet history of northern Eurasia. *Quat. Sci. Rev.* **2004**, *23*, 1229–1271. [CrossRef]
73. Andreev, A.A.; Schirrmeister, L.; Siegert, C.; Bobrov, A.A.; Demske, D.; Seiffert, M.; Hubberten, H.-W. Paleoenvironmental changes in northeastern Siberia during the Late Quaternary—Evidence from pollen records of the Bykovsky Peninsula. *Polarforschung* **2002**, *70*, 13–25.
74. Wetterich, S.; Tumskoy, V.; Rudaya, N.; Andreev, A.A.; Opel, T.; Meyer, H.; Schirrmeister, L.; Hüls, M. Ice Complex formation in arctic East Siberia during the MIS 3 Interstadial. *Quat. Sci. Rev.* **2014**, *84*, 39–55. [CrossRef]
75. Nikolskiy, P.A.; Sulerzhitsky, L.D.; Pitulko, V.V. Last straw versus Blitzkrieg overkill: Climate-driven changes in the Arctic Siberia mammoth population and the Late Pleistocene extinction problem. *Quat. Sci. Rev.* **2011**, *30*, 2309–2328. [CrossRef]
76. Jakobsson, M.; Long, A.; Ingólfsson, Ó.; Kjær, K.H.; Spielhagen, R.F. New insights on Arctic Quaternary climate variability from palaeo-records and numerical modelling. *Quat. Sci. Rev.* **2010**, *29*, 3349–3358. [CrossRef]
77. Jakobsson, M.; Andreassen, K.; Bjarnadóttir, L.R.; Dove, D.; Dowdeswell, J.A.; England, J.H.; Funder, S.; Hjgan, K.; Ingólfsson, Ó.; Jennings, A.; et al. Arctic Ocean glacial history. *Quat. Sci. Rev.* **2014**, *92*, 40–67. [CrossRef]
78. Lambeck, K.; Rouby, H.; Purcell, A.; Sun, Y.; Sambridge, M. Sea level and global ice volumes from the Last Glacial Maximum to the Holocene. *Proc. Nat. Acad. Sci. USA* **2014**, *111*, 15296–15303. [CrossRef]
79. Meyer, H.; Dereviagin, A.Y.; Siegert, C.; Hubberten, H.-W. Paleoclimate studies on Bykovsky Peninsula, North Siberia—Hydrogen and oxygen isotopes in ground ice. *Polarforschung* **2002**, *70*, 37–51.
80. Pitulko, V.V.; Pavlova, E.Y.; Kuzmina, S.A.; Nikolsky, P.A.; Basilyan, A.E.; Tumskoy, V.E.; Anisimov, M.A. Natural–Climatic Changes in the Yana–Indigirka Lowland during the Terminal Kargino Time and Habitat of Late Paleolithic Man in Northern Part of East Siberia. *Dokl. Earth Sci.* **2007**, *417*, 1256–1260. [CrossRef]
81. Pitulko, V.V.; Pavlova, E.Y.; Kuzmina, S.A.; Nikolskiy, P.A.; Basilyan, A.E.; Anisimov, M.A. Landscape-Climatic Changes in Yana Palaeolithic Site Area on Western part of Yana-Indighirka Lowland during Late Pleistocene-Holocene. *Bull. North-East Sci. Cent. Russ. Acad. Sci. Far East Branch* **2013**, *1*, 16–29. (In Russian)
82. Sulerzhitsky, L.D. Features of the radiocarbon chronology of mammoth (mammuthus primigenius) in Siberia and north of east Europe. *Proc. Zool. Inst.* **1995**, *263*, 163–183.
83. Sulerzhitsky, L.D. Features of the radiocarbon chronology of mammoth in Siberia and in the North of East Europe (considered as a substrate for human dispersal). In *Man Occupying the Earth Planet*; Velichko, A.A., Soffer, O.A., Eds.; Institute of Geography RAS: Moscow, Russia, 1997; pp. 184–202.

84. Sulerzhitsky, L.D.; Romanenko, F.A. Age and dispersal of "mammoth" fauna in Asian Polar region (according to radiocarbon data). *Earth's Cryosphere* **1997**, *1*, 12–19.
85. Stuart, A.J. Mammalian extinctions in the late Pleistocene of northern Eurasia and North America. *Biol. Rev.* **1991**, *66*, 453–562. [CrossRef] [PubMed]
86. Boeskorov, G.G.; Baryshnikov, G.F. *Late Quaternary Carnivora of Yakutia*; Nauka Publ.: Moscow, Russia, 2013. (In Russian)
87. Pitulko, V.V.; Nikolsky, P.A.; Girya, E.Y.; Basilyan, A.E.; Tumskoy, V.E.; Koulakov, S.A.; Astakhov, S.N.; Pavlova, E.Y.; Anisimov, M.A. The Yana RHS Site: Humans in the Arctic Before the Last Glacial Maximum. *Science* **2004**, *303*, 52–56. [CrossRef] [PubMed]
88. Pitulko, V.V. Yana B area of the Yana site: Some observations done during the excavations of 2015 through 2018. *Prehist. Archaeol. J. Interdiscip. Stud.* **2019**, *1*, 64–91. (In Russian) [CrossRef]
89. Pitulko, V.; Nikolskiy, P.; Basilyan, A.; Pavlova, E. Human habitation in the Arctic Western Beringia prior the LGM. In *Paleoamerican Odyssey*; Graf, K.E., Ketron, C.V., Waters, M.R., Eds.; Texas A&M University: College Station, TX, USA, 2013; pp. 13–44.
90. Pitulko, V.V.; Pavlova, E.Y.; Nikolsky, P.A.; Ivanova, V.V. Material culture and symbolic behavior of the Upper Paleolithic settlers of Arctic Siberia (with particular reference to the Yana site). *Russ. Archaeol. Yearb.* **2012**, *2*, 33–102. (In Russian)
91. Pitulko, V.V.; Pavlova, E.Y. Upper Palaeolithic Sewing Kit from the Yana Site, Arctic Siberia. *Strat. Plus* **2019**, *1*, 157–224. (In Russian)
92. Basilyan, A.E.; Anisimov, M.A.; Nikolskiy, P.A.; Pitulko, V.V. Wooly mammoth mass accumulation next to the Paleolithic Yana RHS site, Arctic Siberia: Its geology, age, and relation to past human activity. *J. Archaeol. Sci.* **2011**, *38*, 2461–2474. [CrossRef]
93. Nikolskiy, P.A.; Pitulko, V.V. Evidence from the Yana Palaeolithic site, Arctic Siberia, yields clues to the riddle of mammoth hunting. *J. Archaeol. Sci.* **2013**, *40*, 4189–4197. [CrossRef]
94. Pitulko, V.V. Late Pleistocene habitation and adaptations of the Ancient Man across NE Asia. In *Adaptations of Peoples and Cultures to Natural Changes, Social and Man-Caused Transformations*; Derevianko, A.P., Kudelin, A.B., Tishkov, V.A., Eds.; ROSSPEN: Moscow, Russia, 2010; pp. 38–46. (In Russian)
95. Pitulko, V.V.; Pavlova, E.Y.; Ivanova, V.V. Upper Paleolithic art of the Arctic Siberia: Personal adornments from excavations of the Yana site. *Ural. Hist. J.* **2014**, *2*, 6–18. (In Russian)
96. Pitulko, V.V.; Pavlova, E.Y.; Ivanova, V.V.; Nikolskiy, P.A. The Oldest Art of Eurasian Arctic. *Antiquity* **2012**, *86*, 642–659. [CrossRef]
97. Malaurie, J.P. *The Last Kings of Thule: With the Polar Eskimos, as They Face Their Destiny*; English translation of Les derniers rois de Thulé, translated from the French by Adrienne Foulke; Jonathan Cape: London, UK, 1982.
98. Daniels, W.C.; Russell, J.M.; Morrill, C.; Longo, W.M.; Giblin, A.E.; Holland-Stergar, P.; Welker, J.M.; Wen, X.; Hu, A.; Huang, Y. Lacustrine leaf wax hydrogen isotopes indicate strong regional climate feedbacks in Beringia since the last Ice Age. *Quat. Sci. Rev.* **2021**, *269*, 107130. [CrossRef]
99. Stuiver, M.; Grootes, P.M. GISP2 oxygen isotope ratios. *Quat. Res.* **2000**, *53*, 277–283. [CrossRef]
100. Pavlova, E.Y.; Pitulko, V.V. Sartan stadial landscape-climatic conditions in the New Siberian Islands and Yana-Idighirka Lowland. In *VIII All-Russian Quaternary Conference*; Book of Abstracts; RAS South Center: Rostov-on-Don, Russia, 2013; pp. 495–497. (In Russian)
101. Pavlova, E.Y.; Anisimov, M.A.; Pitulko, V.V. Natural environments and climate change during the second half of the Late Pleistocene in Yana-Indighirka lowland and New Siberian islands (as seen from palaeobotanical data). In Proceedings of the 6th All-Russian Quaternary Conference, Novosibirsk, Russia, 19–23 October 2009; Kontorovich, A.E., Volkova, V.S., Khazina, I.V., Khazin, L.B., Eds.; RAS SB Publishers: Novosibirsk, Russia, 2009; pp. 460–464. (In Russian)
102. Pavlova, E.Y.; Anisimov, M.A.; Dorozhkina, M.V.; Pitul'ko, V.V. Traces of ancient glaciation on New Siberia Island (Novosibirskie Islands) and an regional environment in the late Neo-Pleistocene. *Ice Snow* **2010**, *2*, 85–92. (In Russian)
103. Pavlova, E.Y.; Dorozhkina, M.V.; Pitulko, V.V. Materials, techniques and methods for creating high-resolution paleoclimatic records of the Late Neopleistocene and Holocene age. In *Actual Problems of the Modern Palynology, Proceedings of XIV All-Russian Palynological Conference, Moscow, Russia, 5–8 June 2007*; Bolikhovskaya, N.S., Klyuvitkina, T.S., Eds.; Geographical Faculty of Lomonosov Moscow State University: Moscow, Russia, 2017; pp. 239–242. (In Russian)
104. Glushkova, O.Y. Late Pleistocene Glaciations in North-East Asia. In *Quaternary Glaciations. Extent and Chronology: A Closer Look*; Ehlers, J., Gibbard, P.L., Hughes, P.D., Eds.; Elsevier: Amsterdam, The Netherlands, 2011; Voluem 15, pp. 865–875.
105. Kuzmin, Y.V.; Kosintsev, P.A.; Stepanov, A.D.; Boeskorov, G.G.; Cruz, R.J. Chronology and faunal remains of the Khayrgas Cave (Eastern Siberia, Russia). *Radiocarbon* **2017**, *59*, 575–582. [CrossRef]
106. Derevianko, A.P.; Molodin, V.I.; Zenin, V.N.; Leschinsky, S.V.; Maschenko, E.N. *Site du Paleolithique Recent Chestakovo*; IAET SB RAS Press: Novosibirsk, Russia, 2003. (In Russian)
107. Raghavan, M.; Skoglund, P.; Graf, K.E.; Metspalu, M.; Albrechtsen, A.; Moltke, I.; Rasmussen, S.; Stafford, T.W., Jr.; Orlando, L.; Metspalu, E.; et al. Upper Palaeolithic Siberian genome reveals dual ancestry of Native Americans. *Nature* **2014**, *505*, 87–91. [CrossRef]
108. Sitlivy, V.; Medvedev, G.I.; Lipnina, E.A. *Les Civilisations Préhistoriques d'Asie Centrale*; 1. Le Paléolithique de la rive occidentale du lac Baikal; Musées Royaux d'Art et d'Histoire: Brussels, Belgium, 1997.
109. Mochanov, Y.A. *Peopling of North-East Asia: Earliest Stages*; Nauka Publ.: Novosibirsk, Russia, 1977. (In Russian)
110. Ineshin, E.M.; Tetenkin, A.V. *Humans and the Environment in Northern Baikal Siberia during the Late Pleistocene*; Cambridge Scholars: Newcastle upon Tyne, UK, 2017.
111. Dikov, N.N. *Ancient Cultures of North-Eastern Asia*; Nauka Publ.: Moscow, Russia, 1979. (In Russian)

112. Slobodin, S.B. *Archeology of Kolyma and Continental Priokhot'e in Late Pleistocene and Early Holocene*; The U.S. Department of Interior, National Park Service, Shared Beringian Heritage Program, the Government Printing Office: Anchorage, AK, USA, 2014.
113. Dalton, A.S.; Margold, M.; Stokes, C.R.; Tarasov, L.; Dyke, A.S.; Adams, R.S.; Allard, S.; Arends, H.E.; Atkinson, N.; Attig, J.W.; et al. An updated radiocarbon-based ice margin chronology for the last deglaciation of the North American Ice Sheet Complex. *Quat. Sci. Rev.* **2020**, *234*, 106223. [CrossRef]
114. Dyke, A.; Andrews, J.; Clark, P.; England, J.; Miller, G.; Shaw, J.; Veillette, J. The Laurentide and Innuitian ice sheets during the last glacial maximum. *Quat. Sci. Rev.* **2002**, *21*, 9–31. [CrossRef]
115. Margold, M.; Stokes, C.R.; Clark, C.D. Reconciling records of ice streaming and ice margin retreat to produce a palaeogeographic reconstruction of the deglaciation of the Laurentide Ice Sheet. *Quat. Sci. Rev.* **2018**, *180*, 1–30. [CrossRef]
116. Patton, H.; Hubbard, A.; Andreassen, K.; Auriac, A.; Whitehouse, P.L.; Stroeven, A.P.; Shackleton, C.; Winsborrow, M.; Heyman, J.; Hall, A.M. Deglaciation of the Eurasian ice sheet complex. *Quat. Sci. Rev.* **2017**, *169*, 148–172. [CrossRef]
117. Anisimov, M.A.; Ivanova, V.V.; Pushina, Z.V.; Pitulko, V.V. Lagoon deposits of the Zhokhov Island: Their age, formation and importance for paleogeographic reconstructions of the region of Novosibirsk Islands. *Izv. Ross. Akad. Nauk. Ser. Geogr.* **2009**, *5*, 107–119. (In Russian)
118. Degtyarenko, Y.P.; Puminov, A.P.; Blagoveshchenskiy, M.G. Shorelines of the Eastern Arctic Seas in the Late Pleistocene and Holocene. In *Sea and Oceanic Level Fluctuations for 15,000 Years*; Kaplin, P.A., Klige, R.K., Chepalyga, A.L., Eds.; Nauka Publ.: Moscow, Russia, 1982; pp. 179–185. (In Russian)
119. Bauch, H.A.; Mueller-Lupp, T.; Taldenkova, E.; Spielhagen, L.F.; Kassens, H.; Grootes, P.M.; Thiede, J.; Heinemeir, J.; Petryashov, V.V. Chronology of the Holocene transgression at the North Siberian margin. *Glob. Planet. Chang.* **2001**, *31*, 125–139. [CrossRef]
120. Darby, D.A.; Polyak, L.; Bauch, H.A. Past glacial and interglacial conditions in the Arctic Ocean and marginal seas—A review. *Prog. Oceanogr.* **2006**, *71*, 129–144. [CrossRef]
121. Bradley, R.S.; England, J.H. The Younger Dryas and the Sea of Ancient Ice. *Quat. Res.* **2008**, *70*, 1–10. [CrossRef]
122. Taldenkova, E.; Bauch, H.A.; Gottschalk, J.; Nikolaev, S.; Rostovtseva, Y.; Pogodina, I.; Ovsepyan, Y.; Kandiano, E. History of ice-rafting and water mass evolution at the northern Siberian continental margin (Laptev Sea) during Late Glacial and Holocene times. *Quat. Sci. Rev.* **2010**, *29*, 3919–3935. [CrossRef]
123. Schirrmeister, L.; Grosse, G.; Kunitsky, V.; Fuchs, M.C.; Krbetschek, M.; Andreev, A.; Herzschuh, U.; Babyi, O.; Siegert, C.; Meyer, H.; et al. The mystery of Bunge Land (New Siberian Archipelago)—Implications for its formation based on composite palaeo-environmental records, geomorphology, and remote sensing. *Quat. Sci. Rev.* **2010**, *29*, 3598–3614. [CrossRef]
124. Wetterich, S.; Schirrmeister, L.; Andreev, A.A.; Pudenz, M.; Plessen, B.; Meyer, H.; Kunitsky, V.V. Eemian and Late Glacial/Holocene palaeoenvironmental records from permafrost sequences at the Dmitry Laptev Strait (NE Siberia, Russia). *Paleogeogr. Paleoclim. Paleoecol.* **2009**, *79*, 73–95. [CrossRef]
125. Nikolskiy, P.A.; Basilyan, A.E.; Sulerzhitsky, L.D.; Pitulko, V.V. Prelude to the Extinction: Revision of the Achchagyi-Allaikha and Berelyokh mass accumulations of mammoth. *Quat. Int.* **2010**, *219*, 16–25. [CrossRef]
126. Pitulko, V.V.; Basilyan, A.E.; Pavlova, E.Y. The Berelekh Mammoth Graveyard: New Chronological and Stratigraphical Data from the 2009 field season. *Geoarchaeology* **2014**, *29*, 277–299. [CrossRef]
127. Easton, N.A.; MacKay, G. Early Bifaces from the Little John Site (KdVo-6), Yukon Territory, Canada. In *Projectile Point Sequences in Northwestern North America*; Carlson, R.L., Magne, M.P.R., Eds.; Archaeology Press: Burnaby, BC, Canada, 2008; pp. 333–352.
128. Easton, N.A.; MacKay, G.R.; Young, P.B.; Schnurr, P.; Yesner, D.R. Chindadn in Canada? Emergent Evidence of the Pleistocene Transition in Southeast Beringia as Revealed by the Little John Site, Yukon. In *From the Yenisei to the Yukon: Interpreting Lithic Assemblage Variability in Late Pleistocene/Early Holocene Beringia*; Goebel, T., Buvit, I.C., Eds.; Texas A&M University Press: College Station, TX, USA, 2011; pp. 289–307.
129. Gillispie, T.E.; Cook, J.P.; Sattler, R.A.; Younie, A.M. Healy Lake Village: New Data and Analysis from the Chindadn Site. *Alsk. J. Anthropol.* **2014**, *11*, 186–187.
130. Goebel, T.; Powers, W.R.; Bigelow, N. The Nenana Complex of Alaska and Clovis Origins. In *Clovis: Origins and Adaptations*; Bonnichsen, R., Turnmire, K.L., Eds.; Oregon State University: Corvallis, OR, USA, 1991; pp. 49–79.
131. Goebel, T.; Powers, W.R.; Bigelow, N.; Higgs, A.S. Walker Road. In *American Beginnings: The Prehistory and Palaeoecology of Beringia*; West, F.H., Ed.; University of Chicago Press: Chicago, IL, USA, 1996; pp. 356–363.
132. Hoffecker, J.F. Assemblage Variability in Beringia: The Mesa Factor. In *From the Yenisei to the Yukon: Interpreting Lithic Assemblage Variability in Late Pleistocene/Early Holocene Beringia*; Goebel, T., Buvit, I.C., Eds.; Texas A&M University Press: College Station, TX, USA, 2011; pp. 165–178.
133. Powers, W.R.; Hoffecker, J.F. Late Pleistocene settlement in the Nenana Valley, central Alaska. *Am. Antiq.* **1989**, *54*, 263–287. [CrossRef]
134. Waters, M.R.; Stafford, T.W., Jr. Redefining the age of Clovis: Implications for the peopling of the Americas. *Science* **2007**, *315*, 1122–1126. [CrossRef] [PubMed]
135. Waters, M.R.; Stafford, T.W.; Kooyman, B.; Hills, L.V. Late Pleistocene horse and camel hunting at the southern margin of the ice-free corridor: Reassessing the age of Wally's Beach, Canada. *Proc. Nat. Acad. Sci. USA* **2015**, *112*, 4263–4267. [CrossRef]
136. Gavrilov, A.V.; Romanovskii, N.N.; Hubberten, H.-W. Paleogeographic Scenario of the Post-Glacial Transgression on the Shelf of the Laptev Sea. *Earth's Cryosphere* **2006**, *X*, 39–50. (In Russian)

137. Ukraintseva, V.V.; Arslanov, K.A.; Belousova, Z.M.; Ustinov, V.N. The first data on the Early Holocene flora and vegetation of Bolshoi Lyakhovskyi Island (New Siberian archipelago). *Bot. J.* **1989**, *74*, 782–793. (In Russian)
138. Makeyev, V.M.; Arslanov, K.A.; Baranovskaya, O.F.; Ponomareva, D.P.; Kosmodamiansky, A.V.; Tertychnaya, T.V. Stratigraphy, geochronology and paleogeography of Kotelny Island in the Late Pleistocene and Holocene. *Bull. Comm. Quat. Stud.* **1989**, *58*, 58–69. (In Russian)
139. Makeyev, V.M.; Ponomareva, D.P.; Pitulko, V.V.; Chernova, G.M.; Solovyeva, D.V. Vegetation and climate of New Siberian Islands for the past 15 000 years. *Arct. Antarct. Alp. Res.* **2003**, *35*, 28–35. [CrossRef]
140. Bezrodnykh, Y.P.; Veksler, V.S.; Savvaitov, A.S.; Stelle, V.Y. Correlation of Paleogeographic Events of the Late Pleistocene and Holocene in some Arctic regions based on on ^{14}C dating. In *Isotopic-Geochemical Investigations in the Baltic Area and Belorussia*; Viiding, H., Payamyae, P., Eds.; AN ESSR, Institute of Geology: Tallin, Estonia, 1986; pp. 5–12. (In Russian)
141. Kaplina, T.N. Alas Complex of North Yakutia. *Earth's Cryosphere* **2009**, *XII*, 3–17. (In Russian)
142. Kaplina, T.N.; Lozhkin, A.V. The history of vegetation of Primorskaya lowland of Yakutia in the Holocene. In *Evolution of the Environment of the Territory of the USSR during the Late Pleistocene and Holocene*; Velichko, A.A., Spasskaya, I.I., Khotinskiy, N.A., Eds.; Nauka Publ.: Moscow, Russia, 1982; pp. 207–220. (In Russian)
143. Nikolskiy, P.A.; Basilyan, A.E. Svyatoy Nos Cape—The main cross section of the Yana-Indigirka lowland Quaternary deposits. In *Natural History of the Russian Arctic during the Late Pleistocene and Holocene: Paleogeography, Geology and Archeology*; Nikolskiy, P.A., Pitulko, V.V., Eds.; GEOS Publ.: Moscow, Russia, 2004; pp. 5–13. (In Russian)
144. Zimmermann, H.H.; Raschke, E.; Epp, L.S.; Stoof-Leichsenring, K.R.; Schirrmeister, L.; Schwamborn, G.; Herzschuh, U. The History of Tree and Shrub Taxa on Bol'shoy Lyakhovsky Island (New Siberian Archipelago) since the Last Interglacial Uncovered by Sedimentary Ancient DNA and Pollen Data. *Genes* **2017**, *8*, 273. [CrossRef]
145. Vartanyan, S.L.; Arslanov, K.A.; Karhu, J.; Possnert, G.; Sulerzhitsky, L.D. Collection of radiocarbon dates on the mammoths (Mammuthus primigenius) and other genera of Wrangel Island, northeast Siberia, Russia. *Quat. Res.* **2008**, *70*, 51–59. [CrossRef]
146. Kremenetski, C.V.; Sulerzhitsky, L.D.; Hantemirov, R. Holocene history of the northern range limits of some trees and shrubs in Russia. *Arct. Antarct. Alp. Res.* **1998**, *30*, 317–333. [CrossRef]
147. Binney, H.A.; Willis, K.J.; Edwards, M.E.; Bhagwat, S.A.; Anderson, P.M.; Andreev, A.A.; Blaauw, M.; Damblon, F.; Haesaerts, P.; Kienast, F.; et al. The distribution of late-Quaternary woody taxa in northern Eurasia: Evidence from a new macrofossil database. *Quat. Sci. Rev.* **2009**, *28*, 2445–2464. [CrossRef]
148. Binney, H.; Edwards, M.; Macias-Fauria, M.; Lozhkin, A.; Anderson, P.; Kaplan, J.O.; Andreev, A.; Bezrukova, E.; Blyakharchuk, T.; Jankovska, V.; et al. Vegetation of Eurasia from the last glacial maximum to present: Key biogeographic patterns. *Quat. Sci. Rev.* **2017**, *157*, 80–97. [CrossRef]
149. Pitulko, V.V. Ancient Humans in Eurasian Arctic Ecosystems: Environmental Dynamics and Changeability of Subsistence Models. *World Archaeol.* **1999**, *30*, 421–436. [CrossRef]
150. Romanovskii, N.N.; Hubberten, H.-W.; Gavrilov, A.V.; Tumskoy, V.E.; Tipenko, G.S.; Grigoriev, M.N.; Siegert, C. Thermokarst and land-ocean interactions, Laptev Sea region, Russia. *Permafr. Periglac. Proc.* **2000**, *11*, 137–152. [CrossRef]
151. Pitulko, V.V. *Zhokhovskaya Stoyanka*; Dmitriy Bulanin: St. Petersburg, Russia, 1998. (In Russian)
152. Pitulko, V.V.; Pavlova, E.Y. Geoarchaeology, Age and Chronology of the Zhokhov Site. *Vestn. St. Petersburg Univ. Hist.* **2022**, *67*, 1253–1295. [CrossRef]
153. Pitulko, V.V. Terminal Pleistocene/Early Holocene Occupation in North East Asia and the Zhokhov Assemblage. *Quat. Sci. Rev.* **2001**, *20*, 267–275. [CrossRef]
154. Pitulko, V.V.; Ivanova, V.V.; Kasparov, A.K.; Pavlova, E.Y. Reconstructing Prey Selection, Hunting Strategy and Seasonality of the Early Holocene frozen site in the Siberian High Arctic: A Case Study on the Zhokhov Site faunal remains, De Long Islands. *Environ. Archaeol.* **2015**, *20*, 120–157. [CrossRef]
155. Pitulko, V.V.; Kasparov, A.K. Archaeological dogs from the Early Holocene Zhokhov site in the Eastern Siberian Arctic. *J. Archaeol. Sci. Rep.* **2017**, *13*, 491–515. [CrossRef]
156. Pitulko, V.V.; Kuzmin, Y.V.; Glascock, M.D.; Pavlova, E.Y.; Grebennikov, A.V. «They came from the ends of the earth»: Long-distance exchange of obsidian in the High Arctic during the Early Holocene. *Antiquity* **2019**, *93*, 28–44. [CrossRef]
157. Gerasimova, M.M.; Astakhov, S.N.; Velichko, A.A. *The Palaeolithic Man, His Material Culture and Natural Environment: Illustrated Catalog of Paleoanthropological Finds in Russia and Adjacent Territories*; Nestor-Istoriia Publ.: Saint Petersburg, Russia, 2007.
158. Fu, Q.; Li, H.; Moorjani, P.; Jay, F.; Slepchenko, S.M.; Bondarev, A.A.; Johnson, P.L.F.; Aximu-Petri, A.; Prüfer, K.; de Filippo, C.; et al. Genome sequence of a 45,000-year-old modern human from western Siberia. *Nature* **2014**, *514*, 445–450. [CrossRef] [PubMed]
159. Lee, E.J.; Merriwether, D.A.; Kasparov, A.K.; Khartanovich, V.I.; Nikolskiy, P.A.; Shidlovskiy, F.K.; Gromov, A.V.; Chikisheva, T.A.; Chasnyk, V.G.; Timoshin, V.B.; et al. A genetic perspective of prehistoric hunter-gatherers in the Siberian Arctic: Ancient DNA analysis of human remains from 8000 years ago. *J. Archaeol. Sci. Rep.* **2018**, *17*, 943–949. [CrossRef]
160. Lu, H.Y.; Yi, S.W.; Xu, Z.W.; Zhou, H.X.; Zeng, L.; Zhu, F.Y.; Feng, H.; Dong, L.N.; Zhuo, H.X.; Yu, K.F.; et al. Chinese deserts and sand fields in Last Glacial Maximum and Holocene Optimum. *Chin. Sci. Bull.* **2013**, *58*, 2775–2783. [CrossRef]
161. Yang, X.; Jiang, W.; Yang, S.; Kong, Z.; Luo, Y. Vegetation and climate changes in the western Chinese Loess Plateau since the Last Glacial Maximum. *Quat. Int.* **2015**, *372*, 58–65. [CrossRef]

162. Pitulko, V.V.; Pavlova, E.Y. Late Pleistocene and Early Holocene Paleoenvironments, Human Settlement and Adaptations in the East Siberian Arctic. *Strat. Plus* **2023**, *1*, 193–228. (In Russian) [CrossRef]
163. Pavlov, P.Y. The Paleolithic of Northeastern Europe: New Data. *Archaeol. Ethnogr. Anthropol. Eurasia* **2008**, *33*, 33–45. (In Russian) [CrossRef]
164. Clark, P.U.; Dyke, A.S.; Shakun, J.D.; Carlson, A.E.; Clark, J.; Wohlfarth, B.; Mitrovica, J.X.; Hostetler, S.W.; McCabe, A.M. The last glacial maximum. *Science* **2009**, *325*, 710–713. [CrossRef]
165. Hoffecker, J.F.; Hoffecker, I.T. Technological complexity and the global dispersal of modern humans. *Evol. Anthropol.* **2017**, *26*, 285–299. [CrossRef]
166. Hoffecker, J.F.; Hoffecker, I.T. The Structural and Functional Complexity of Hunter-Gatherer Technology. *J. Archaeol. Method Theory* **2018**, *25*, 202–225. [CrossRef]
167. Mellars, P. The Impossible Coincidence. A Single-Species Model for the Origins of Modern Human Behavior in Europe. *Evol. Anthropol.* **2005**, *14*, 12–27. [CrossRef]
168. Derevianko, A.P.; Shun'kov, M.V.; Kozlikin, M.B.; Fedorchekno, A.Y.; Pavlenok, G.D.; Belousova, N.E. Early Upper Paleolithic Bone Needle from the Main Chamber of Denisova Cave. In *Problems of Archaeology, Ethnography and Anthropology of Siberia and Neighboring Territories*; IAET SB RAS Press: Novosibirsk, Russia, 2016; Volume XXII, pp. 72–75. (In Russian)
169. Shalagina, A.; Baumann, M.; Kolobova, K.; Krivoshapkin, A. Bone Needles from Upper Paleolithic Complexes of the Strashnaya Cave (North-Western Altai). *Teor. I Prakt. Arkheologicheskikh Issled.* **2018**, *21*, 89–98. (In Russian) [CrossRef]
170. Rasmussen, S.O.; Bigler, M.; Blockley, S.P.; Blunier, T.; Buchardt, S.L.; Clausen, H.B.; Cvijanovic, I.; Dahl-Jensen, D.; Johnsen, S.J.; Fischer, H.; et al. A stratigraphic framework for abrupt climatic changes during the Last Glacial period based on three synchronized Greenland ice-core records: Refining and extending the INTIMATE event stratigraphy. *Quat. Sci. Rev.* **2014**, *106*, 14–28. [CrossRef]
171. Brigham-Grette, J.; Lozhkin, A.V.; Anderson, P.M.; Glushkova, O.Y. Paleoenvironmental conditions in western Beringia before and during the last glacial maximum. In *Entering America Northeast Asia and Beringia before the Last Glacial Maximum*; Madsen, D.B., Ed.; The University of Utah Press: Salt Lake City, UT, USA, 2004; pp. 29–62.
172. Oswalt, W.H. *An Anthropological Analysis of Food-Getting Technology*; John Wiley & Sons: New York, NY, USA, 1976.
173. Vereschagin, N.K. Berelekh mammoth "graveyard". *Proc. Zool. Inst.* **1977**, *72*, 5–50. (In Russian)
174. Alfimov, A.V.; Berman, D.I. Beringian climate during the Late Pleistocene and Holocene. *Quat. Sci. Rev.* **2001**, *20*, 127–134. [CrossRef]
175. Kuzmina, S.A.; Sher, A.V.; Edwards, M.E.; Haile, J.; Yan, E.V.; Kotov, A.V.; Willerslev, E. The late Pleistocene environment of the Eastern West Beringia based on the principal section at the Main River, Chukotka. *Quat. Sci. Rev.* **2011**, *30*, 2091–2106. [CrossRef]

quaternary

MDPI

Article

Foxes in Retrospect—Unraveling Human-Fox Relationships through Fox Tooth Ornaments in the Swabian Jura

Flavia Venditti [1,*], Madison J. McCartin [2], Melanie-Larisa Ostermann [2], Nicholas J. Conard [1,3] and Sibylle Wolf [1,3,*]

1 Department of Early Prehistory and Quaternary Ecology, University of Tübingen, Schloß Hohentübingen, Burgsteige 11, 72070 Tübingen, Germany; nicholas.conard@uni-tuebingen.de
2 Institute for Archaeological Sciences, University of Tübingen, Hölderlinstrasse 12, 72074 Tübingen, Germany; mjmccartin@ucdavis.edu (M.J.M.); laraostermann20@gmail.com (M.-L.O.)
3 Senckenberg Centre for Human Evolution and Palaeoenvironment at the University of Tübingen, Hölderlinstrasse 12, 72074 Tübingen, Germany
* Correspondence: flavia.venditti@uni-tuebingen.de (F.V.); sibylle.wolf@ifu.uni-tuebingen.de (S.W.)

Abstract: Personal ornaments play an important role in our understanding of human cultural and behavioral change during the Upper Paleolithic, providing insights into intangible aspects of human cultural behavior. Some ornament forms are better studied than others, and fox tooth ornaments, despite their frequent occurrence and broad spatiotemporal span, are relatively under-addressed. Here we present the first comprehensive study of 40 perforated fox teeth recovered from four cave sites in southwestern Germany. This region's rich record of symbolic representations, as well as evidence of long-standing human–fox relationships, make the Swabian Jura an ideal case study for investigations of fox tooth ornaments. By applying a holistic approach, including geometric morphometrics and traceology coupled with experimental archaeology, we show that fox teeth were mostly perforated by bifacial scraping and grooving and were worn as ornaments. We discuss the role of foxes within human socio-symbolic and paleoenvironmental systems during the Upper Paleolithic of the Swabian Jura, and we contextualize our results within the broader context of sites across Europe during the Upper Paleolithic. The data we provide are in line with general trends observed across the continent and offer insight into the role of foxes during the Upper Paleolithic, especially regarding human subsistence, cultural expression, and ornament production.

Keywords: personal ornamentation; traceology; geometric morphometrics; Upper Paleolithic; Swabian Jura; foxes (*Vulpes vulpes* and *Vulpes lagopus*)

Citation: Venditti, F.; McCartin, M.J.; Ostermann, M.-L.; Conard, N.J.; Wolf, S. Foxes in Retrospect—Unraveling Human-Fox Relationships through Fox Tooth Ornaments in the Swabian Jura. *Quaternary* **2023**, *6*, 50. https://doi.org/10.3390/quat6030050

Academic Editor: Juan Manuel López García

Received: 16 August 2023
Revised: 2 September 2023
Accepted: 6 September 2023
Published: 21 September 2023

1. Introduction

In the Swabian Jura (southwestern Germany), Upper Paleolithic symbolic culture is characterized by ivory figurines, musical instruments, and personal ornamentation [1–9]. Although small, ornaments are particularly informative, as they are numerous, well-preserved, diverse in form, and occur throughout the entirety of the Upper Paleolithic. Their hypothesized role as social and symbolic objects also lends insight onto intangible aspects of human cultural behavior, such as personal identity, group affiliation, language, movement, trade, and social status [8,10–17].

During the Upper Paleolithic, ornaments were produced from a variety of organic and inorganic raw materials, including ivory, tooth, bone, shell, fossil, and stone. Those made from ivory, shells, and cervid teeth are well studied [8,10,11,18–30], while fox tooth ornaments, primarily made from red (*Vulpes vulpes*) and arctic fox (*Vulpes lagopus*) canines, are comparatively under-addressed.

Fox tooth ornaments first appear during the Châtelperronian and persist throughout the entire Upper Paleolithic at sites spanning the Iberian Peninsula to the East European Plain (see Supplementary Materials) [13,17,31]. Most assemblages are small, usually less than twenty specimens (Table S1), although exceptional cases include sites like Pavlov I (Czechia) and Sunghir (Russia), each represented by over 200 fox tooth ornaments in association with Gravettian burials [10,32–34].

Despite their abundance and wide geographic span, detailed information on fox tooth ornaments is limited (e.g., on species and tooth selection, metric data, production methods, use-wear, taphonomy). This is especially disappointing, given their potential to illuminate not only the socio-symbolic aspects of ornamentation, but also the broader relationship between humans, foxes, and their shared environment. Unlike other ornament forms, animal teeth come from a living organism and are, therefore, directly related to local ecology and human subsistence practices. Fox tooth ornaments can thus be linked to the broader role of foxes on the landscape (e.g., paleoecology, human–animal relationships), in human diets (e.g., hunting methods, nutrition), and as a raw material (e.g., fur, bone tools, ornamentation).

This study presents an assemblage of Upper Paleolithic fox tooth ornaments (n = 40) from four sites in the Swabian Jura: Hohle Fels (n = 12), Geißenklösterle (n = 12), Brillenhöhle (n = 3), and Hohlenstein-Stadel (n = 13) (Supplementary Materials and Figures 1 and 2). This region's rich record of symbolic representation, as well as evidence of long-standing human–fox relationships [35–39], make the Swabian Jura the ideal location to explore the production and use of these ornaments and their relevance for broader human–fox relationships in the Upper Paleolithic.

In this study, we explore fox tooth ornaments from a diachronic and synchronic perspective by analyzing ornaments from the entire Upper Paleolithic sequence (Aurignacian to Magdalenian) and by considering intra- and inter-ornament variability and the influence of culture versus technological preferences. In particular, we aim to:

1. investigate diachronic changes in technology, production, and function and how these changes relate to the methods of the perforation and mode of use among the three cultures (Aurignacian, Gravettian, and Magdalenian) and within the four analyzed sites in the Swabian Jura,
2. explore the potential for species preferences between red fox and arctic fox within the analyzed sites and discuss the results in the framework of fox ecology and behavior, and
3. contextualize the results from the Swabian Jura within a broader context, drawing from an extensive literature review of fox tooth ornaments from sites across Europe during the Upper Paleolithic.

To do so, we apply three analytical methods to obtain a more holistic view of fox exploitation and ornament production, use, and discard. Geometric morphometrics facilitates the reliable differentiation of red versus arctic fox in isolated teeth, allowing us to investigate species selection and preference, fox paleoecology, and the interaction between humans and foxes in the Swabian Jura. This is complemented by traceological analysis coupled with experimental archaeology, in which microscopic observations of the perforations are used to identify manufacturing techniques and use-wear traces.

Foxes in the Swabian Jura

During the Late Pleistocene, climatic and environmental conditions influenced the distribution and composition of fox species on the landscape and their subsequent availability to hunter-gatherer populations [40,41]. At the start of the Upper Paleolithic, both red and arctic foxes were sympatric in Central Europe. Subsequent cooling associated with the Last Glacial Maximum forced red foxes to retreat into glacial refugia, such as the Iberian Peninsula [42,43]. Sympatric populations appeared again in Central Europe during the Late Glacial, but Allerød warming later pushed arctic foxes northward. A final period of sympatry is recorded during the Younger Dryas, followed by the northern retreat of arctic foxes in response to Holocene warming, after which both species attained their modern distributions [42].

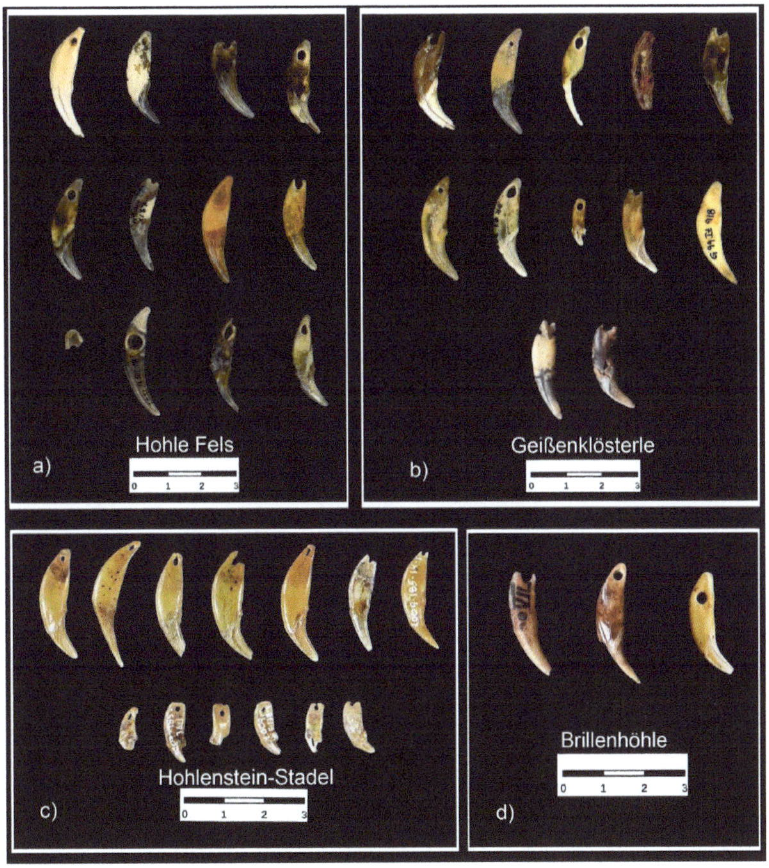

Figure 1. Perforated fox teeth analyzed in this study and their geographic location. Items from left to right. (**a**) Upper row: ID 461, ID 631.1, ID 1278.1, ID 1295; middle row: ID 1517.1, ID 1137, ID 1228.2, ID 964.1; lower row: ID 1227.1, ID 1742, ID 2099.1, ID 1713. (**b**) Upper row: ID 1B, ID 1A, ID 391, ID 89, ID 369; middle row: ID 286, ID 38F, ID 658.2, ID 226, ID 918; lower row: ID 487, ID 1511. (**c**) Upper row: ID 1.0380, ID 2.0380, ID 3.0380, ID 4.0380, ID. 5.0380, ID 141-1, ID 14-1; lower row: ID 33-1, ID 36-1, ID 167-1, ID 178-1, ID 9000-1, ID 9001-1. (**d**) ID 130, ID 131, ID 132.

In the Swabian Jura, both red and arctic foxes are found at numerous archaeological sites throughout the Paleolithic, generally accounting for 1–5% of NISP. Their abundance increased over time compared to other carnivores and large herbivores. A similar increase is recorded in hare, likely due to common acquisition methods, such as traps and snares, which elevated the catch rate of small game [44,45]. Juvenile foxes (<1 year) are rare, only accounting for ~2% of fox NISP. Most come from the Middle Paleolithic of Hohlenstein-Stadel (n = 17) [46], reflecting natural deaths from denning animals. The comparative dearth of juveniles from Upper Paleolithic contexts (Vogelherd Aurignacian, n = 1 [35]; Geißenklösterle Gravettian, n = 4 [39]; Vogelherd Magdalenian, n = 1 [35]; Langmahdalde Magdalenian, n = 6 [47]) may indicate issues with preservation or excavation techniques, that caves were not used as fox dens, or that humans did not intentionally hunt juveniles [44].

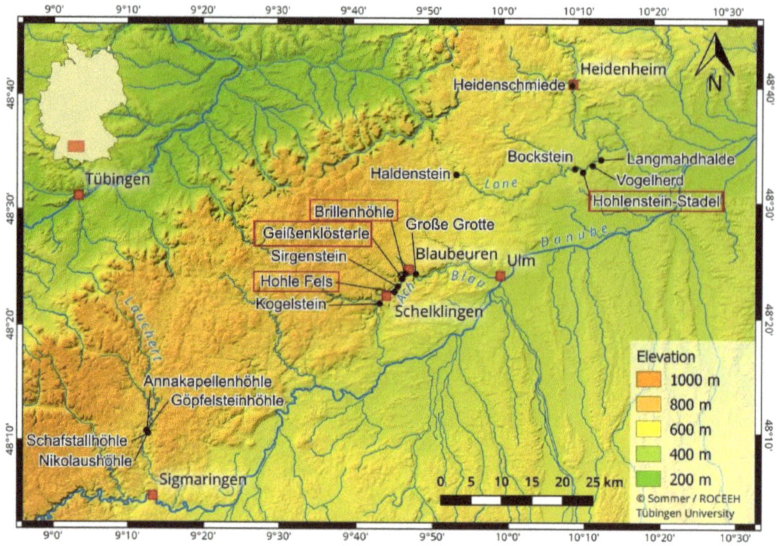

Figure 2. Map of the Swabian Jura with the location of important Paleolithic cave sites from the Ach, Lone, and Lauchert valleys. Four sites outlined in red provide Upper Paleolithic fox tooth pendants and are the focus of this study: Hohle Fels (1) Geißenklösterle (2), Brillenhöhle (3), Hohlenstein-Stadel (4). (Map provided by Christian Sommer/ROCEEH (2019): A map collection of the Paleolithic of the Swabian Jura (Version 1.0.0) (Data set). (modified by F.V.).

Evidence of fox butchery is rare compared to other taxa in the Swabian Jura [44]. During the Middle Paleolithic, a single phalanx from Bockstein exhibits a circular cut mark associated with skinning [48]. During the Aurignacian, cut marks on two mandibles at Vogelherd [49,50] and a fifth metacarpal at Geißenklösterle [39] also indicate skinning, while cut marks on a rib and ulna at Hohle Fels are from defleshing [37]. During the Gravettian, only two cut marks are present on fox remains at Hohle Fels, one on a pelvis and the other on a femur, both related to defleshing [37]. Lastly, during the Magdalenian, one cut-marked ulna is recorded from Hohle Fels [51] and one cut-marked mandible is present at Langmahdhalde [52], indicating defleshing and skinning, respectively. In addition to cut marks, human tooth marks are also recorded on three Aurignacian and Gravettian fox elements from Hohle Fels [53]. Combined, these traces clearly establish foxes were used for both fur and food, as well as a raw material for the 40 fox tooth ornaments discussed in this study.

Human–fox relationships also extended beyond human predation, as evidenced by isotopic studies of fox paleoecology, which suggest some foxes were commensal to humans [38,44,54,55]. Prior to the Upper Paleolithic, foxes inhabited expected niches as either rodent specialists or commensals to large carnivores (for an overview on fox ecology and behavior, see the Supplementary Materials). However, during the Aurignacian, a new fox niche emerged, characterized by a diet of primarily reindeer with a smaller proportion of rodents and other large game. The likely explanation is that some foxes became commensal to humans to take advantage of food waste, perhaps in response to increasing human population density [38,55]. By engaging in this synanthropic commensal behavior, foxes were able to access new, more reliable food resources as well as protection from large carnivores who were deterred by humans. Humans also benefited, gaining more frequent access to foxes for food, fur, and raw materials, as well as a means of food disposal, which indirectly protected humans from large carnivores [54]. This relationship lasted throughout the Aurignacian and Gravettian before dogs and wolves largely took over the niche during the Magdalenian, although some foxes remained commensal [38,55].

2. Materials and Methods

2.1. Archaeological and Comparative Materials

The complete assemblage includes 40 perforated fox tooth ornaments from the Aurignacian, Gravettian, and Magdalenian horizons of four sites in the Swabian Jura: Hohle Fels, Geißenklösterle, Brillenhöhle, and Hohlenstein-Stadel (Table 1). Most are canines (n = 32), although incisors (n = 4) and premolars (n = 2) are also present as well as some indeterminate root fragments (n = 2).

Table 1. List of studied perforated fox tooth ornaments from the Swabian Jura. QU = quadrant, ID = find ID, GH = geological horizon, AH = archaeological horizon, Morph = morphological species identification, GM = geometric-morphometric species identification from this study, [1] = Kölbl and Conard [5].

Culture	QU	ID	GH	AH	Species ID (Morph)	Species ID (GM)	Tooth
Hohle Fels:							
NA (Upper Paleolithic)	28	964.1	1ka-7	0/I-IV	*Vulpes* sp.	*Vulpes vulpes*	Canine
Gravettian	58	1295	3c	IIc	*Vulpes lagopus* [1]	*Vulpes lagopus*	Canine
Gravettian	79	1228.29	3c-3d	IIc-IId	*Vulpes* sp.	*Vulpes lagopus*	Canine
Gravettian	66d	461	3bl	IIb	*Vulpes* sp. [1]	*Vulpes vulpes*	Canine
Aurignacian-Gravettian	76	1227.1	5	IIe	*Vulpes* sp. ?	NA	Root fragment
Aurignacian-Gravettian	59	2099.1	5-7	IIe-IV	*Vulpes lagopus* [1]	NA	Canine
Aurignacian	24	1137	7	IV	*Vulpes* sp.	Indeterminate	Canine
Aurignacian	26	631.1	7a	Va	*Vulpes* sp.	*Vulpes lagopus*	Canine
Aurignacian	69	1713	7	IV	*Vulpes lagopus* [1]	*Vulpes vulpes?*	Canine
Aurignacian	89	1517.1	7	IV	*Vulpes* sp.	*Vulpes lagopus*	Canine
Aurignacian	24	1278.1	7	IV	*Vulpes* sp.	*Vulpes lagopus*	Canine
Aurignacian	65	1742	7	IV	*Vulpes* sp.	NA	Canine
Geißenklösterle:							
NA (likely Gravettian)	22	1A	disturbed	disturbed	*Vulpes* sp.	*Vulpes vulpes* ?	Canine
NA (likely Gravettian)	22	1B	disturbed	disturbed	*Vulpes* sp.	*Vulpes lagopus*	Canine
Gravettian	67	658.2	7	It	*Vulpes* sp.	NA	Incisor
Gravettian	98	391	7	It	*Vulpes* sp.	*Vulpes vulpes*	Canine
Gravettian	99	918	7	It	*Vulpes* sp.	*Vulpes lagopus* ?	Canine
Gravettian	100	89	5b	Ir	*Vulpes* sp. [1]	*Vulpes* sp.	Canine
Gravettian	140	226	7	It	*Vulpes* sp.	*Vulpes lagopus*	Canine
Gravettian	98	286	7	It	*Vulpes* sp.	NA	Canine
Gravettian	98	387	7	It	*Vulpes lagopus* [1]	NA	Canine
Aurignacian	59	369	16	IIIb	*Vulpes* sp. [1]	*Vulpes vulpes*	Canine
Aurignacian	58	487	15	III	*Vulpes* sp.	NA	Canine
Aurignacian	67	1511	15	IIIa	*Vulpes* sp.	NA	Canine
Hohlenstein-Stadel:							
Aurignacian	NA	1.0380	19./20. Meter, 6. Hieb	NA	*Vulpes* sp.	*Vulpes lagopus*	Canine
Aurignacian	NA	2.0380	19./20. Meter, 6. Hieb	NA	*Vulpes* sp.	*Vulpes vulpes*	Canine
Aurignacian	NA	3.0380	19./20. Meter, 6. Hieb	NA	*Vulpes* sp.	*Vulpes vulpes*	Canine
Aurignacian	NA	4.0380	19./20. Meter, 6. Hieb	NA	*Vulpes* sp.	*Vulpes lagopus*	Canine
Aurignacian	NA	5.0380	19./20. Meter, 6. Hieb	NA	*Vulpes* sp.	*Vulpes lagopus*	Canine
Aurignacian or Magdalenian	NA	36-1	backdirt	NA	*Vulpes* sp.	NA	Incisor
Aurignacian or Magdalenian	NA	90001-1	backdirt	NA	*Vulpes* sp.	NA	Incisor
Aurignacian or Magdalenian	NA	9000-1	backdirt	NA	*Vulpes* sp.	NA	Incisor
Aurignacian or Magdalenian	NA	14-1	backdirt	NA	*Vulpes* sp.	*Vulpes* sp.	Canine
Aurignacian or Magdalenian	NA	141-1	backdirt	NA	*Vulpes* sp.	*Vulpes* sp.	Canine
Aurignacian or Magdalenian	NA	178-1	backdirt	NA	*Vulpes* sp.	NA	Premolar
Aurignacian or Magdalenian	NA	33-1	backdirt	NA	*Vulpes* sp.	NA	Premolar
Aurignacian or Magdalenian	NA	167-1	backdirt	NA	*Vulpes* sp. ?	NA	Root fragment
Brillenhöhle:							
Gravettian	NA	132	NA	VII	*Vulpes* sp.	NA	Canine
Gravettian	P7	131	NA	VII	*Vulpes* sp.	NA	Canine
Gravettian	NA	130	NA	VII	*Vulpes* sp.	NA	Canine

The ornaments from Hohle Fels are stored at the University of Tübingen as part of a larger study of the faunal assemblage. Most of the ornaments from Geißenklösterle are also stored at the University of Tübingen, while two others are on display at the Landesmuseum Württemberg Stuttgart alongside the ornaments from Brillenhöhle. The Hohlenstein-Stadel ornaments are stored and curated at the Museum Ulm, some of which are on display.

The Hohle Fels assemblage includes 12 total ornaments, which come from the Aurignacian layers Vb-IId (n = 8), the Gravettian layers IIe-IIb (n = 3), and a profile collapse spanning multiple Upper Paleolithic horizons (n = 1) [3,5,53,56–59].

Geißenklösterle hosts 12 total ornaments, which come from the Aurignacian layers IIIb-III (n = 3), the Gravettian layers It-Ir (n = 7), and from disturbed contexts that are likely Gravettian (n = 2) [60,61]. Here, we publish one of the Aurignacian ornaments (ID 1511) for the first time. Two additional Gravettian ornaments are known from Kölbl and Conard [5] (Kat. 75, ID 359) and the Geißenklösterle database (ID 786); however, these two objects could not be located for this study and are not included in the total count.

For Brillenhöhle, the assemblage includes three Gravettian ornaments from layer VII (Riek, 1973).

Lastly, the Hohlenstein-Stadel assemblage includes 13 total ornaments. Those excavated in association with the lion man figurine (n = 5) are considered Aurignacian [61–64], while those excavated from the backdirt (n = 8) cannot be assigned to a specific cultural period [63,65]. The latter likely belong to the Aurignacian, although the Magdalenian cannot be excluded.

The assemblage we present here is likely incomplete. It is likely that excavators in the early 20th century overlooked smaller finds due to their coarser excavation methods. Excavations are also ongoing at Brillenhöhle and Hohle Fels, and we can expect to discover more animal tooth pendants from these sites in the future.

For the geometric morphometric analysis, our sample includes a total of 24 Aurignacian and Gravettian ornaments from Hohle Fels (n = 10), Geißenklösterle (n = 7), and Hohlenstein-Stadel (n = 7) (Table S2). To establish a comparative baseline, we also curated 42 modern, unmodified, adult fox canines (Table S3) from the comparative zooarchaeological collection of the University of Tübingen alongside two canines from the private collection of Chris Baumann (University of Tübingen). The comparative sample includes 13 upper and 9 lower arctic fox canines (Banks Island, Canada) as well as 9 upper and 11 lower red fox canines (southwestern Germany).

2.2. Geometric Morphometric Methods

Micro-computed tomography (μCT) scanning of the archaeological sample was carried out with a Nikon XT H 320 μCT scanner at isotopic voxel length between 20 and 61 μ from the Terrestrial Palaeoclimatology working group at the University of Tübingen. For the comparative sample, 16 arctic and 16 red fox canines were scanned with the same machine, while 6 arctic and 4 red fox canines were scanned with a GE Phoenix v | tome | x s240 μCT scanner at an isotopic voxel length between 15 and 55 μ from the Paleoanthropology High Resolution CT Laboratory at the University of Tübingen.

Our geometric morphometric analysis follows protocols established by Benazzi et al. [66] and Röding et al. [67] for the cervical outline of hominin molars with some modifications for fox canines. As the cervical outline lacks homologous fixed points and a clear start or end point, landmark collection does not fall into the usual concept of fixed or semi-landmarks, thus necessitating additional steps during data collection for the removal of orientation, location, and absolute size from the landmark coordinates.

After scanning, we post-processed and sectioned the μCT images in Avizo 9.2 (Visualization Sciences Group), orienting each tooth to establish a best-fit plane along the cervical outline. We then virtually sectioned the images along this outline, removing the portion below the cervical plane, so that only the crown remained. Next, in Rhinoceros 6 (Robert McNeel and Associates), we translated the best-fit plane into an x-y coordinate system for landmark selection. To ensure consistent positioning, each tooth was rotated so that the lingual surface was parallel to the right side of the z-axis. Additionally, the area centroid of the cervical outline was calculated, upon which the outline was centered and translated to the Cartesian coordinate position (10,10,0). After positioning, sixteen landmark positions along the cervical outline were generated through equiangular spaced radial vectors from the centroid. The coordinates at the intersection of the vector with the cervical outline served as pseudo-landmarks and were recorded clockwise starting from the middle-upper landmark. Following established methods [68,69], all left canines were mirrored and treated as right canines. However, it should be noted that fluctuating asymmetry in dentition may increase noise when left and right teeth are combined [67].

Before analyzing the landmark data, we tested for intra-observer error to evaluate the alignment, reliability of landmark positions, and the relative reproducibility of individual landmarks. This was calculated as the percent error of the Euclidean distance between the configuration centroid and repeated landmark measurements. To accomplish this, we performed each step in Aviso and Rhinoceros five times for a random sample with a time gap of five hours, one day, one week, and one month after the first analysis. The randomization was performed by assigning a number to each used sample, and a random number generator was then used to select a sample for error measurements. Afterward, for each repeated landmark set, the centroid and the 16 Euclidean distances between the centroid and each landmark were calculated using unscaled landmark data. We then calculated the percent error for each landmark as well as the average and standard deviation.

After data collection, we performed multivariate statistical analyses in the program R (R Development Core Team 2011, version 4.1.2, © 2021) using multiple packages: Geomorph [70,71], ggplot2 [72], ggpubr [73], and ggrepel [74]. Analyses were conducted separately for upper and lower canines. For some teeth, the cervical outline was absent or unclear due to surface modifications or poor preservation. We considered five ornaments too incomplete for further analysis (ID 391, ID 89, ID 1227.1, ID 14-1, ID 141-1). However, for teeth with less severe damage, we attempted to reconstruct the cervical outline with the function estimate.missing in geomorph (method = "TPS"). The method was successful for a CN45 upper canine, ID 1A, ID 226, and ID 2.0380 but unsuccessful in the case of the CN99 upper canine. The former were included in subsequent analyses, while the latter was removed from the statistical sample.

We analyzed the landmark sets with two Principal Component Analyses (PCA)—a Shape PCA (gm.prcomp function) and a Form PCA (prcom function) [70]. The Shape PCA only utilizes the scaled and superimposed landmark coordinates, whereas the Form PCA, defined as shape considered with size, uses the logarithmized centroid size added as a variable. To satisfy the assumptions of our PCAs, we estimated Cook's distance for each individual to ensure our dataset did not include outliers or influential specimens, and no canines fell over the recommended cut-off value [75]. Shape changes along the PCs were visualized as landmark configurations at ±2 standard deviations [67].

2.3. Traceological Methods

Traceological analyses were carried out on the complete ornament assemblage in the Material Culture Laboratory (hereafter MCL) at the University of Tübingen. We performed the analysis using three microscopes to identify manufacture traces, perforation techniques, and use-wear traces. While we analyzed the manufacturing traces on broken and complete perforations, observations at higher magnification in search of use-wear traces were only carried out on teeth with intact and well-preserved perforations.

Following well-established protocols for microwear analyses of material culture [76–78], we combined observations at low and high magnification. Macroscopic traces were observed with an Olympus SZX7 stereomicroscope with magnification ranging from 8× to 56× and equipped with an LED ring light source in combination with a Hirox HRX-01, a 3D digital microscope covering a magnification range of 20×–2500×. Observations at higher magnification were performed with an Olympus BX53M metallographic microscope in transmitted light with magnification up to 500×, alongside the aforementioned 3D digital microscope.

Before analysis, the perforations of the best-preserved teeth were cleaned with a cotton swab immersed in demineralized water. For the description of macro and micro traces, we used established nomenclature [78–81]. When describing the location of traces, we used the dial plate system [82], in which specific "hours" on the dial plate correspond to a sector, i.e., sector 1 corresponds to hours 12 to 1, sector 2 to hours 1 to 2, and so on.

We compared the technological and use-wear traces observed on the archeological specimens with a reference collection of 28 perforated fox teeth, which were produced for this study (see Supplementary Materials). Additionally, comparisons were made with previous microwear studies of animal tooth ornaments from other prehistoric and ethnographic contexts [25,82,83,83–88].

3. Results

3.1. Geometric-Morphometric Results

Our novel approach to differentiate red versus arctic fox individuals from loose canines was ultimately very successful. Of the total assemblage, we identified 17 (71%) ornaments as either red (n = 10) or arctic fox (n = 7) individuals. Two ornaments (8%) were classified as indeterminate, and five (21%) were excluded due to damage affecting the visibility of the cervical outline (Table 2). Of the arctic foxes, six correspond to the Aurignacian and four to the Gravettian. Of the red foxes, five correspond to the Aurignacian and two to the Gravettian.

Table 2. Species classification of the archaeological ornaments after geometric-morphometric analyses.

Site	Red Fox	Arctic Fox	Indeterminate	Unsuitable for Analysis
Geißenklösterle (n = 7)	2	3	0	2
Hohle Fels (n = 10)	3	4	2	1
Hohlenstein-Stadel (n = 7)	2	3	0	2
Total (n = 24)	7	10	2	5

The statistical analyses revealed that, while both the Shape and Form PCAs exhibited significant distinction and little overlap between species, the Shape PCA is best suited for this analysis. The Form PCA, which considers the size of the cervical outline, reveals more variability between the archaeological ornaments than the comparative sample.

Error measurements fall under the acceptable 5% deviation between the five repeated measurements (Figure S7). Landmark 2 exhibited the greatest error (1.3%), where the error measurements fell under 0.4%. This may indicate landmark 2 was more susceptible to variations in outline placement in Avizo or in subsequent orientation in Rhinoceros. However, the overall low deviation between samples indicates our method is suitable for fox canines.

3.1.1. Upper Canines

The upper canine assemblage includes the archaeological specimens ID 369, ID 226, ID 1137, ID 964.1, ID 461, ID 1228, and ID 2.0380 alongside the modern comparative upper canines (n = 21).

The Shape PCA (Figure 3a) cumulatively explains 84% of variance in the first two PCs, with PC1 and PC2 explaining 64% and 20% of variance, respectively. Red and arctic fox canines mainly differ across PC1, where arctic foxes plot more positively and red foxes plot more negatively, although a small overlap occurs between 0.00 and 0.01. On PC2, red foxes exhibit a narrower range than arctic foxes and fall mostly within the arctic fox range.

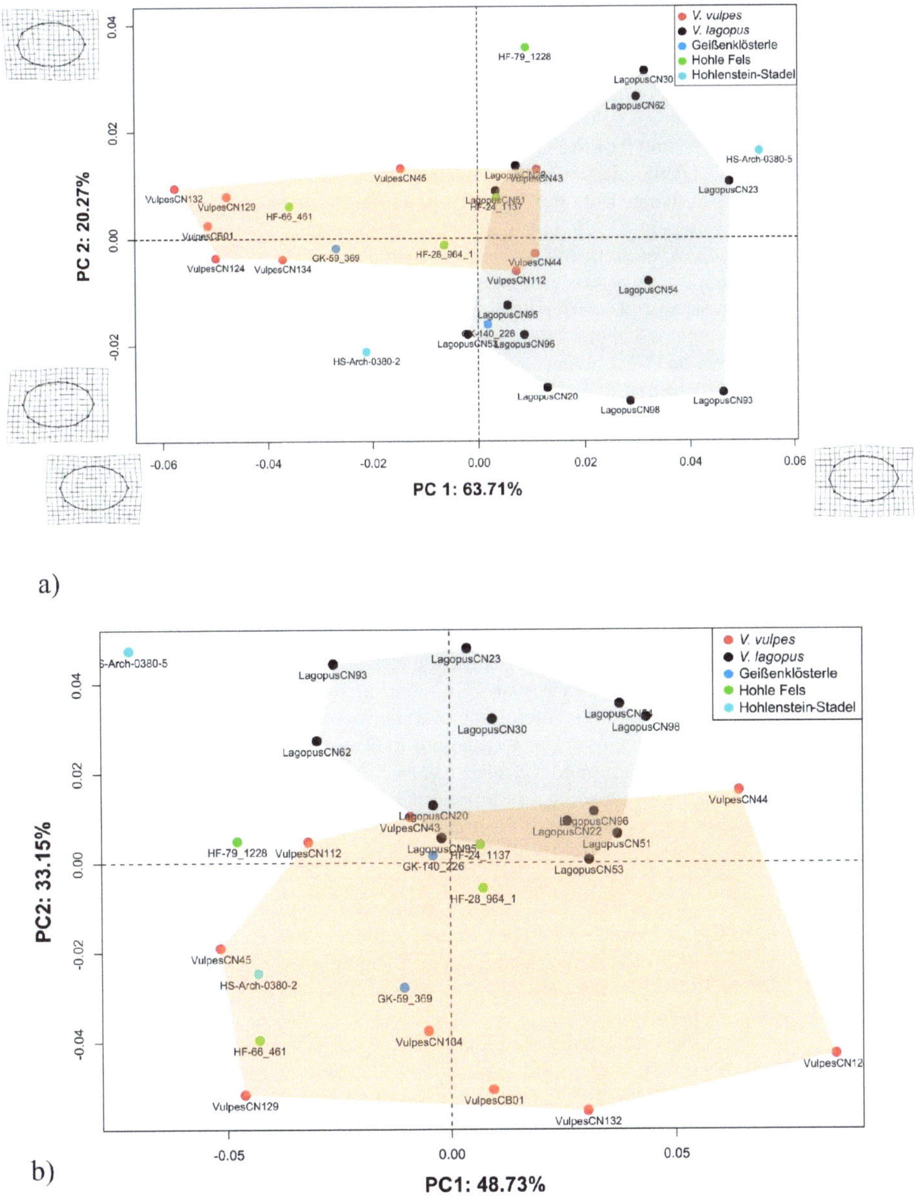

Figure 3. GM: Shape (**a**) and Form (**b**) PCA of upper canines and pendants.

Although most ornaments plot clearly within the arctic or red fox groups, some require further consideration. ID 1137 plots in an area of overlap along both PC1 and PC2, so a clear species assignment was not possible. Three pendants plot outside the distribution of the recent upper canines for both species: ID 1228, ID 2.0380, and ID 5.0380. Although clear species classifications were not possible, ID 1228 is likely an arctic fox, given the more positive PC2 value, and ID 5.0380 is likely also an arctic fox, given the more positive PC1 value.

The Form PCA (Figure 3b) cumulatively explains 9% of variance in the first two PCs, with PC1 and PC2 explaining 49% and 33% of variance, respectively. Similar to the Shape PCA, a distinction between red and arctic foxes is apparent. While red fox presents a broader range than arctic fox in PC1, the primary separation is visible in PC2, where arctic fox plots more positively and red fox plots more negatively. Areas of overlap occur between −0.04 and 0.04 in PC1 and between 0.00 and 0.01 in PC2.

Although size is considered in the Form PCA, species classifications do not fundamentally change from the Shape PCA; however, some specimens require further consideration: Ornament ID 226 plots within the arctic fox range in the Shape PCA but plots slightly outside of the arctic fox and within the red fox range on PC2 in the Form PCA. This indicates that ID 226 can be viewed as a larger arctic fox canine. As in the Shape PCA, ID 1137 plots in an area of overlap between species, inhibiting a clear species assignment. ID 1228 still plots outside of both groups, but within the red fox range on PC1 and in an overlapping area on PC2; however, a clear assignment was not possible. ID 2.380 plots within the red fox distribution on both PC1 and PC2; considering the Shape PCA, this tooth was tentatively identified as a red fox canine. Lastly, ID 5.0380 plots outside of the red and arctic fox values on PC1 and within the arctic fox values on PC2, indicating that the size of this tooth falls outside of the range of recent arctic foxes.

3.1.2. Lower Canines

The lower canine assemblage includes the archaeological specimens ID 1A, ID 1B, ID 918, ID 1278.1, ID 631, ID 1295, ID 1713, ID 1517, ID 1.0380, ID 3.0380, and ID 4.0380 alongside the modern comparative lower canines (n = 22).

The Shape PCA (Figure 4a) cumulatively explains 76% of variance in the first two PCs, with PC1 and PC2 explaining 60% and 16% of variance, respectively. The species are primarily distinguished on PC1, where arctic foxes plot more positively and red foxes more negatively with an area of overlap between −0.01 and 0.01. PC2 exhibits similar ranges between groups and is thus a poor indicator of species.

Most ornaments exhibit clear adherence to either red or arctic fox, although some specimens are more ambiguous. ID 1A and ID 1713 both plot within the red fox group, although overlap with the arctic fox group along PC1 makes this identification tentative. ID 1B and ID 4.0380 both plot outside the distribution of recent fox canines, but their more positive PC1 values indicate these individuals correspond to arctic foxes.

The Form PCA (Figure 4b) mostly confirms the identifications presented in the Shape PCA and explains 84% of variance in the first two PCs, with PC1 and PC2 explaining 59% and 25% of variance, respectively. The greatest species separation occurs along PC2 with arctic fox plotting more positively and red fox more negatively with some overlap between −0.01 and 0.01.

Several ornaments plot outside the defined species groups, indicating the size of cervical outlines varied more in the archaeological sample than the modern sample. In the Shape PCA, ID 1A is tentatively classified as a red fox given its more negative PC2 value, which is supported by the Form PCA's PC1 value. Although ID 1713 plots within the red fox group, the potential for overlap is high, making the classification as red fox tentative. ID 981 overlaps with the red fox range along PC1; however, the location within the arctic fox group of the Shape PCA combined with the more positive PC2 values of the Form PCA suggests a tentative identification as arctic fox. For ID 1B, ID 1295, ID 1.0380, and ID 4.0380, more positive PC2 values indicate they belong to arctic foxes. Lastly, the more negative PC2 values of ID 3.0380 indicates this individual is a red fox.

3.2. Traceological Results

We present the traceological results by site, focusing on root modifications to determine perforation techniques. Additionally, we pay special attention to the identification of potential macro and microwear traces related to the use of the teeth as pendants or as ornaments sewn onto clothes or other goods.

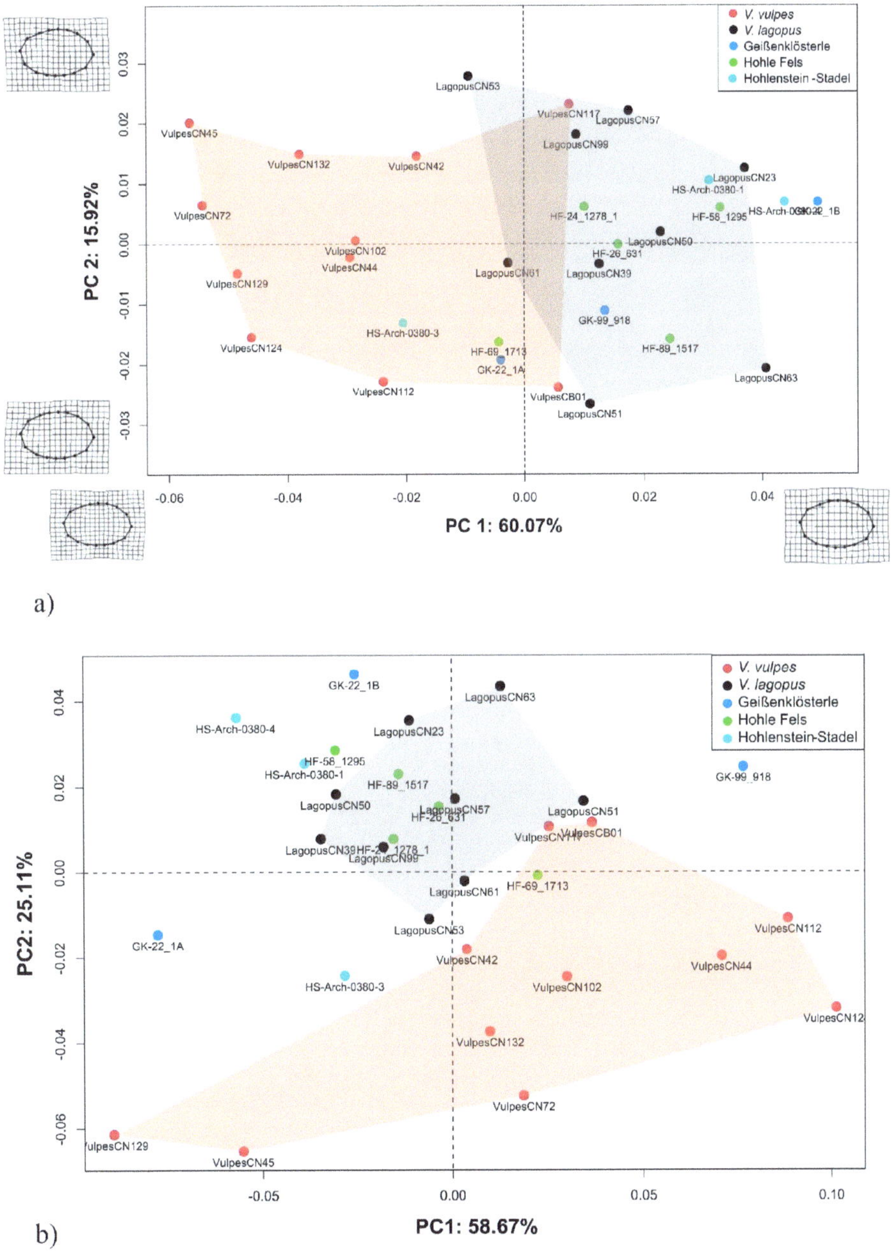

Figure 4. GM: Shape (**a**) and Form (**b**) PCA of lower canines and pendants.

3.2.1. Hohle Fels

Among the 12 fox canines analyzed at Hohle Fels, seven exhibit complete perforations, while three are broken at the tip of the root and one fragment only represents the upper part of the eyelet. The twelfth tooth is incompletely perforated. Overall, 11 teeth were suitable for the macro and microscopic analysis (Table 3).

Table 3. List of archeological fox teeth analyzed in this study from Hohle Fels.

Site	Culture	Find Number (ID)	Tooth	Integrity of Perforation	Technique of Perforation	Use Traces
Hohle Fels	Aurignacian	1517.1	Canine	Complete	Scraping + grooving + drilling	Used: rounding/ smoothing
Hohle Fels	Aurignacian	1137	Canine	Fragmentary	Scraping	Not diagnostic
Hohle Fels	Aurignacian	631.1	Canine	Complete	Probably scraping + widening	Not diagnostic
Hohle Fels	Aurignacian	1713	Canine	Complete	Scraping + grooving + widening	Used: rounding/smoothing/Polish and striae
Hohle Fels	Aurignacian	1278.1	Canine	Complete (broke during analysis)	Scraping + grooving	Used: rounding/ smoothing
Hohle Fels	Aurignacian	1742	Canine	Complete	Drilling	Used: rounding/ smoothing
Hohle Fels	Aurignacian-Gravettian	2099.1	Canine	Complete	Scraping + grooving	Possibly used
Hohle Fels	Gravettian	1295	Canine	Complete	Scraping + grooving+ possibly widening	Used: rounding/smoothing/Polished
Hohle Fels	Gravettian	1228.29	Canine	Perforation unaccomplished	Carving + Scraping + grooving	Not Used
Hohle Fels	Aurignacian-Gravettian	1227.1	Root fragment	Fragmentary	NA	NA
Hohle Fels	Gravettian	461	Canine	Complete	Scraping + drilling	Used: rounding/ smoothing
Hohle Fels	NA	964.1	Canine	Fragmentary	Scraping + grooving	Not diagnostic

Under the microscopes, the assemblage shows different degrees of preservation: most teeth have shiny, damaged surfaces and two exhibit "very poor" preservation. Only two teeth are considered "well preserved".

Manufacturing Traces

At Hohle Fels, we identified two manufacturing techniques: (1) scraping/grooving and (2) drilling.

1. Scraping/grooving: nine teeth show traces of scraping or grooving, which we define as isolated groups of close, narrow striations, and/or deep grooves (Figure 5a–c). Technological deformations in sectors 1 and 6 of five teeth indicate deep gouging for the purpose of creating an opening in the root (Figure 5a). In three cases, we observed specular lateral deformations in sectors 3 and 9, which we interpret as the result of the widening and smoothing of the perforations through a semi-circular motion with a borer or pointed lithic tool (Figure 5c,d and comparison in the Supplementary Materials and Figure S2g–i). This scraping/grooving technique was always performed bifacially and resulted in elongated perforations on the vertical axis of the root (Figure 5a,c). Interestingly, tooth ID 1228 exhibits bilateral whittling/carving traces complemented by scraping, but it was abandoned during early manufacture before achieving a complete perforation, providing some insight into the order of operations (Figure 5e).

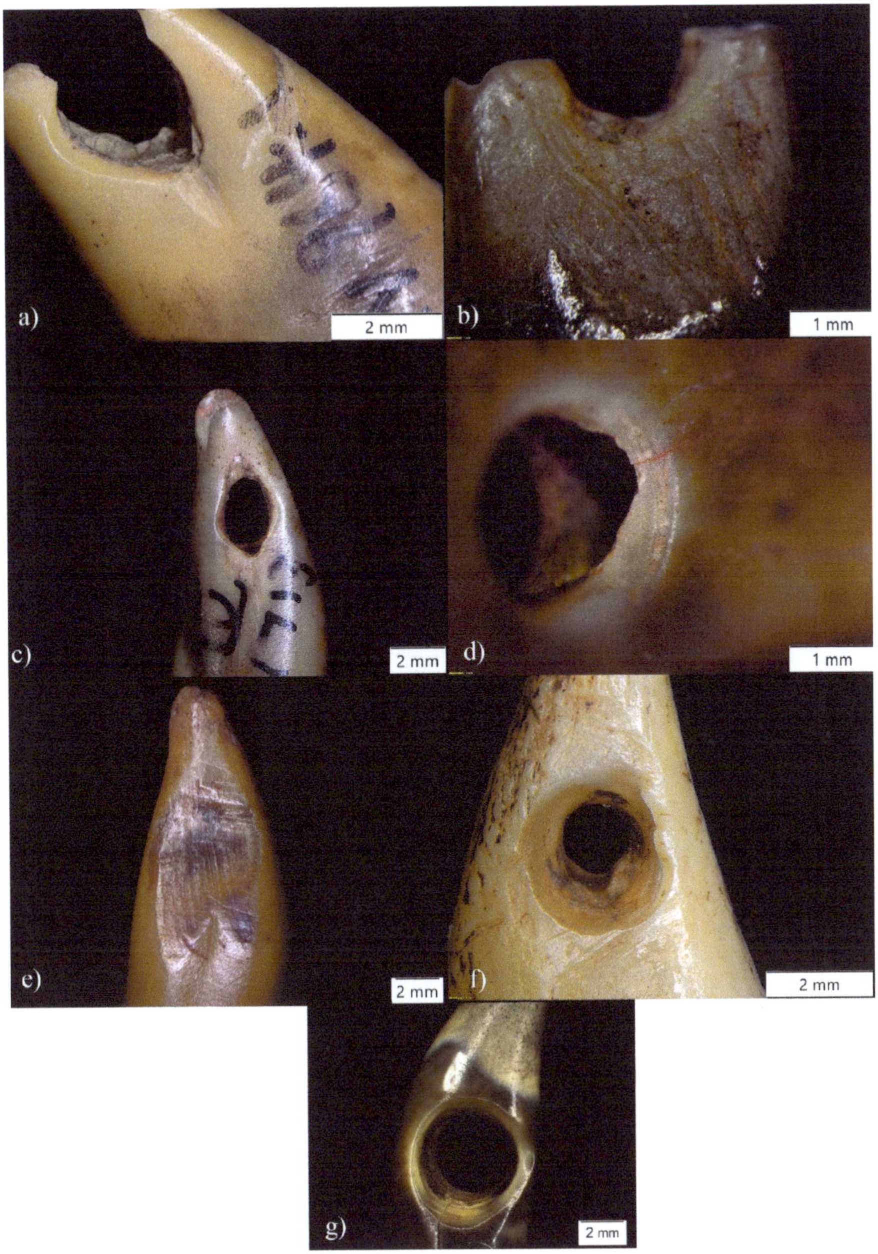

Figure 5. Overview of technological traces recorded on fox teeth from Hohle Fels. (**a**) ID 964, grooving marks; (**b**) ID 1137, scraping and grooving marks; (**c**) ID 1713, scraping and grooving marks and evidence of inner widening of the hole (notice the lateral half-moon enlargement); (**d**) ID 1517.1, circular striations by semi-rotational drilling; (**e**) ID 1228.2, whittling/carving marks and scraping traces; (**f**) ID 461, drilling traces; (**g**) ID 1742, drilling traces.

2. Drilling: Two teeth (ID 1742 and 461) exhibit clear concentric striations on the inner walls of their perforations, characteristic of the rotating action of a borer or a pointed

lithic tool. According to our reference collection, the perforation of tooth ID 461 was created by bifacial free hand drilling (Figure 5f). Tooth ID 1742 was probably bifacially perforated with the use of a bow drill or a hafted hand drill (Figure 5g). The difference between the two techniques is indicated by the high regularity of the drilling traces and the asymmetrical shape of the perforation, which indicates bow drilling. These features may result from the difficulty of controlling a bow drill at high speeds (see for comparison Figure S1h). Only the root surface of ID 461 appears to be prepared by scraping prior to drilling.

Use-Wear Related Traces

All seven teeth with complete perforations showed very round and smooth perforations, specimens ID 1742 and ID 1713 in particular. We found clear microwear traces associated with use on two specimens. On tooth ID 1295 and ID 1713, we recorded smooth and polished areas in sectors 1, 2, and 12 with linear traces in the form of long, shallow, and straight striations running parallel to the longitudinal axis of the root (Figure 6a–d, see Figure S6e for comparison).

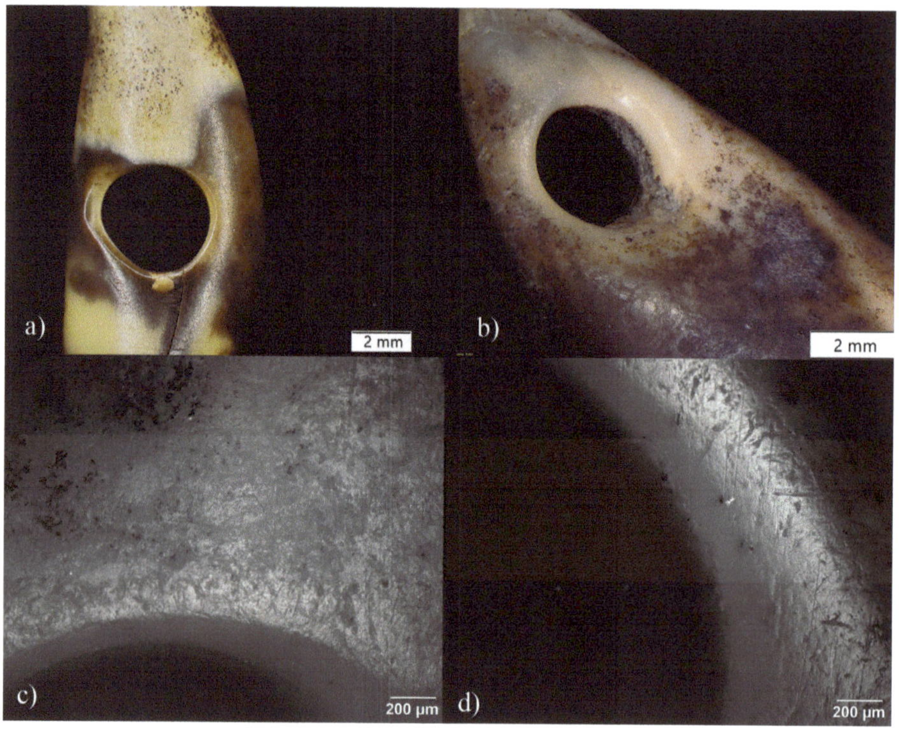

Figure 6. Overview of use-related traces recorded on fox teeth from Hohle Fels. (**a**) ID 1742, high rounding and smoothing; (**b**) ID 1295, rounding and smoothing; (**c**) ID 1295, rough polish developed in sector 12; (**d**) ID 1713, smooth polish with parallel long and narrow striations developed in sector 3.

3.2.2. Geißenklösterle

At Geißenklösterle, fox canines are represented by 12 total specimens divided as follows: five teeth with complete perforations, five with broken perforations, one non-human modified tooth, and another modified tooth with a broken root displaying no perforation (Table 4). The state of preservation is variable: two specimens exhibit very shiny surfaces, one is affected by patina, and three have calcite encrustation around the holes, while the remainder are in medium-good state of preservation.

Table 4. List of archeological fox teeth analyzed in this study from Geißenklösterle.

Site	Culture	Find Number (ID)	Tooth	Integrity of Perforation	Technique of Perforation	Use Traces
Geißenklösterle	Aurignacian	487	Canine	Fragmentary	Scraping + drilling/ lateral carving	Not eligible
Geißenklösterle	Aurignacian	1511	Canine	Fragmentary	Possibly scraped + widening	Not eligible
Geißenklösterle	Aurignacian	369	Canine	Fragmentary	Grooving	Not diagnostic
Geißenklösterle	most likely Gravettian	1A	Canine	Fragmentary	Gouging + widening	Not diagnostic
Geißenklösterle	most likely Gravettian	1B	Canine	Complete	Scraping + drilling	Not diagnostic
Geißenklösterle	Gravettian	391	Canine	Complete	Grooving + Carving	Used: rounding/polish/striae (lateral sides)
Geißenklösterle	Gravettian	918	Canine		Non-human modified	/
Geißenklösterle	Gravettian	658.2	Incisor (?)	Complete	Light scraping + possibly widening	Used: rounding/smoothing/ Polish (apex)
Geißenklösterle	Gravettian	89	Canine	Not preserved	NA	NA
Geißenklösterle	Gravettian	226	Canine	Fragmentary	Scraping + grooving	Not diagnostic
Geißenklösterle	Gravettian	387	Canine	Complete	Possibly scraping + widening	Used: rounding/polish/striae (lateral sides)
Geißenklösterle	Gravettian	286	Canine	Complete	Scraping/grooving + widening	Not diagnostic

Manufacturing Traces

As in Hohle Fels, at Geißenklösterle we identified two main production techniques: (1) scraping/grooving/gouging and (2) drilling.

1. Scraping/grooving/gouging. Scraping and grooving appeared in eight cases, sometimes combined with gouging. In all cases, this technique was performed bifacially. Scraping is represented by a bundle of close and narrow, long and short striations, while grooving is attested by deep isolated long grooves (Figure 7a–c). In the case of ID 391, the perforation was opened by bifacial gouging and superficial carving of the root, whose traces are evident in sectors 12 and 6 (Figure 7d). The perforations associated with this technique are elongated or sub-oval.
2. Drilling. Drilling is attested at Geißenklösterle from only two teeth, both showing traces of scraping to prepare the surface of the root. Clear circular and concentric drilling traces are visible at low and high magnification. The irregularity of the perforations suggests the use of free hand drills (Figure 7e,f). Interestingly, the root of ID 487 is worked in two ways, suggesting two different attachment methods. The root was previously scraped and subsequently drilled in order to obtain a circular hole, which later broke (Figure 7f,g). Underneath the perforation, the root is carved on both lateral sides to create two deep incisions, which might have served to secure a thread (Figure 7f,g). Unfortunately, we could not establish whether the perforation broke ab antiquo or was a result of post-depositional events. Moreover, we were unable to perform observations at higher magnification in search of use-related traces on the carved lateral sides, as the piece could not leave the museum. We thus cannot establish whether the root was carved after the perforation broke to re-facilitate attachment or if it was a stylistic choice.

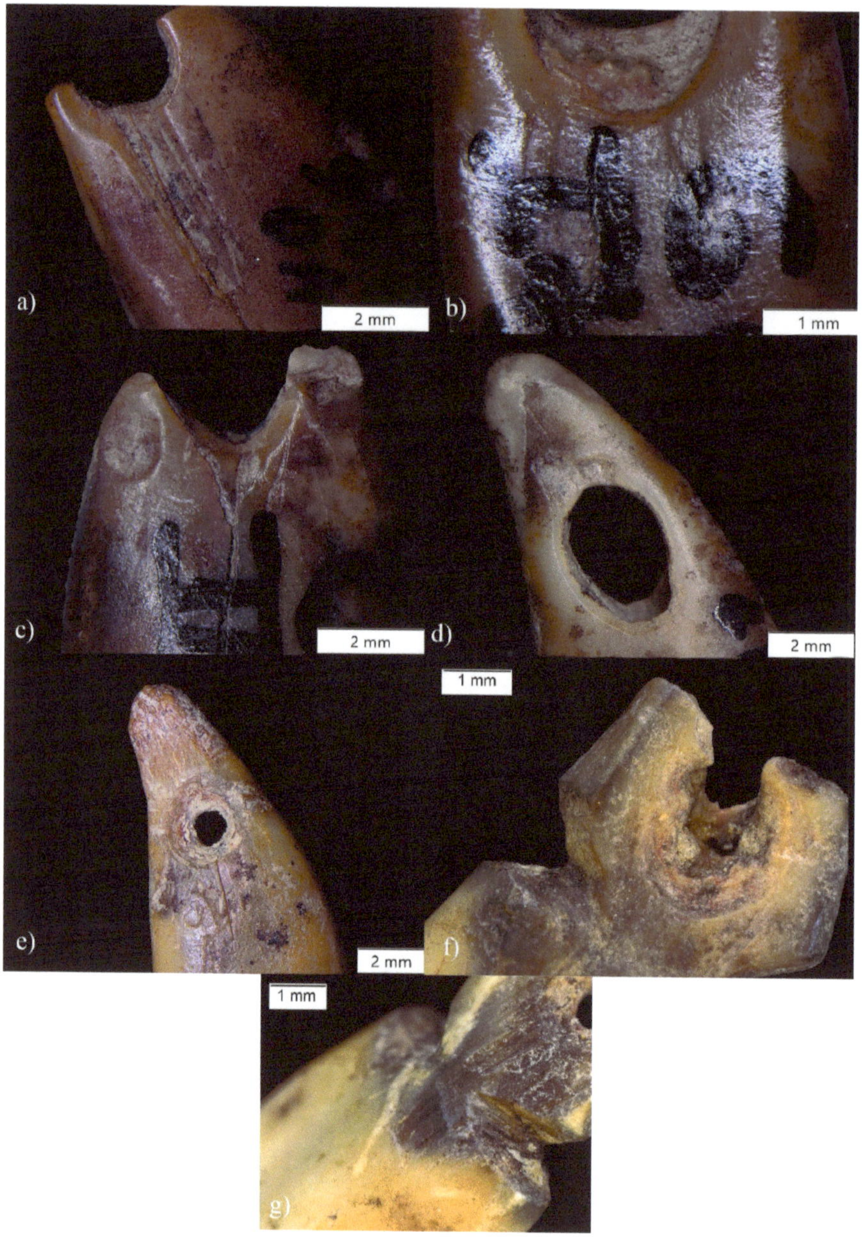

Figure 7. Overview of technological traces recorded on fox teeth from Geißenklösterle. (**a**) ID 226, grooving marks; (**b**) ID 658.2 scraping marks; (**c**) ID 360, grooving mark; (**d**) ID 391, carving traces in sector 12; (**e**) ID 1b, drilling traces; (**f**) ID 487, drilling traces and lateral carved notches; (**g**) ID 487, close up of scraping marks.

Use-Wear Related Traces

Out of four complete perforations, three showed polish and shallow long striations on the internal rim. In particular, ID 658.2 showed a small, polished area in sector 12 with a developed rounding of the internal rim (Figure 8a) while ID 391 and ID 76 exhibited a smooth polish on sectors 3 and 9 associated with striations (Figure 8b,c).

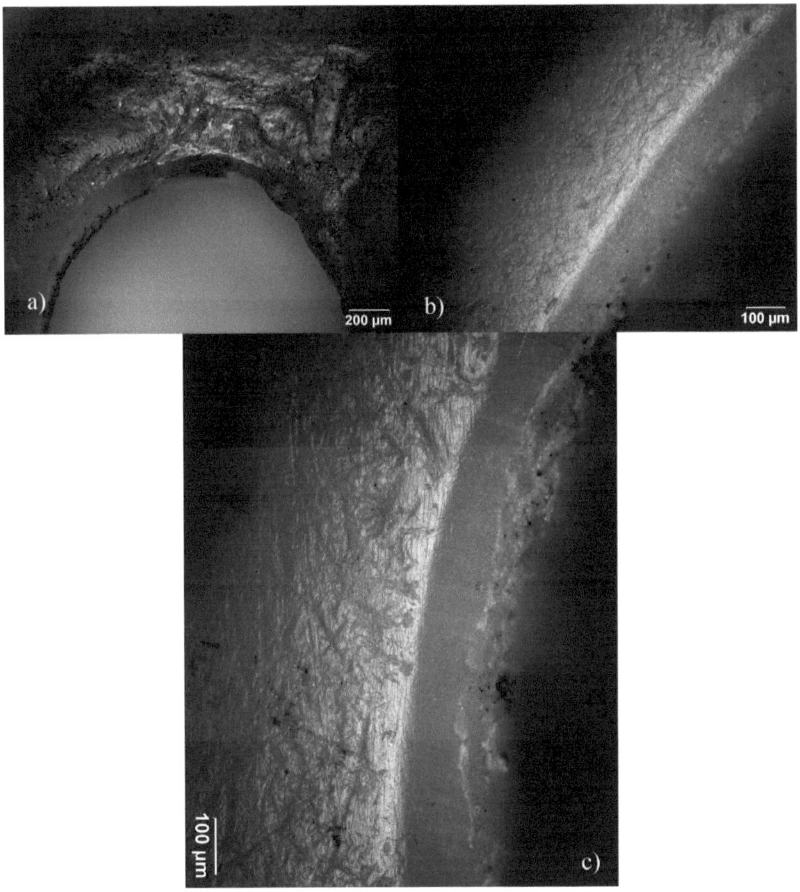

Figure 8. Overview of use-related traces recorded on fox teeth from Geißenklösterle. (**a**) ID 658.2, polish developed in sector 12; (**b**) ID 38F, smooth polish developed in sector 9; (**c**) ID 391, smooth polish with short shallow striations developed in sector 9.

3.2.3. Hohlenstein-Stadel

Perforated fox teeth are represented at Hohlenstein-Stadel by 13 specimens including seven canines, three incisors, and two premolars. A root fragment showing a complete perforation could not be assigned to a specific tooth type (Table 5). The state of preservation of the entire sample is good. Most teeth displayed complete perforations, and only four were broken in the upper part of the eyelet. Unfortunately, 10 teeth showed varnish around the perforations, and some had traces of glue inside the holes, which hampered observation at high magnification.

Table 5. List of archeological fox teeth analyzed in this study from Hohlenstein-Stadel.

Site	Culture	Find Number (ID)	Tooth	Integrity of Perforation	Technique of Perforation	Use Traces
Hohlenstein-Stadel	Aurignacian	1.0380	Canine	Complete	Scraping + widening	Most likely used: rounding
Hohlenstein-Stadel	Aurignacian	2.0380	Canine	Complete	Scraping + carving	Most likely used: rounding
Hohlenstein-Stadel	Aurignacian	3.0380	Canine	Complete	Scraping	Most likely used: rounding
Hohlenstein-Stadel	Aurignacian	4.0380	Canine	Fragmentary	Scraping + grooving	Not diagnostic
Hohlenstein-Stadel	Aurignacian	5.0380	Canine	Complete	Scraping	Not diagnostic
Hohlenstein-Stadel	Aurignacian or Magdalenian	14.1	Canine	Fragmentary	Grooving	Not diagnostic
Hohlenstein-Stadel	Aurignacian or Magdalenian	141.1	Canine	Fragmentary	Scraping + possibly widening	Not diagnostic
Hohlenstein-Stadel	Aurignacian or Magdalenian	167.1	Root fragment	Complete	Possibly abrasion/ scraping + widening	Most likely used: rounding
Hohlenstein-Stadel	Aurignacian or Magdalenian	178.1	Premolar	Complete	Possibly abrasion/ scraping + widening	Most likely used: rounding
Hohlenstein-Stadel	Aurignacian or Magdalenian	33.1	Premolar	Complete	Carving + widening	Most likely used: rounding
Hohlenstein-Stadel	Aurignacian or Magdalenian	36.1	Incisor (?)	Complete	Possibly abrasion/ scraping + widening	Used: rounding/ smoothing
Hohlenstein-Stadel	Aurignacian or Magdalenian	9000.1	Incisor (?)	Complete	Possibly abrasion/ scraping + possibly widening	Most likely used: rounding
Hohlenstein-Stadel	Aurignacian or Magdalenian	9001.1	Incisor (?)	Fragmentary	Possibly abrasion/ scraping + possibly widening	Not diagnostic

Manufacturing Traces

Only one technique was attested at Hohlenstein-Stadel for perforating fox teeth: scraping. All 13 teeth were longitudinally scraped to flatten and thin the root, the marks of this preparation being visible as long and irregular striations (Figure 9a,b,d,e). In several cases, the extension of the striations exceeded the area of the hole, reaching the enamel. Bifacial longitudinal scraping was performed until perforations were achieved, resulting in perforations with an elongated form (Figure 9b,d,e). However, we recognized traces corresponding to the widening of the hole, likely to achieve a more rounded and regular shape (Figure 9c).

It is interesting to notice that the five canines likely associated with the Aurignacian horizon show a common singular technological trait: the upper part of the root is strongly bent, almost to form a step between the end of the perforation and the apex of the root (Figure 9d–f). This was due to the deep and strong scraping resulting in an almost carved surface on the apical side of the perforation, well visible in tooth ID 2.380 (Figure 9e).

Hohlenstein-Stadel is the only site in the Swabian Jura to report perforated fox premolars. Interestingly, two premolars and three incisors show flat facets around the perforation in association with striations, which differ from those on the canines in being wider, more regular, deeper, and in an oblique direction (Figure 9g,h). In those cases, abrasion against a coarse surface might better explain these traces than scraping with a lithic tool. However, they are subsequently worked following the same scheme, namely by flattening the surface until achieving the perforation and later widening the hole with bifacial semi-rotation.

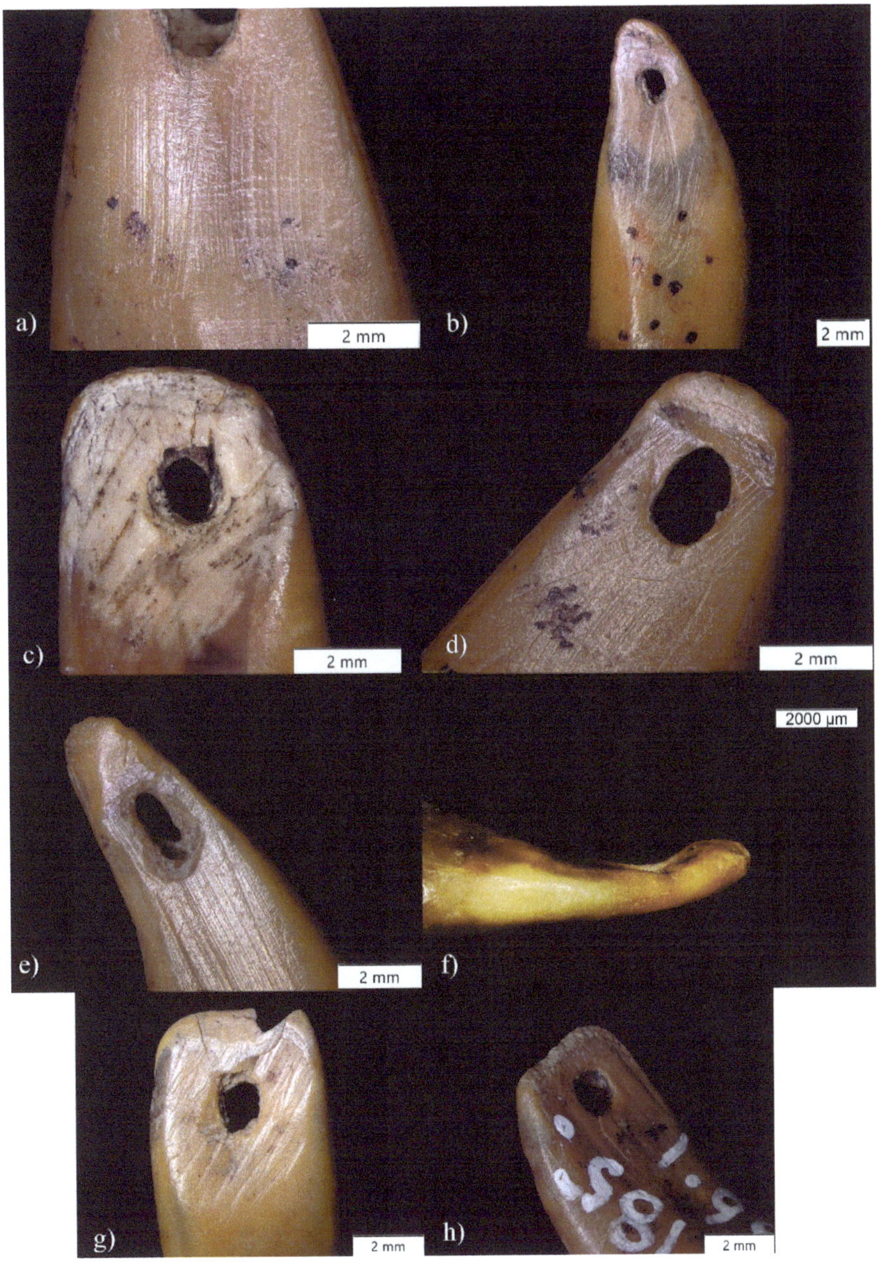

Figure 9. Overview of technological traces recorded on fox teeth from Hohlenstein-Stadel. (**a**) ID 3.380, scraping marks; (**b**) ID 1.380, scraping and grooving marks; (**c**) ID 178.1, widening by semi-rotational drilling; (**d**) ID 3.380, deep scraping marks resulting in a bent upper root; (**e**) ID 2.380, striations and carving marks; (**f**) ID 33.1, deep carving resulting in a bent upper root; (**g**) ID 167.1, abrasive/scraping marks; (**h**) ID 36.1, flat facet and abrasive/scraping marks.

Use-Wear Related Traces

Among the teeth with complete perforations, five showed a well-developed rounding of their internal and external rim, especially localized in sectors 12, 3, and 9 (Figure 10a–d). On five better-preserved teeth subjected to observation at higher magnification, we did not identify any diagnostic use-related microwear. The rest of the assemblage exhibited traces of varnish and glue and, consequently, it was not possible to perform more detailed microscopic observations.

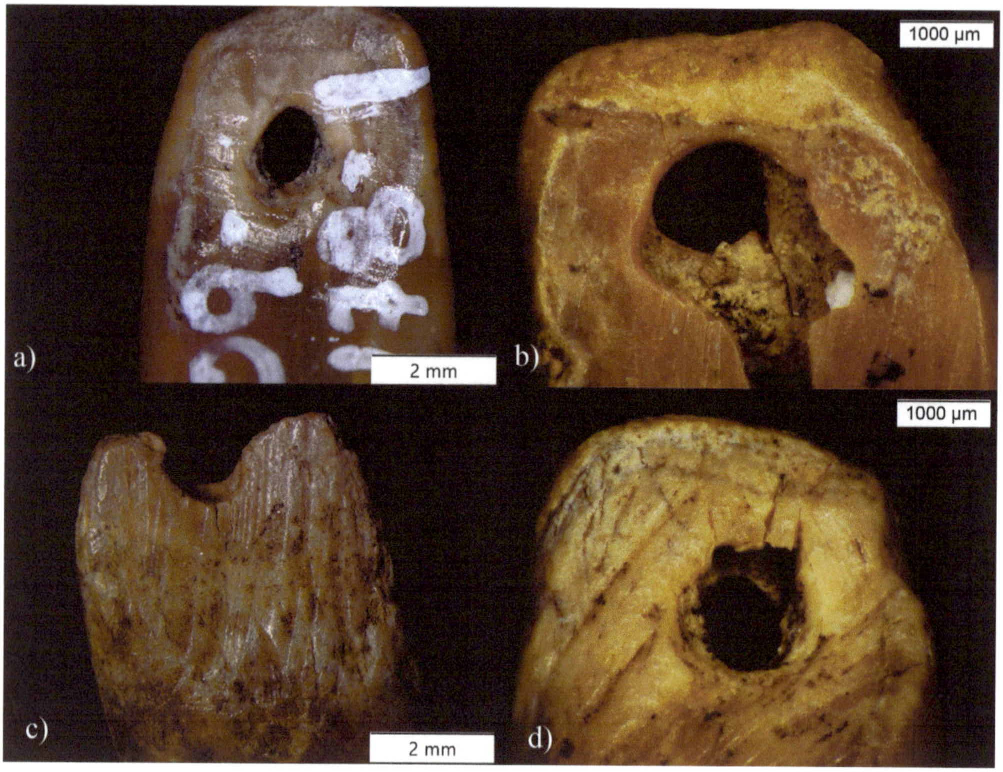

Figure 10. Overview of use-related traces recorded on fox teeth from Hohlenstein-Stadel. (**a**) ID 178.1, rounding and smoothing of the inner and outer rim; (**b**) ID 36.1, rounding and smoothing of the rim; (**c**) ID 9001.1 rounding of the rim; (**d**) ID 178.1, rounding of the inner wall of the perforation.

3.2.4. Brillenhöhle

At Brillenhöhle, three perforated canines come from the Gravettian. Two show complete perforations and good preservation of the surfaces, while one is broken on the apical side of the root (Table 6). We could only make macroscopic observations of these teeth and were unable to perform observations at higher magnification because the specimens could not leave the museum.

Manufacturing Traces

The perforations of all three ornaments were achieved by scraping/grooving. However, scraping was not performed as a preliminary thinning treatment, but rather to directly achieve the perforation. On ID 130, the remaining lower part of the perforation in sector 6 is V-shaped, suggesting an original elongated hole made by grooving/carving (Figure 11a). Specimen ID 132 exhibited unifacial short grooves and a depression on the upper part of the hole produced by deeply grooving the root. On the other face, the perforation showed traces of widening for regularizing the shape of the hole (Figure 11b). Tooth ID 131 carries a few technological traces and exhibits a round and smooth perforation. Due to the presence of encrusted soil particles on the inner wall, it was not possible to ascertain if the perforation was made by drilling or by semi-rotation for widening the hole. No clear rotational traces are visible even if the roundness of the hole suggests a complete rotational movement. A depression in the upper part of the root might suggest a scraping/grooving activity (Figure 11c).

Table 6. List of archeological fox teeth analyzed in this study from Brillenhöhle.

Site	Culture	Find Number (ID)	Tooth	Integrity of Perforation	Technique of Perforation	Use Traces on Perforation
Brillenhöhle	Gravettian	132	Canine	Complete	Grooving/carving + widening	Most likely used: rounding/smoothing
Brillenhöhle	Gravettian	131	Canine	Complete	Scraping/grooving + widening or drilling	Most likely used: rounding/smoothing
Brillenhöhle	Gravettian	130	Canine	Fragmentary	Scraping + grooving	Not diagnostic

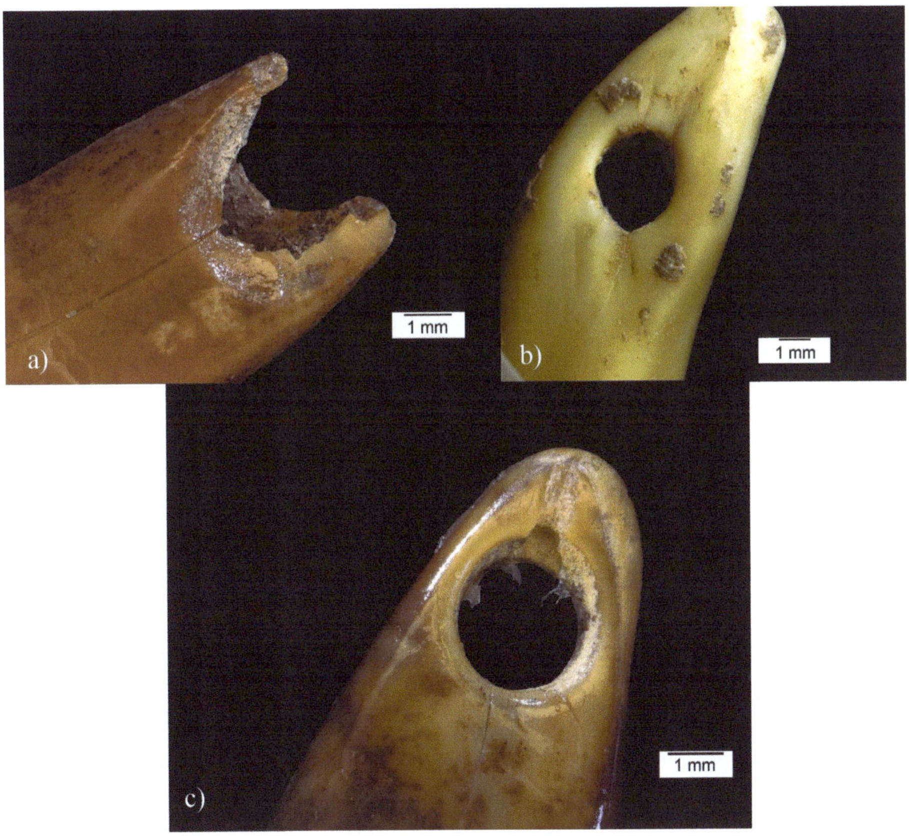

Figure 11. Overview of technological traces recorded on fox teeth from Brillenhöhle. (**a**) ID 130, carving marks, (**b**) ID 132, grooving marks; (**c**) ID 131, grooving mark in the upper perforation.

4. Discussion

The use of animal teeth as ornaments during the Upper Paleolithic provides valuable insights into the creativity, symbolism, and cultural practices of our ancestors, giving us a glimpse into their way of life and how they interacted with the natural and animal world. Indeed, the use of animal remains as a source of food, as well as a raw material (e.g., teeth for ornaments, fur for clothes, bones for tools), evidences an intimate relationship between humans and animals. This is particularly true for animals like foxes, which exhibit behaviors showing their charisma, adaptiveness, and ability to influence their surroundings and shape human communities. The human–fox relationship—today, as in the past—is often characterized by the commensal behavior of foxes and their coexistence with humans and their ability to thrive in human-constructed environments [38,44,54,55,89–91]. The result of this relationship is still observed in urban and suburban environments today, where foxes often live in close proximity to humans [92,93].

This human–fox relationship also existed in prehistory, as evidenced by the frequent occurrence of fox bones and teeth in the archaeological record, many of which were used to produce ornaments [55,94]. Although foxes are not the most common taxa in the archeological record of the Swabian Jura [55], their perforated teeth are a particularly informative class of artifacts with the potential to shed light on the role foxes played in past human societies.

4.1. Production of Fox Tooth Ornaments in the Swabian Jura

The results presented in this study show that canines were the preferred tooth type for ornamentation made from foxes in the Swabian Jura. In our assemblage, Hohlenstein-Stadel exhibits ornaments made from other teeth, namely incisors (one of which is also present at Geißenklösterle) and premolars, albeit in lesser quantities than canines. Based on the available published data, only a handful of other sites share this variability in tooth choice, one of which is the Magdalenian site of Petersfels (Germany), a few hundred kilometers south of the Ach and Lone valleys in the Swabian Jura (see Table S1) [95–97]. Given the uncertain context of some of the Hohlenstein-Stadel ornaments as either Aurignacian or Magdalenian, it is possible that the use of premolars and incisors indicates a Magdalenian origin for these five ornaments, like those represented at Petersfels.

From a traceological point of view, we noticed that five Hohlenstein-Stadel canines (Aurignacian) were processed by scraping with a lithic tool, while the two premolars and three incisors (Aurignacian or Magdalenian) were possibly abraded to achieve the perforations. This difference might suggest a deliberate cultural choice in technique. However, the other two canines from the uncertain context also show traces in line with the five Aurignacian canines, suggesting the chosen technique was more dependent on tooth type and likely morphology rather than a cultural or individual preference. Tooth morphology also influences the practical aspects of perforation. Canines, for example, are larger with a greater root surface area than other teeth, making ornament production easier and perhaps providing a specific aesthetic function regarding their shape, size, and luster.

Despite the likely morphological influence on production technique, some of the Hohlenstein-Stadel ornaments may still belong to the Magdalenian, given their uncertain stratigraphic context. However, a cultural explanation for tooth preference is complicated by the broader record of fox tooth ornaments throughout the Upper Paleolithic. Ornaments made from fox incisors and premolars are found at numerous sites across Europe in the Magdalenian but also the Gravettian [34,86,94–96,98,99] (Table S1). Retracing the context of the Hohlenstein-Stadel ornaments, as well as traceological comparison with other ornament assemblages, may thus illuminate the potential hunter-gatherer networks that influenced tooth choice in the Swabian Jura.

Regarding the perforation techniques, our four sites show a common trend: the preferred technique was bifacial scraping and grooving combined with the widening of the hole to regularize the perforation. In most of the analyzed teeth, the scraping and grooving technique was not performed to thin the root as a surface preparation, but rather to make the perforation itself. Only two specimens (one from Hohle Fels and one from Geißenklösterle) were instead scraped before being rotationally drilled. In one case, we recorded deep gouging around the perforation (ID 2.0380), while in a few others, the root was scraped and carved producing a hollowed facet around the hole (ID 1228, ID 33-1).

According to our experimental reproductions, scraping and grooving was an efficient method of perforation that did not require any prior surface treatment. For teeth from larger animals with thicker roots, this might not be the case. By analyzing perforated teeth from other animal species at Hohle Fels and Geißenklösterle, we noticed that horse incisors always exhibit a scraped surface before the perforation is achieved by drilling (F.V., personal communication). This reflects the need to adjust perforation techniques according to the constraints of the raw material.

In our assemblage, we did not notice any correlation between a specific technique with a particular cultural horizon or site. This might reflect our singular focus on fox teeth. Comparing the perforated teeth from different animal species at these sites (for an overview of numbers and percentages see Table S6) may highlight variations in techniques revealing a tooth functional-dependent choice, rather than cultural or individual preferences. However, it is worth noting that the five canines assigned to the Aurignacian at Hohlenstein-Stadel show the same technological trait of bent upper roots (Figure 9d,f). This signature is a distinct marker for the site and strengthens the attribution of the teeth to the same Aurignacian horizon. Moreover, we suggest here the possibility that these five teeth were worked by the same person using a distinctive and repetitive technical gesture, which was not recorded at the other sites.

Another aspect to discuss is if and how the teeth were worn. Microtraces on teeth are always very ambiguous because of their naturally shiny surfaces and animal life history, especially considering the various post-depositional processes that can also impact preservation. However, macro traces such as rounding and smoothing localized on the inner and outer rim of perforations are a clear marker of use. More difficult is to reconstruct when teeth served as hanging pendants or when they were sewn to clothes or other objects. In our assemblage, we considered all well-preserved teeth with complete perforations to be most likely used, based on the presence of developed and localized rounding and smoothing of the perforations (Tables 3–6). For the teeth exhibiting use-related polishes and striations, we do not feel confident in assigning a specific attachment mode or binding method. However, the presence of striations recorded on some specimens may be related to the use of abrasive strings like nettle or other rough vegetal/animal materials. Strings treated with colorant such as ochre can also develop smoother polish and bundles of striations. Indeed, we recorded traces of red pigment in seven teeth from Hohle Fels and one from Geißenklösterle. Other than ID 1713 (Figure 5c), which shows reddish traces inside the inner rim in association with rounding, polish, and striations likely from ochred strings, we interpreted most red coloration as post-depositional, given its presence within recent fractures, in the cavity of the root, and/or mixed with the adhering sediment.

4.2. Selection of Red versus Arctic Fox

The relationship between humans and a specific fox species—red or arctic—is often impossible to determine in the archaeological record, given the nondiagnostic nature of isolated faunal elements. Differentiation in loose teeth is particularly tricky (see Supplementary Materials). Although some teeth are distinguishable based on size, canines are more ambiguous [100], and prior attempts at metric differentiation [101] exhibit notable flaws that decrease the reliability of species determinations [39,102–105]. For this reason, a geometric morphometric approach is advantageous and a potential improvement on existing morphological and metric methods.

With the help of our modern comparative sample, we successfully determined the species of 17 (71%) ornaments from our archaeological sample. Although some ornaments plotted outside the recent fox ranges, their position within the PCAs still shows tendencies toward a specific classification. The larger range of the archaeological sample may indicate noise in landmark placements, which is expected in archaeological specimens due to post-depositional damage resulting in alterations to the cervical outline. Other factors, such as variation between individuals, sexes, and populations also impact crown heights.

Our results show that both red and arctic foxes were used to produce ornaments in the Swabian Jura. On a site level, there is no clear preference for species; however, the total assemblage includes slightly more arctic than red foxes. Between cultural periods, there is also no strong preference for species—the Aurignacian is represented by five red foxes and six arctic foxes, while the Gravettian includes two red foxes and four arctic foxes.

Concerning the broader Upper Paleolithic record of fox tooth ornaments, few studies provide species-specific identifications. Our literature review revealed limited information from Aurignacian assemblages, but more from the Gravettian and the Magdalenian. In the Gravettian, two notable sites report high numbers of perforated arctic fox teeth: 42 at the Czech site of Dolní Věstonice I (alongside 31 indeterminate) and 41 at the Russian site of Zaraysk (Table S1) [99,106,107]. For the Magdalenian, 16 perforated arctic fox teeth are reported from Petersfels (alongside 25 indeterminate), at least 5 come from Kniegrotte [95–97], and 162 were recovered from Wilczyce in Poland (Table S1) [86]. Few sites report perforated red fox canines, with Dolni Věstonice I being the only clear example from our literature review [99,106,107].

Despite the overwhelming abundance of perforated arctic fox canines, both arctic and red foxes were available to hunter-gatherers across Europe. In the Swabian Jura, red and arctic foxes were even found within the same layers at sites like Geißenklösterle [39]. Although both species were likely sympatric during the Upper Paleolithic (see Supplementary Materials), their abundance may have varied. In modern sympatric contexts, arctic foxes are observed in smaller numbers [108–111]; if the same were true for the Paleolithic, hunter-gatherers would have encountered red foxes more frequently. As acquisition methods like traps and snares were non-selective, the ratio of captured animals would thus reflect this natural abundance. However, setting traps at known fox dens, perhaps dens used by a single species, may have enabled selective acquisition (see Supplementary Materials).

It is thus possible that the relative scarcity of arctic foxes on the landscape increased their value as a motif for ornamentation and influenced their preferential exploitation across Europe. However, local paleoenvironments and faunal assemblages should be considered. Although foxes were likely sympatric in a broad sense, our modern understanding of interspecific fox relationships may be a poor analog for the Paleolithic. Additionally, climate and environmental conditions varied significantly throughout this period, sometimes within a short timeframe, potentially influencing the degree of species overlap. The resolution of the archaeological record may also limit our ability to parse moments in time when conditions were more or less hospitable to a particular species. A closer look at the ratio of red and arctic foxes in the broader faunal assemblages would further illuminate their nuanced relationship within the local paleoenvironment. Biomolecular techniques, such as aDNA and ZooMS [112,113], would also increase the reliability of these distinctions.

In the Swabian Jura, some information on the ratio of foxes in the broader faunal assemblages is available [39,46,51,55]. As with the ornaments, there is no overwhelming preference for arctic versus red fox. This near-equal distribution suggests the ornament ratio likely reflects the hunted ratio and perhaps the ratio of animals on the landscape. Butchery traces indicate foxes were used as food and for their fur. Concerning the former, other species, namely ungulates, provided a far greater contribution to hunter-gatherer diet. However, fox fur may have served a unique purpose. Arctic foxes are uniquely valued amongst modern arctic hunter-gatherer populations for their small size and bright, white color [114,115]. This then may be a convincing argument for the relative overabundance of arctic fox canines elsewhere in Europe—an aesthetic or functional preference for their fur.

A final consideration is the method of identification. Arctic fox canines, due to their small size, are far more recognizable with simple morphological and metric characteristics. Those in the area of overlap may go unidentified, or are identified as arctic fox by default, and few specimens are large enough to be confidently identified as red fox. Unfortunately, few studies report their methods of species determination and the commonly applied metric method [101] may be biased towards arctic foxes [39]. Future application of geometric-morphometric methods, like those applied in this study, would help to clarify this point and allow further investigation of the other factors influencing red versus arctic fox abundance in the Upper Paleolithic.

4.3. Fox Tooth Ornaments in the Broader European Context

We report the data from our literature review of fox tooth ornaments across the European Upper Paleolithic in Table S1. These ornaments first appear in small numbers during the Châtelperronian at sites such as La Grotte du Renne [20,116]. Increasing in the Aurignacian, fox tooth ornaments dominate various German, French, and Russian sites [10]. The richest French sites are those of Isturitz, Saint-Jean-de-Verges, and La Quina, showing 13, 14, and 28 perforated canines, respectively [25,117,118]. In Russia, Kostenki hosts at least 37 pierced fox teeth from the Initial and Early Upper Paleolithic layers, the richest for the Aurignacian [119]. Comprehensive studies on the investigation of perforation techniques are generally rare, but at Isturitz fox canines were first thinned by fine abrasion and then bifacially scraped to open a perforation. Deep carving of the root followed by scraping and bifacial semi-rotational perforation is also attested [25].

Foxes continued to be an important ornament resource during the Gravettian, especially in Central and Eastern Europe. Hunter-gatherers also began to exploit a greater variety of tooth types, including incisors and premolars in addition to canines. In Czechia, two sites in Moravia recorded a high number of fox teeth used as pendants. At Pavlov I, a total of 284 perforated fox teeth were discovered, among which 146 are incisors, 105 are canines, and 33 are premolars [94]. The nearby site of Dolní Věstonice I reports 73 perforated teeth, of which 42 are canines [106,107]. Moving eastward, in Russia, the site of Zaraysk exhibits only 32 incisors and 9 premolars for a total of 41 teeth, all assigned to arctic fox and found together as a "necklace" [99]. The funerary site of Sunghir in Russia offers the best glimpse into the complex worldview of some of the first anatomically modern humans in Europe. Within the several burials, a large number of fox tooth ornaments (>300) were discovered in situ and in close association with human remains, offering a snapshot of how personal ornaments were worn by this group of hunter-gatherers [17,120]. Numerous perforated fox teeth were found on the forehead of an adult male (Sunghir 1), >40 on the cranial vault and >250 on the belt of another individual (Sunghir 2), one in association with an isolated cranium (Sunghir 5), and at least two found in association with a final individual (Sunghir 10). In addition to heavy adornment, the individuals were also covered with a large quantity of red ochre and other grave goods [17].

Foxes are also well-represented in Magdalenian ornament assemblages from Germany and France with 41 perforated teeth from Petersfels [95–97] and 45 from Gönnersdorf in Germany [31,121,122] as well as at least 35 from the Magdalenian layers of Isturitz [85] and La Vache [123], 12 from La Madeleine [124], and 13 from Bruniquel [125,126] in France. Although canines were the preferred tooth type, Petersfels also exhibits several incisors and premolars. With 162 fox teeth, the Polish site of Wilczyce is the richest with 149 pierced incisors alongside 13 canines. These were all acquired from the upper jaws of a minimum of 31 arctic foxes and may represent the remains of a necklace associated with a neonatal burial [86].

These data show that the Upper Paleolithic sites of the Swabian Jura are in line with the general trends observed across Europe concerning the use of fox teeth as ornaments. Their exploitation does not represent a unique signature or a key cultural marker for the region, but a shared and well-established behavior amongst European Upper Paleolithic groups. Canines are the preferred tooth type in both the Swabian Jura and the whole of Europe, while incisors and premolars are less frequently used, although their presence might represent a cultural preference or individual choice within certain groups of hunter-gatherers. The techniques of perforation used in the analyzed sites compare well with other European sites; although, scraping seems more common to achieve the perforation itself rather than as a surface preparation method. The ubiquitous presence of perforated fox teeth throughout the Upper Paleolithic in Central Europe attests to the important role and intimate relationship this animal established with prehistoric human communities. Although rare, evidence of fox butchery in the Swabian Jura suggests they were hunted for food, for their furs, and as a raw material for ornamentation. The diversity of resources provided by foxes along with their commensal behavior likely granted them a special significance in the ecological, dietary, functional, and symbolic realms of Upper Paleolithic hunter-gatherers.

5. Conclusions

Our results provide insight into the role of foxes during the Upper Paleolithic, especially regarding human subsistence, cultural expression, and ornament production.

Focusing on the latter, we cannot overlook the fact that foxes were one of many species exploited for ornamentation in the Swabian Jura. Other taxa whose teeth were used as ornaments include various herbivores like horses, reindeer and red deer, as well as carnivores such as wolves, bears, and lions (Table S6). However, like the other species, foxes had their own unique relationship with humans. As previously discussed, their opportunistic nature likely drew foxes to human food waste, bringing both species closer together in commensal relationships, which are attested to throughout the Upper Paleolithic. Once connected, foxes played at least three key roles: as a resource, as a symbol, and as a companion animal [127]. The latter two are less tangible in the archaeological record, while study of material culture derived from foxes allows us to illuminate the economic and symbolic significance of foxes in the Upper Paleolithic.

Although butchery traces on fox remains are rare, cut marks suggest foxes were likely consumed as food in the Swabian Jura. Other marks indicate their use as a raw material, particularly for their fur, a valuable resource for clothing and other sewn goods [53,55]. The striking color and waterproof quality of their pelts likely played a role in the selection and use of foxes for this resource [121]. While direct evidence is lacking, fox bones may have also been used for tools (e.g., awls). Otherwise, fox tooth ornaments are the best example of the use of foxes as a raw material.

The study of fox tooth ornaments helps us to unravel the symbolic importance of foxes among Upper Paleolithic hunter-gatherers. This is because ornaments are social objects, capable of reflecting otherwise unsaid aspects of the wearer, conveying messages on ethnic, gender, social, and/or personal identity [14,15,128,129]. We can thus assume foxes, both as a raw material and as a species, had a symbolic value in the lives of the hunter-gatherers who wore their teeth.

Despite their clear symbolic importance, it is surprising that foxes are rarely depicted in parietal or figurative art, other than a cave engraving at the Aurignacian site of Altxerri B Cave in Spain [130]. In comparison, ungulates and larger predators, such as lions or bears, are regularly depicted. This appears to change, however, during the Pre-Pottery Neolithic and Natufian, when foxes increase in abundance, are commonly depicted in relief (e.g., at Göbekli Tepe, Turkey) [131], and are buried alongside humans (e.g., at Uyun al-Hammam, Jordan) [132]. A continued strengthening of the human–fox relationship is also evidenced in the Neolithic [131,133], with the first possible evidence of fox domestication in the Early-Middle Bronze Age [134].

To conclude, there is no doubt that the relationship between humans and foxes was significant and started as early as the Initial Upper Paleolithic. The effect of this long-lasting link still persists today in the folklore and mythology of many cultures, which depict foxes as beautiful, clever, cunning, and magical beings [135,136].

Supplementary Materials: The following supporting information can be downloaded at: https://www.mdpi.com/article/10.3390/quat6030050/s1, Figure S1: Overview of experimental technological traces by drilling, Figure S2: Overview of experimental technological traces by scraping/grooving, Figure S3: Overview of experimental technological traces by scraping or abrading and drilling, Figure S4: Example of various ornament replicas and modes of attachment during the experimentation, Figure S5: Pre- and post-use macrographs of perforations, Figure S6: Micrographs of polish recorded on the perforations and corresponding location, Figure S7: Error margins for the 16 landmarks placed along the cervical outline, Table S1: Summary of the literature review on Upper Paleolithic fox tooth ornaments, Table S2: List of archaeological fox canines included in the geometric-morphometric analysis, Table S3: List of modern fox canines included in the geometric-morphometric analysis, Table S4: Experimental fox canine replicas and related technological results, Table S5: Experimental fox canine replicas with wear information and related recorded traces, Table S6: Perforated animal teeth from the Swabian Jura (Hohle Fels, Geißenklösterle, Brillenhöhle, Bockstein, Hohlenstein-Stadel, and Vogelherd). Ornaments unidentifiable to species and ornaments from uncertain stratigraphic contexts are not included [137–248].

Author Contributions: Conceptualization, F.V., M.J.M. and S.W.; Methodology, F.V. and M.-L.O.; Formal analysis, F.V. and M.-L.O.; Investigation, F.V. and M.-L.O.; Data curation, F.V., M.J.M. and M.-L.O.; Writing—original draft preparation, F.V., M.J.M. and M.-L.O.; Review and editing, F.V., M.J.M., N.J.C., M.-L.O. and S.W.; Visualization, F.V. and M.-L.O.; Project administration, F.V., M.J.M. and S.W.; Funding acquisition, S.W.; Resources, N.J.C. and S.W. All authors have read and agreed to the published version of the manuscript.

Funding: This research was funded by the Senckenberg Center for Human Evolution and Palaeoenvironment at the University of Tübingen.

Data Availability Statement: All relevant data are within the manuscript and its Supplementary Materials files.

Acknowledgments: We are grateful to Ran Barkai for inviting us to contribute to this Special Issue. We thank Chris Baumann for providing the recent fox canines from his private collection, Alexandros F. Karakostis and Gabriel Ferreira for their support during CT-scans, and Julia Zastrow for her help during the GM-data collection. We would also like to thank Viviane Bolin and Nadine Rabovsky from Museum Ulm, Fabian Haack from Landesmuseum Württemberg Stuttgart and the Archäologisches Landesmuseum Baden-Württemberg, Zentrales Fundarchiv Rastatt for their assistance and for providing access to the teeth under curation analyzed in this work. Thanks are also due to our colleague Rudolf Walter for his active contribution to the experimental program. We are finally grateful to the three anonymous reviewers for their constructive comments which contributed to improve this manuscript.

Conflicts of Interest: The authors declare no conflict of interest.

References

1. Riek, G. *Die Eiszeitjägerstation Am Vogelherd Im Lonetal*; Akademische Verlagsbuchhandlung Franz F. Heine: Tübingen, Germany, 1934.
2. Hahn, J. *Kraft Und Aggression: Die Botschaft Der Eiszeitkunst Im Aurignacien Süddeutschlands?* Archaeologica Venatoria: Tübingen, Germany, 1986; Volume 7, ISBN 978-3-921618-24-0.
3. Conard, N.J. Palaeolithic Ivory Sculptures from Southwestern Germany and the Origins of Figurative Art. *Nature* **2003**, *426*, 830–832. [CrossRef]
4. Conard, N.J. A Female Figurine from the Basal Aurignacian of Hohle Fels Cave in Southwestern Germany. *Nature* **2009**, *459*, 248–252. [CrossRef] [PubMed]
5. Kölbl, S.; Conard, N.J. (Eds.) *Eiszeitschmuck-Status Und Schönheit*; Museumsheft; Urgeschichtliches Museum Blaubeuren: Blaubeuren, Germany, 2003; ISBN 1617-2655.
6. Conard, N.J.; Malina, M. Schmuck Und Vielleicht Auch Musik Am Vogelherd Bei Niederstotzingen-Stetten Ob Lontal, Kreis Heidenheim. *Archäol. Ausgrabungen Baden-Württemberg 2005* **2006**, 21–25.
7. Conard, N.J.; Malina, M.; Münzel, S.C. New Flutes Document the Earliest Musical Tradition in Southwestern Germany. *Nature* **2009**, *460*, 737–740. [CrossRef] [PubMed]
8. Wolf, S. *Schmuckstücke: Die Elfenbeinbearbeitung Im Schwäbischen Aurignacien*; Tübingen Monographien zur Urgeschichte; Kerns Verlag: Tübingen, Germany, 2015; ISBN 978-3-935751-21-6.
9. Dutkiewicz, E. *Zeichen: Markierungen, Muster Und Symbole Im Schwäbischen Aurignacien*; Tübingen Kerns Verlag: Tübingen, Germany, 2021; ISBN 3-935751-34-6.
10. White, R. Technological and Social Dimensions of "Aurignacian-Age" Body Ornaments Across Europe. In *Before Lascaux: The Complex Record of the Early Upper Paleolithic*; Knecht, H., Pike-Tay, A., White, R., Eds.; CRC Press: Boca Raton, FL, USA, 1993; pp. 277–300.

11. White, R. Systems of Personal Ornamentation in the Early Upper Palaeolithic: Methodological Challenges and New Observations. In *Rethinking the Human Revolution: New Behavioural and Biological Perspectives on the Origin and Dispersal of Modern Humans*; McDonald Institute Monographs; Mellars, P., Boyle, K., Bar-Yosef, O., Stringer, C., Eds.; McDonald Institute for Archaeological Research: Cambridge, UK, 2007; pp. 287–302, ISBN 1-902937-46-5.
12. McBrearty, S.; Brooks, A.S. The Revolution That Wasn't: A New Interpretation of the Origin of Modern Human Behavior. *J. Hum. Evol.* **2000**, *39*, 453–563. [CrossRef]
13. Vanhaeren, M.; d'Errico, F. Aurignacian Ethno-Linguistic Geography of Europe Revealed by Personal Ornaments. *J. Archaeol. Sci.* **2006**, *33*, 1105–1128. [CrossRef]
14. Kuhn, S.L.; Stiner, M.C. Paleolithic Ornaments: Implications for Cognition, Demography and Identity. *Diogenes* **2007**, *54*, 40–48. [CrossRef]
15. Zilhão, J. The Emergence of Ornaments and Art: An Archaeological Perspective on the Origins of "Behavioral Modernity". *J. Archaeol. Res.* **2007**, *15*, 1–54. [CrossRef]
16. D'Errico, F.; Vanhaeren, M. Earliest Personal Ornaments and Their Significance for the Origin of Language Debate. In *The Cradle of Language*; Studies in the Evolution of Language; Sands, B., Güldemann, T., Eds.; Oxford University Press: Oxford, UK, 2009; pp. 204–218.
17. Trinkaus, E.; Buzhilova, A.P. Diversity and Differential Disposal of the Dead at Sunghir. *Antiquity* **2018**, *92*, 7–21. [CrossRef]
18. White, R. Beyond Art: Toward an Understanding of the Origins of Material Representation in Europe. *Annu. Rev. Anthropol.* **1992**, *21*, 537–564. [CrossRef]
19. White, R. Substantial Acts: Meaning in Upper Paleolithic Representation. In *Beyond Art: Pleistocene Image and Symbol*; Wattis Symposium Series in Anthropology; Conkey, M.W., Soffer, O., Stratmann, D., Jablonski, N.C., Eds.; University of California Press: San Francisco, CA, USA, 1997; pp. 93–121.
20. White, R. Personal Ornaments from the Grotte Du Renne at Arcy-Sur-Cure. *Athena Rev.* **2001**, *2*, 41–46.
21. Stiner, M.C. "Standardization" in Upper Paleolithic Ornaments at the Coastal Sites of Riparo Mochi and Üçagızlı Cave. In *The Chronology of the Aurignacian and of the Transitional Technocomplexes: Dating, Stratigraphies, Cultural Implications*; Trabalhos de Arqueologia; Zilhão, J., D'errico, F., Eds.; Instituto Português de Arqueologia: Lisboa, Portugal, 2003.
22. Stiner, M. Finding a Common Bandwidth: Causes of Convergence and Diversity in Paleolithic Beads. *Biol. Theory* **2014**, *9*, 51–64. [CrossRef]
23. Heckel, C. Physical Characteristics of Mammoth Ivory and Their Implications for Ivory Work in the Upper Paleolithic. *Mitteilungen Der Ges. Für Urgesch.* **2009**, *18*, 71–91.
24. Cristiani, E.; Farbstein, R.; Miracle, P. Ornamental Traditions in the Eastern Adriatic: The Upper Palaeolithic and Mesolithic Personal Adornments from Vela Spila (Croatia). *J. Anthropol. Archaeol.* **2014**, *36*, 21–31. [CrossRef]
25. White, R.; Normand, C. Early and Archaic Aurignacian Personal Ornaments from Isturitz Cave: Technological and Regional Perspectives. *Palethnol. Archéol. Et Sci. Hum.* **2015**, *7 | 2015*. Available online: http://journals.openedition.org/palethnologie/789 (accessed on 12 September 2023). [CrossRef]
26. Wolf, S.; Conard, N.J. Personal Ornaments of the Swabian Aurignacian. *Palethnol. Archéol. Et Sci. Hum.* **2015**, *7 | 2015*. Available online: http://journals.openedition.org/palethnologie/897 (accessed on 12 September 2023). [CrossRef]
27. Wolf, S.; Heckel, C. Ivory Ornaments of the Aurignacian in Western Europe: Case Studies from France and Germany. *L'Anthropologie* **2018**, *122*, 348–373. [CrossRef]
28. Peresani, M.; Forte, M.; Quaggiotto, E.; Colonese, A.C.; Romandini, M.; Cilli, C.; Giacobini, G. Marine and Freshwater Shell Exploitation in the Early Upper Paleolithic: Re-Examination of the Assemblages from Fumane Cave (NE Italy). *PaleoAnthropology* **2019**, *2019*, 64–81.
29. Stiner, M.C.; Kuhn, S.L.; Güleç, E. Early Upper Paleolithic Shell Beads at Üçağızlı Cave I (Turkey): Technology and the Socioeconomic Context of Ornament Life-Histories. *J. Hum. Evol.* **2013**, *64*, 380–398. [CrossRef]
30. Tátá, F.; Cascalheira, J.; Marreiros, J.; Pereira, T.; Bicho, N. Shell Bead Production in the Upper Paleolithic of Vale Boi (SW Portugal): An Experimental Perspective. *J. Archaeol. Sci.* **2014**, *42*, 29–41. [CrossRef]
31. Álvarez-Fernández, E. Los Objetos de Adorno-Colgantes Del Paleolítico Superior y Del Mesolítico En La Cornisa Cantábrica y En El Valle Del Ebro: Una Visión Europea. Ph.D. Dissertation, Universidad de Salamanca, Salamanca, Spain, 2006.
32. Trinkaus, E.; Svoboda, J.A.; Wojtal, P.; Fišákova, M.N.; Wilczyński, J. Human Remains from the Moravian Gravettian: Morphology and Taphonomy of Additional Elements from Dolní Věstonice II and Pavlov I. *Int. J. Osteoarchaeol.* **2010**, *20*, 645–669. [CrossRef]
33. Trinkaus, E.; Buzhilova, A.P.; Mednikova, M.B.; Dobrovolskaya, M.V. *The People of Sunghir: Burials, Bodies, and Behavior in the Earlier Upper Paleolithic*; Human Evolution Series; Oxford University Press: Oxford, UK, 2014; ISBN 978-0-19-938105-0.
34. Wojtal, P.; Wilczyński, J.; Bocheński, Z.M.; Svoboda, J.A. The Scene of Spectacular Feasts: Animal Remains from Pavlov I South-East, the Czech Republic. *Quat. Int.* **2012**, *252*, 122–141. [CrossRef]
35. Niven, L. *The Palaeolithic Occupation of Vogelherd Cave: Implications for the Subsistence Behavior of Late Neanderthals and Early Modern Humans*; Tübingen Publications in Prehistory; Kerns Verlag: Tübingen, Germany, 2006; ISBN 978-3-935751-04-9.
36. Kitagawa, K.; Krönneck, P.; Conard, N.J.; Münzel, S. Exploring Cave Use and Exploitation Among Cave Bears, Carnivores and Hominins in the Swabian Jura, Germany. *J. Taphon.* **2012**, *10*, 439–461.

37. Conard, N.J.; Kitagawa, K.; Krönneck, P.; Böhme, M.; Münzel, S. The Importance of Fish, Fowl and Small Mammals in the Paleolithic Diet of the Swabian Jura, Southwestern Germany. In *Vertebrate Paleobiology and Paleoanthropology*; Springer Science: Dordrecht, The Netherlands, 2013; pp. 173–190, ISBN 978-94-007-6765-2.
38. Baumann, C.; Bocherens, H.; Drucker, D.G.; Conard, N.J. Fox Dietary Ecology as a Tracer of Human Impact on Pleistocene Ecosystems. *PLoS ONE* **2020**, *15*, e0235692. [CrossRef] [PubMed]
39. Münzel, S.C. Die Jungpleistozäne Großsäugerfauna Aus Dem Geißenklösterle. In *Geißenklösterle II. Chronostratigraphie, Paläoumwelt und Subsistenz im Mittel- und Jungpaläolithikum der Schwäbischen Alb*; Conard, N.J., Bolus, M., Münzel, S.C., Eds.; Kerns Verlag: Tübingen, Germany, 2019; p. 416.
40. Hofreiter, M.; Stewart, J. Ecological Change, Range Fluctuations and Population Dynamics during the Pleistocene. *Curr. Biol.* **2009**, *19*, R584–R594. [CrossRef]
41. Pushkina, D.; Raia, P. Human Influence on Distribution and Extinctions of the Late Pleistocene Eurasian Megafauna. *J. Hum. Evol.* **2008**, *54*, 769–782. [CrossRef]
42. Sommer, R.; Benecke, N. Late-Pleistocene and Early Holocene History of the Canid Fauna of Europe (Canidae). *Mamm. Biol.* **2005**, *70*, 227–241. [CrossRef]
43. Sommer, R.S.; Nadachowski, A. Glacial Refugia of Mammals in Europe: Evidence from Fossil Records. *Mammal Rev.* **2006**, *36*, 251–265. [CrossRef]
44. Baumann, C.; Wong, G.L.; Starkovich, B.M.; Münzel, S.C.; Conard, N.J. The Role of Foxes in the Palaeolithic Economies of the Swabian Jura (Germany). *Archaeol. Anthropol. Sci.* **2020**, *12*, 1–17. [CrossRef]
45. Stiner, M.C.; Munro, N.D.; Surovell, T.A. The Tortoise and the Hare: Small-Game Use, the Broad-Spectrum Revolution, and Paleolithic Demography. *Curr. Anthropol.* **2000**, *41*, 39–79. [CrossRef]
46. Kitagawa, K. Exploring Hominins and Animals in the Swabian Jura: Study of the Paleolithic Fauna from Hohlenstein-Stadel. Ph.D. Thesis, Universität Tübingen, Tübingen, Germany, 2014.
47. Wong, G.L.; Starkovich, B.M.; Drucker, D.G.; Conard, N.J. New Perspectives on Human Subsistence during the Magdalenian in the Swabian Jura, Germany. *Archaeol. Anthropol. Sci.* **2020**, *12*, 217. [CrossRef]
48. Krönneck, P. Die Pleistozäne Makrofauna Des Bocksteins (Lonetal-Schwäbische Alb) Ein Neuer Ansatz Zur Rekonstruktion Der Paläoumwelt. Ph.D. Thesis, Universität Tübingen, Tübingen, Germany, 2012.
49. Boger, U.; Starkovich, B.M.; Conard, N.J. New insights gained from the faunal material recovered during the latest excavations at Vogelherd Cave. *Mitteilungen Ges. Urgesch.* **2014**, *23*, 57–81.
50. Niven, L. From Carcass to Cave: Large Mammal Exploitation during the Aurignacian at Vogelherd, Germany. *J. Hum. Evol.* **2007**, *53*, 362–382. [CrossRef] [PubMed]
51. Napierala, H.; Münzel, S.C.; Conard, N.J. Die Fauna Des Magdalénien Vom Hohle Fels. In *Das Magdalénien des Hohle Fels. Chronologische Stellung, Lithische Technologie und Funktion der Rückenmesser*; Tübinger Monographien zur Urgeschichte; Kerns Verlag: Tübingen, Germany, 2014; pp. 275–317.
52. Wong, G.L.; Starkovich, B.M.; Conard, N.J. Human Subsistence and Environment during the Magdalenian at Langmahdhalde: Evidence from a New Rock Shelter in the Lone Valley, Southwest Germany. *Mitteliungen Der Ges. Für Urgesch.* **2017**, *26*, 103–123.
53. Camarós, E.; Münzel, S.C.; Cueto, M.; Rivals, F.; Conard, N.J. The Evolution of Paleolithic Hominin–Carnivore Interaction Written in Teeth: Stories from the Swabian Jura (Germany). *J. Archaeol. Sci. Rep.* **2016**, *6*, 798–809. [CrossRef]
54. Baumann, C. The Paleo-Synanthropic Niche: A First Attempt to Define Animal's Adaptation to a Human-Made Micro-Environment in the Late Pleistocene. *Archaeol. Anthropol. Sci.* **2023**, *15*, 63. [CrossRef]
55. Baumann, C.; Starkovich, B.M.; Drucker, D.G.; Münzel, S.C.; Conard, N.J.; Bocherens, H. Dietary Niche Partitioning among Magdalenian Canids in Southwestern Germany and Switzerland. *Quat. Sci. Rev.* **2020**, *227*, 106032. [CrossRef]
56. Conard, N.J.; Malina, M. Die Ausgrabungen Am Hohle Fels Bei Schelklingen, Alb-Donau-Kreis. *Archäol. Ausgrabungen Baden-Württemberg* **2007**, *2006*, 17–21.
57. Conard, N.J.; Malina, M. New Evidence for the Origins of Music from the Caves of the Swabian Jura. *Orient Archäol.* **2008**, *22*, 13–22.
58. Conard, N.J.; Malina, M. Die Ausgrabungen 2004 in Den Frühen Jungpaläolithischen Schichten Des Hohle Fels Bei Schelklingen, Alb-Donau-Kreis. pro Ano. *Archäol. Ausgrabungen Baden-Württemberg* **2005**, *2004*, 17–21.
59. Conard, N.J.; Janas, A.; Malina, M. Vielfältige Funde Aus Dem Aurignacien Und Ein Bemalter Stein Aus Dem Magdalénien Vom Hohle Fels Bei Schelklingen. *Archäol. Ausgrabungen Baden-Württemberg* **2014**, *2013*, 58–63.
60. Hahn, J. (Ed.) *Die Geißenklösterle-Höhle Im Achtal Bei Blaubeuren I: Fundhorizontbildung Und Besiedlung Im Mittelpaläolithikum Und Im Aurignacien*; Forschungen und Berichte zur Vor- und Frühgeschichte in Baden-Württemberg; Theiss Verlag: Stuttgart, Germany, 1988; ISBN 3-8062-0794-1.
61. Hahn, J. Besiedlung Und Tätigkeit Im Aurignacien Des Hohlenstein-Stadels. In *Die altsteinzeitliche Elfenbeinstatuette aus der Höhle Stadel im Hohlenstein bei Asselfingen, Alb-Donau-Kreis*; Fundberichte aus Baden-Württemberg: Baden-Württemberg, Germany, 1989; Volume 14, pp. 97–118.
62. Hahn, J. Nachgrabungen Im Hohlen Felsen Bei Schelklingen, Alb-Donau-Kreis. *Archäol. Korresp.* **1977**, *7*, 241–248.
63. Wolf, S. Schmuck aus den neuen Ausgrabungen. In *Löwenmensch und mehr: Die Ausgrabungen 2008–2013 in Den Altsteinzeitlichen Schichten der Stadel-Höhle im Hohlenstein (Lonetal), Gemeinde Asselfingen, Alb-Donau-Kreis*; Forschungen und Berichte zur Archäologie in Baden-Württembrg; Kind, C.-J., Ed.; Reichert Verlag: Wiesbaden, Germany, 2019; pp. 125–128.

64. Reinhardt, B.; Wehrberger, K. *Der Löwenmensch*; Ulmer Museum: Ulm, Germany, 2005; ISBN 3-928738-04-6.
65. Kind, C.-J.; Ebinger-Rist, N.; Wolf, S.; Beutelspacher, T.; Wehrberger, K. The Smile of the Lion Man. Recent Excavations in Stadel Cave (Baden-Württemberg, South-Western Germany) and the Restoration of the Famous Upper Palaeolithic Figurine-Das Lächeln Des Löwenmenschen. Neue Ausgrabungen in Der Stadel-Höhle (Baden-Württemberg, Südwestdeutschland) Und Die Restaurierung Der Berühmten Jungpaläolithischen Figur. *Quartär–Int. Jahrb. Zur Erforsch. Des Eiszeitalt. Und Der Steinzeit* **2014**, *61*, 129–145.
66. Benazzi, S.; Fornai, C.; Buti, L.; Toussaint, M.; Mallegni, F.; Ricci, S.; Gruppioni, G.; Weber, G.W.; Condemi, S.; Ronchitelli, A. Cervical and Crown Outline Analysis of Worn Neanderthal and Modern Human Lower Second Deciduous Molars. *Am. J. Phys. Anthropol.* **2012**, *149*, 537–546. [CrossRef] [PubMed]
67. Röding, C.; Zastrow, J.; Scherf, H.; Doukas, C.; Harvati, K. Crown Outline Analysis of the Hominin Upper Third Molar from the Megalopolis Basin, Peloponnese, Greece. In *Ancient connections in Eurasia*; Kerns Verlag: Tübingen, Germany, 2021.
68. Bauer, C.C.; Benazzi, S.; Darlas, A.; Harvati, K. Geometric Morphometric Analysis and Internal Structure Measurements of the Neanderthal Lower Fourth Premolars from Kalamakia, Greece. *Quat. Int.* **2018**, *497*, 14–21. [CrossRef]
69. Harvati, K.; Bauer, C.C.; Grine, F.E.; Benazzi, S.; Ackermann, R.R.; van Niekerk, K.L.; Henshilwood, C.S. A Human Deciduous Molar from the Middle Stone Age (Howiesons Poort) of Klipdrift Shelter, South Africa. *J. Hum. Evol.* **2015**, *82*, 190–196. [CrossRef]
70. Adams, D.; Collyer, M.; Kaliontzopoulou, A.; Baken, E. Geomorph: Geometric Morphometric Analyses of 2D and 3D Landmark Data 2021. Available online: https://cran.r-project.org/web/packages/geomorph/index.html (accessed on 30 February 2022).
71. Baken, E.K.; Collyer, M.L.; Kaliontzopoulou, A.; Adams, D.C. Geomorph v4.0 and GmShiny: Enhanced Analytics and a New Graphical Interface for a Comprehensive Morphometric Experience. *Methods Ecol. Evol.* **2021**, *12*, 2355–2363. [CrossRef]
72. Wickham, H. *Ggplot2: Elegant Graphics for Data Analysis*; Springer: New York, NY, USA, 2016.
73. Kassambara, A. Ggpubr: "ggplot2" Based Publication Ready Plots 2023. Available online: https://cran.r-project.org/web/packages/ggpubr/index.html (accessed on 21 May 2023).
74. Slowikowski, K. Ggrepel: Automatically Position Non-Overlapping Text Labels with "Ggplot2". R Package Version 0.9.1. 2021. Available online: https://CRAN.R-Project.Org/Package=ggrepel (accessed on 21 May 2023).
75. Bollen, K.A.; Jackman, R.W. Regression Diagnostics: An Expository Treatment of Outliers and Influential Cases. *Sociol. Methods Res.* **1985**, *13*, 510–542. [CrossRef]
76. Keeley, L.H. *Experimental Determination of Stone Tool Uses: A Microwear Analysis*; University of Chicago Press: Chicago, IL, USA, 1980; ISBN 978-0-226-42889-5.
77. Semenov, S.A. *Prehistoric Technology: An Experimental Study of the Oldest Tools and Artefacts from Traces of Manufacture and Wear*; Cory, Adams & Mackay: London, UK, 1964; ISBN 978-0-389-20571-5.
78. Van Gijn, A.L. *The Wear and Tear of Flint. Principles of Functional Analysis Applied to Dutch Neolithic Assemblages*; Analecta Praehistorica Leidensia; Leiden University: Leiden, The Netherlands, 1990; ISBN 978-90-73368-02-6.
79. Sidéra, I.; Legrand, A. Tracéologie Fonctionnelle Des Matières Osseuses: Une Méthode. *Bull. De La Soc. Préhist. Française* **2006**, *103*, 291–304. [CrossRef]
80. Buc, N. Experimental Series and Use-Wear in Bone Tools. *J. Archaeol. Sci.* **2011**, *38*, 546–557. [CrossRef]
81. Vaughan, P. *Use-Wear Analysis of Flaked Stone Tools*; University of Arizona Press: Tucson, AZ, USA, 1985; ISBN 978-0-8165-0861-7.
82. Osipowicz, G.; Piličiauskienė, G.; Orłowska, J.; Piličiauskas, G. An Occasional Ornament, Part of Clothes or Just a Gift for Ancestors? The Results of Traceological Studies of Teeth Pendants from the Subneolithic Sites in Šventoji, Lithuania. *J. Archaeol. Sci. Rep.* **2020**, *29*, 102130. [CrossRef]
83. Mărgărit, M.; Boroneanț, A.; Bonsall, C. Stone and Osseous Adornments in the Mesolithic and Early Neolithic of the Iron Gates. *Open Archaeol.* **2021**, *7*, 779–797. [CrossRef]
84. Chauvière, F.-X.; Castel, J.-C. Le Statut Du Renard à Combe Saunière (Sarliac-Sur-l'Isle, Dordogne) et Dans Le Solutréen de l'Est Aquitain. In *Petits Animaux et Sociétés Humaines. Du Complément Alimentaire aux Ressources Utilitaires. XXIVè Rencontres Internationales D'archéologie et D'histoire d'Antibes, Antibes: Edition APDCA*; Brugal, J.P., et Desse, J., Eds.; APDCA: Antibes, France, 2004; Volume 24, pp. 389–402.
85. Mereau, A.-L. Les dents animales perforées au Magdalénien: Nouvelles perspectives fonctionnelles. Etude de quatre sites pyrénéens: Isturitz, le Mas d'Azil, Arudy et Gourdan. Ph.D. Thesis, Archéologie et Préhistoirferee. Universite de Paris 1-Pantheon Sorbonne, Paris, France, 2012.
86. Winiarska-Kabacińska, M. Animal Perforated Teeth Deposit from Magdalenian Site at Wilczyce (Southeastern Poland). What They Were Used for? A Use-Wear Case Study. *Cuad. De Prehist. Y Arqueol. De La Univ. De Granada* **2017**, *27*, 73–82.
87. Falci, C.G.; Cuisin, J.; Delpuech, A.; Van Gijn, A.; Hofman, C.L. New Insights into Use-Wear Development in Bodily Ornaments through the Study of Ethnographic Collections. *J. Archaeol. Method Theory* **2019**, *26*, 755–805. [CrossRef]
88. D'Errico, F.; Martí, A.P.; Wei, Y.; Gao, X.; Vanhaeren, M.; Doyon, L. Zhoukoudian Upper Cave Personal Ornaments and Ochre: Rediscovery and Reevaluation. *J. Hum. Evol.* **2021**, *161*, 103088. [CrossRef] [PubMed]
89. Audet, A.M.; Robbins, C.B.; Larivière, S. Alopex Lagopus. *Mamm. Species* **2002**, *713*, 1–10. [CrossRef]
90. Hulme-Beaman, A.; Dobney, K.; Cucchi, T.; Searle, J.B. An Ecological and Evolutionary Framework for Commensalism in Anthropogenic Environments. *Trends Ecol. Evol.* **2016**, *31*, 633–645. [CrossRef]
91. Larivière, S.; Pasitschniak-Arts, M. Vulpes Vulpes. *Mamm. Species* **1996**, *537*, 1–11. [CrossRef]

92. Walter, T.; Zink, R.; Laaha, G.; Zaller, J.G.; Heigl, F. Fox Sightings in a City Are Related to Certain Land Use Classes and Sociodemographics: Results from a Citizen Science Project. *BMC Ecol.* **2018**, *18*, 50. [CrossRef]
93. Scott, D.M.; Berg, M.J.; Tolhurst, B.A.; Chauvenet, A.L.; Smith, G.C.; Neaves, K.; Lochhead, J.; Baker, P.J. Changes in the Distribution of Red Foxes (*Vulpes vulpes*) in Urban Areas in Great Britain: Findings and Limitations of a Media-Driven Nationwide Survey. *PLoS ONE* **2014**, *9*, e99059. [CrossRef]
94. Wojtal, P.; Svoboda, J.A.; Roblíčková, M.; Wilczyński, J. Carnivores in the Everyday Life of Gravettian Hunters-Gatherers in Central Europe. *J. Anthropol. Archaeol.* **2020**, *59*, 101171. [CrossRef]
95. Albrecht, G.; Berke, H.; Poplin, F. *Naturwissenschaftliche Untersuchungen an Magdalénien-Inventaren Vom Petersfels, Grabungen 1974–1976*; Verlag Archaeologica Venatoria: Tübingen, Germany, 1983; ISBN 3-921618-17-7.
96. Albrecht, G.; Berke, H.; Burkert, W.; Haas-Campen, S.; Hahn-Weishaupt, A. Aufsätze Die Funde Vom Petersfels in Der Städtischen Sammlung Engen Im Hegau. *Fundberichte Aus Baden-Württemberg* **1994**, *19*, 1–62.
97. Peters, E.; Toepfer, V. Der Abschluß Der Grabungen Am Petersfels Bei Engen Im Badischen Hegau. *Prähist. Z.* **1932**, *23*, 155–199.
98. Händel, M.; Simon, U.; Einwögerer, T.; Neugebauer-Maresch, C. New Excavations at Krems-Wachtberg–Approaching a Well-Preserved Gravettian Settlement Site in the Middle Danube Region. *Quartär* **2009**, *56*, 187–196.
99. Trusov, A.B.; Zhitenev, B.C. о Па ааo Co. In *o, aaa, a*; Russian Academy of Sciences in Moscow: Moscow, Russia, 2008; pp. 427–434.
100. Szuma, E.; Germonpré, M. Size of the Lower Carnassial in the Arctic and the Red Fox from Late Pleistocene in Belgium Compared to Other Ancient and Extant Populations. *Mammal Res.* **2020**, *65*, 127–139. [CrossRef]
101. Poplin, F. *Les Grands Vertébrés de Gönnersdorf: Fouilles 1968*; Frantz Steiner Verlag GMBH: Wiesbaden, Germany, 1976; ISBN 3-515-02007-1.
102. Harris, S. Age Determination in the Red Fox (*Vulpes vulpes*)—An Evaluation of Technique Efficiency as Applied to a Sample of Suburban Foxes. *J. Zool.* **1978**, *184*, 91–117. [CrossRef]
103. Szuma, E. Geographic Variation of Tooth and Skull Sizes in the Arctic Fox *Vulpes* (*Alopex*) *Lagopus*. *Ann. Zool. Fenn.* **2008**, *45*, 185–199. [CrossRef]
104. Szuma, E. Geography of Sexual Dimorphism in the Tooth Size of the Red Fox *Vulpes vulpes* (Mammalia, Carnivora). *J. Zool. Syst. Evol. Res.* **2008**, *46*, 73–81. [CrossRef]
105. Chevallier, C.; Gauthier, G.; Berteaux, D. Age Estimation of Live Arctic Foxes *Vulpes lagopus* Based on Teeth Condition. *Wildl. Biol.* **2017**, *2017*, 1–6. [CrossRef]
106. Lázničková-Galetová, M. Perforated Animal Teeth. In *Dolní Věstonice II: Chronostratigraphy, Paleoethnology, Paleoanthropology*; Svoboda, J.A., Ed.; Academy of Sciences of the Czech Republic, Institute of Archaeology: Brno, Czech Republic, 2016.
107. Svoboda, J.A. *DOLNÍ VĚSTONICE-PAVLOV. Place: South Moravia. Time: 30 Thousand Years Ago*; Regionální Muzeum Mikulové: Mikulové, Czech Republic, 2018; ISBN 978-80-85088-36-6.
108. Macpherson, A.H. *The Dynamics of Canadian Arctic Fox Populations*; Canadian Wildlife Service Report Series; Department of Indian Affairs and Northern Development Ottawa: Ottawa, ON, Canada, 1969; ISBN 0069-0031.
109. Hersteinsson, P.; Macdonald, D.W. Interspecific Competition and the Geographical Distribution of Red and Arctic Foxes *Vulpes vulpes* and *Alopex lagopus*. *Oikos* **1992**, *64*, 505–515. [CrossRef]
110. Pulliainen, E. *Alopex lagopus* (Linnaeus, 1758)-Eisfuchs. In *Handbuch der Säugetiere Europas. Raubsäuger I*; AULA Verlag: Wiesbaden, Germany, 1993; pp. 195–214.
111. Wandeler, A.I.; Lüps, P. *Vulpes vulpes* (Linnaeus, 1758)–Rotfuchs. *Handb. Der Säugetiere Eur.* **1993**, *1*, 139–193.
112. Essel, E.; Zavala, E.I.; Schulz-Kornas, E.; Kozlikin, M.B.; Fewlass, H.; Vernot, B.; Shunkov, M.V.; Derevianko, A.P.; Douka, K.; Barnes, I.; et al. Ancient Human DNA Recovered from a Palaeolithic Pendant. *Nature* **2023**, *618*, 328–332. [CrossRef] [PubMed]
113. Buckley, M.; Collins, M.J. Collagen Survival and Its Use for Species Identification in Holocene-Lower Pleistocene Bone Fragments from British Archaeological and Paleontological Sites. *Antiqua* **2011**, *1*, e1. [CrossRef]
114. Hatt, G.; Taylor, K. Arctic Skin Clothing in Eurasia and America an Ethnographic Study. *Arct. Anthropol.* **1969**, *5*, 3–132.
115. Monchot, H.; Gendron, D. Fox Exploitation by the Paleoeskimo at the Tayara Site, Nunavik. *Arct. Anthropol.* **2011**, *48*, 15–32. [CrossRef]
116. Vanhaeren, M.; d'Errico, F.; Julien, M.; Mourer-Chauviré, C.; Lozouet, P. Les Objets de Parure. In *Le Châtelperronien de la grotte du Renne (Arcy-sur-Cure, Yonne, France). Les fouilles d'André Leroi-Gourhan (1949–1963)*; PALEO; Julien, M., David, F., Girard, M., Roblin-Jouve, A., Eds.; Société des Amis du Musée National de Préhistoire et de la Recherche Archéologique: Les Eyzies-de-Tayac, France, 2019; pp. 259–285.
117. Vézian, J.; Vézian, J. Les Gisements de La Grotte de Saint-Jean-de-Verges (Ariège). *Gall. Préhist.* **1966**, *9*, 93–130. [CrossRef]
118. Granger, J.-M.; Lévêque, F. Parure Castelperronienne et Aurignacienne: Étude de Trois Séries Inédites de Dents Percées et Comparaisons. *Comptes Rendus De L'acad. Des Sci.-Ser. IIA-Earth Planet. Sci.* **1997**, *325*, 537–543. [CrossRef]
119. Sinitsyn, A.A. Figurative and decorative art of Kostenki: Chronological and cultural differentiation. In *L'art Pléistocène dans le Monde/Pleistocene art of the World/Arte Pleistoceno en el Mundo, Actes du Congrès IFRAO, Tarascon-sur-Ariège, Septembre 2010, Symposium «Art Mobilier Pléistocène»*; N° spécial de Préhistoire, Art et Sociétés, Bulletin de la Société Préhistorique Ariège-Pyrénées, LXV–LXVI, 2010–2011, CD; Clottes, J., (dir.); Société préhistorique Ariège-Pyrénées: Tarascon-sur-Ariège, France, 2012; pp. 1339–1359.

120. Bahder, O.N. Das Zweite Grab in Der Paläolithischen Siedlung Sungir Im Mittleren Rußland. *Quartär–Int. Jahrb. Zur Erforsch. Des Eiszeitalt. Und Der Steinzeit* **1970**, *21*, 103–104.
121. Street, M.; Turner, E. *The Faunal Remains from Gönnersdorf*; Verlag des Römisch-Germanischen Zentralmuseums: Mainz, Germany, 2013; ISBN 978-3-7954-2765-8.
122. Álvarez-Fernández, E. Jet Pearls and Personal Adornment on Fox and Red Deer Teeth of the Magdalenian Age at Gonnersdorf and Andernach-Martinsberg-2 (Neuwied, Rheinland Pfalz, Germany). *Zephyrus* **1999**, *52*, 79–106.
123. Alix, C. Les Dents Percées Ou Encochées. In *La Grotte de la Vache (Ariège): Fouilles Romain robert*; Les occupations du Magdalénien; Réunion des Musées Nationaux: Paris, France, 2003; pp. 368–390.
124. Vanhaeren, M.; d'Errico, F. La Parure de l'enfant de La Madeleine (Fouilles Peyrony). Un Nouveau Regard Sur l'enfance Au Paléolithique Supérieur. *PALEO Rev. D'archéol. Préhist.* **2001**, *13*, 201–240. [CrossRef]
125. Ladier, E.; Welté, A.-C.; Lambert, G.-N. Les objets de parure de la vallée de l'Aveyron. Le Courbet, Bruniquel-Montastruc et autres abris; documents inédits ou retrouvés. *Paléo Rev. D'archéol. Préhist.* **1994**, *6*, 197–231. [CrossRef]
126. Ladier, E.; Welté, A.-C. Les Objets de Parure de La Vallée de l'Aveyron, Fontales, Abris de Bruniquel (Plantade, Lafaye, Gandil). *Paléo Rev. D'archéol. Préhist.* **1993**, *5*, 281–317. [CrossRef]
127. Griffith, J. Of Beasts and Men: What Does The Fox Say? Cardiff University SHARE EJournal Blog (Of Beasts & Men Series). *Cardiff Univ. SHARE Ejournal Blog* **2018**. Available online: https://cushareejournal.wordpress.com/2018/11/05/of-animals-and-men-what-does-the-fox-say/ (accessed on 12 September 2023).
128. D'Errico, F.; Henshilwood, C.; Lawson, G.; Vanhaeren, M.; Tillier, A.-M.; Soressi, M.; Bresson, F.; Maureille, B.; Nowell, A.; Lakarra, J. Archaeological Evidence for the Emergence of Language, Symbolism, and Music–an Alternative Multidisciplinary Perspective. *J. World Prehist.* **2003**, *17*, 1–70. [CrossRef]
129. Vanhaeren, M. Speaking with Beads: The Evolutionary Significance of Personal Ornaments. In *From Tools to Symbols: From Early Hominids to Modern Humans*; D'Errico, F., Backwell, L., Eds.; Wits University Press: Johannesburg, South Africa, 2005; pp. 525–554. ISBN 978-1-86814-637-6.
130. González-Sainz, C.; Ruiz-Redondo, A.; Garate-Maidagan, D.; Iriarte-Avilés, E. Not Only Chauvet: Dating Aurignacian Rock Art in Altxerri B Cave (Northern Spain). *J. Hum. Evol.* **2013**, *65*, 457–464. [CrossRef] [PubMed]
131. Peters, J.; Schmidt, K. Animals in the Symbolic World of Pre-Pottery Neolithic Göbekli Tepe, South-Eastern Turkey: A Preliminary Assessment. *Anthropozoologica* **2004**, *39*, 179–218.
132. Maher, L.A.; Stock, J.T.; Finney, S.; Heywood, J.J.; Miracle, P.T.; Banning, E.B. A Unique Human-Fox Burial from a Pre-Natufian Cemetery in the Levant (Jordan). *PLoS ONE* **2011**, *6*, e15815. [CrossRef]
133. Reshef, H.; Anton, M.; Bocquentin, F.; Vardi, J.; Khalaily, H.; Davis, L.; Bar-Oz, G.; Marom, N. Tails of Animism: A Joint Burial of Humans and Foxes in Pre-Pottery Neolithic Motza, Israel. *Antiquity* **2019**, *93*, e28. [CrossRef]
134. Grandal-d'Anglade, A.; Albizuri, S.; Nieto, A.; Majó, T.; Agustí, B.; Alonso, N.; Antolín, F.; López, J.B.; Moya, A.; Rodríguez, A. Dogs and Foxes in Early-Middle Bronze Age Funerary Structures in the Northeast of the Iberian Peninsula: Human Control of Canid Diet at the Sites of Can Roqueta (Barcelona) and Minferri (Lleida). *Archaeol. Anthropol. Sci.* **2019**, *11*, 3949–3978. [CrossRef]
135. Johnson, T.W. Far Eastern Fox Lore. *Asian Folk. Stud.* **1974**, *33*, 35–68. [CrossRef]
136. Van Deusen, K. *Kiviuq: An Inuit Hero and His Siberian Cousins*; McGill-Queen's Press-MQUP: Montreal, QC, Canada, 2009; Volume 54.
137. Fraas, O. Der Hohlenstein Und Der Höhlenbär. *Jahresh. Ver. Vaterländische Naturkunde Württemberg* **1862**, *18*, 156–188.
138. Schmidt, R.R. *Die Diluviale Vorzeit Deutschlands. With Contributions by E. Koken and J. Schliz*; E. Schweizerbartsche Verlagsbuchhandlung Nägele: Stuttgart, Germany, 1912.
139. Bolus, M.; Conard, N.J. 100 Jahre Robert Rudolf Schmidts, Die Diluviale Vorzeit Deutschlands'. *Mitteilungen Ges. Urgesch.* **2012**, *21*, 63–89.
140. Conard, N.J.; Bolus, M. Radiocarbon Dating the Appearance of Modern Humans and Timing of Cultural Innovations in Europe: New Results and New Challenges. *J. Hum. Evol.* **2003**, *44*, 331–371. [CrossRef] [PubMed]
141. Conard, N.J.; Bolus, M. The Swabian Aurignacian and Its Place in European Prehistory. In *Towards a definition of the Aurignacian*; Trabalhos de Arqueologia; Instituto Português de Arqueologia: Lisboa, Portugal, 2006.
142. Conard, N.J.; Dippon, G.; Goldberg, P. Chronostratigraphy and Archeological Context of the Aurignacian Deposits at Geißenklösterle. In *The Chronology of the Aurignacian and of the Transitional Technocomplexes: Dating, Stratigraphies, Cultural implications*; Trabalhos de Arqueologia; Instituto Português de Arqueologia: Lisboa, Portugal, 2003.
143. Conard, N.J. The Demise of the Neanderthal Cultural Niche and the Beginning of the Upper Paleolithic in Southwestern Germany. In *Neanderthal Lifeways, Subsistence and Technology: One Hundred Fifty Years of Neanderthal Study*; Vertebrate Paleobiology and Paleontology Series; Conard, N.J., Richter, J., Eds.; Springer: Dordrecht, The Netherlands, 2011; pp. 223–240. ISBN 978-94-007-0414-5.
144. Miller, C. Geoarchaeology and the Interpretation of the Past in the Caves of Swabia: Current and Future Research Priorities. In *Human Origin Sites and the World Heritage Convention in Eurasia*; Nuria, S., Ed.; UNESCO Publishing: Paris, France, 2015; Volume 2, ISBN 978-92-3-100109-3.
145. Fraas, O. Resultate Der Ausgrabungen Im Hohlefels Bei Schelklingen. *Jahresh. Ver. Vaterländische Naturkunde Württemberg* **1872**, *28*, 21–36.

146. Conard, N.J.; Uerpmann, H.-P. Die Ausgrabungen 1997 Und 1998 Im Hohle Fels Bei Schelklingen, Alb-Donau-Kreis. *Archäol. Ausgrabungen Baden-Württemberg* **1998**, 47–52.
147. Bataille, G.; Conard, N.J. Burin-Core Technology in Aurignacian Horizons IIIa and IV of Hohle Fels Cave (Southwestern Germany): Die Stichelkern-Technologie Der Aurignacien-Horizonte IIIa Und IV Der Hohle Fels-Höhle (Südwestdeutschland). *Quartär–Int. Jahrb. Erforsch. Eiszeitalt. Und Steinzeit* **2018**, *65*, 7–49.
148. Taller, A.; Conard, N.J. Das Gravettien Der Hohle Fels-Höhle Und Seine Bedeutung Für Die Kulturelle Evolution Des Europäischen Jungpaläolithikums: The Gravettian of Hohle Fels Cave and Its Implications for the Cultural Evolution of the European Upper Palaeolithic. *Quartär–Int. Jahrb. Erforsch. Eiszeitalt. Steinzeit* **2016**, *63*, 89–123.
149. Housley, R.A.; Gamble, C.S.; Street, M.; Pettitt, P. Radiocarbon Evidence for the Lateglacial Human Recolonisation of Northern Europe. In Proceedings of the Prehistoric Society; Cambridge University Press: Cambridge, UK, 1997; Volume 63, pp. 25–54.
150. Taller, A. *Das Magdalénien des Hohle Fels: Chronologische Stellung, lithische Technologie und Funktion der Rückenmesser*; Kerns Verlag: Tübingen, Germany, 2014; ISBN 9783935751209.
151. Dutkiewicz, E.; Wolf, S.; Floss, H.; Conard, N.J. Les objets en ivoire du Jura Souabe. The ivory objects of the Swabian Jura. *L'Anthropologie* **2018**, *122*, 447–468. [CrossRef]
152. Velliky, E.C.; Porr, M.; Conard, N.J. Ochre and Pigment Use at Hohle Fels Cave: Results of the First Systematic Review of Ochre and Ochre-Related Artefacts from the Upper Palaeolithic in Germany. *PLoS ONE* **2018**, *13*, e0209874. [CrossRef]
153. Wolf, S.; Conard, N.J.; Floss, H.; Dapschauskas, R.; Velliky, E.; Kandel, A. The Use of Ochre and Painting During the Upper Paleolithic of the Swabian Jura in the Context of the Development of Ochre Use in Africa and Europe. *Open Archaeol.* **2018**, *4*, 185–205. [CrossRef]
154. Higham, T.; Basell, L.; Jacobi, R.; Wood, R.; Ramsey, C.B.; Conard, N.J. Testing Models for the Beginnings of the Aurignacian and the Advent of Figurative Art and Music: The Radiocarbon Chronology of Geißenklösterle. *J. Hum. Evol.* **2012**, *62*, 664–676. [CrossRef] [PubMed]
155. Richter, D.; Waiblinger, J.; Rink, W.J.; Wagner, G.A. Thermoluminescence, Electron Spin Resonance and14C-Dating of the Late Middle and Early Upper Palaeolithic Site of Geißenklösterle Cave in Southern Germany. *J. Archaeol. Sci.* **2000**, *27*, 71–89. [CrossRef]
156. Reimer, P.J.; Austin, W.E.N.; Bard, E.; Bayliss, A.; Blackwell, P.G.; Bronk Ramsey, C.; Butzin, M.; Cheng, H.; Edwards, R.L.; Friedrich, M.; et al. The IntCal20 Northern Hemisphere Radiocarbon Age Calibration Curve (0–55 Cal kBP). *Radiocarbon* **2020**, *62*, 725–757. [CrossRef]
157. Bronk Ramsey, C. Bayesian analysis of radiocarbon dates. *Radiocarbon* **2009**, *51*, 337–360. [CrossRef]
158. Floss, H. L'art Mobilier Aurignacien Du Jura Souabe et Sa Place Dans l'art Paléolithique - Die Kleinkunst Des Aurignacien Auf Der Schwäbischen Alb Und Ihre Stellung in Der Paläolithischen Kunst. In Proceedings of the Les chemins de l'Art aurignacien en Europe - Das Aurignacien und die Anfänge der Kunst in Europa, Aurignac, France, 16–18 September 2005; Floss, H., Rouquerol, N., Eds.; Musée-forum d'Aurignac: Aurignac, France, 2007; p. 476.
159. Floss, H. The Oldest Portable Art: The Aurignacian Ivory Figurines from the Swabian Jura (Southwest Germany). *Palethnol. Archéol. Sci. Hum.* **2015**. [CrossRef]
160. Münzel, S.C. Subsistence Patterns in the Gravettian of the Ach Valley. A Former Tributary of the Danube in the Swabian Jura. In *The Gravettian along the Danube*; The Dolní Věstonice Studies; Svoboda, J.A., Sedláčková, L., Eds.; Archeologický ústav AV ČR Brno: Brno-střed, Czech Republic, 2004; pp. 71–85.
161. Riek, G. *Das Paläolithikum Der Brillenhöhle Bei Blaubeuren*; Müller & Gräff: Stuttgart, Germany, 1973; ISBN 978-3-87532-055-8.
162. Tafelmaier, Y. „Verdamp Lang Her": Eiszeitliche Jäger Und Sammler in Südwestdeutschland: Einblicke in Aktuelle Forschungen Zur Altsteinzeit Am Landesamt Für Denkmalpflege. *Denkmalpfl. Baden-Württemberg Nachrichtenblatt Landesdenkmalpflege* **2023**, 62–71. [CrossRef]
163. Conard, N.J.; Bolus, M. Radiocarbon Dating the Late Middle Paleolithic and the Aurignacian of the Swabian Jura. *J. Hum. Evol.* **2008**, *55*, 886–897. [CrossRef]
164. Orschiedt, J. Secondary Burial in the Magdalenian: The Brillenhöhle (Blaubeuren, Southwest Germany). *Paléo Rev. d'archéol. Préhist.* **2002**, *14*, 241–256. [CrossRef]
165. Sala, N.; Conard, N.J. Taphonomic Analysis of the Hominin Remains from Swabian Jura and Their Implications for the Mortuary Practices during the Upper Paleolithic. *Quat. Sci. Rev.* **2016**, *150*, 278–300. [CrossRef]
166. Beck, D. *Das Mittelpaläolithikum Des Hohlenstein-Stadel Und Bärenhöhle-Im Lonetal*; Habelt: Bonn, Germany, 1999.
167. Wetzel, R. Der Hohlestein Im Lonetal: Dokumente Alteuropäischer Kulturen Vom Eiszeitalter Bis Zur Völkerwanderung. *Mitteilungen Ver. Naturwissenschaft Math. Ulm* **1961**, *27*, 21–75.
168. Beutelspacher, T.; Ebinger-Rist, N.; Kind, C.J. Neue Funde Aus Der Stadelhöhle Im Hohlenstein Bei Asselfingen. *Archäol. Ausgrabungen Baden-Württemberg* **2011**, *2010*, 65–70.
169. Bolus, M.; Conard, N.J.; Kandel, A.W. Grabungen Vor Dem Hohlenstein Im Lonetal, Gemeinden Bissingen Und Asselfingen, Alb-Donau-Kreis. *Archäol. Ausgrabungen Baden-Württemberg* **1998**, 40–47.
170. Gamble, C. Hunting Strategies in the Central European Palaeolithic. *Proc. Prehist. Soc.* **1979**, *45*, 35–52. [CrossRef]
171. Wolf, S.; Conard, N.J.; Kind, C.-J. Schmuck Aus Dem Aurignacien von Der Schwäbischen Alb Im Vergleich Mit Inventaren Aus Dem Lahntal Und Dem Rheinland. *Archäol. Korresp.* **2013**, *43*, 295–313.

172. Wehrberger, K. *Begleitbuch zur Ausstellung Die Rückkehr des Löwenmenschen-Geschichte, Mythos, Magie: Ulmer Museum, 15. November 2013–9. Juni 2014: Ein Projekt des Ulmer Museums und des Landesamts für Denkmalpflege im Regierungspräsidium Stuttgart*; Thorbecke: Ostfildern, Germany, 2013; ISBN 978-3-7995-0509-3.
173. Bartolini-Lucenti, S.; Madurell-Malapeira, J. Unraveling the Fossil Record of Foxes: An Updated Review on the Plio-Pleistocene *Vulpes* Spp. from Europe. *Quat. Sci. Rev.* **2020**, *236*, 106296. [CrossRef]
174. Kurtén, B. *Pleistocene Mammals of Europe*; Weidenfeld and Nicolson: London, UK, 1968; ISBN 978-1-138-52995-3.
175. Voigt, D. Red Fox. In *Wild Furbearer Management and Conservation in North America*; Novak, M., Baker, J., Obbard, M., Malloch, B., Eds.; Ontario Trappers Association Publishers: North Bay, ON, Canada, 1987.
176. Von Königswald, W. *Lebendige Eiszeit. Klima Und Tierwelt Im Wandel*; Theiss Verlag: Darmstadt, The Netherlands, 2002; ISBN 978-3-89678-691-3.
177. Nowak, R.M. *Walker's Mammals of the World*; Fifth; The Johns Hopkins University Press: Baltimore, MD, USA, 1991; ISBN 978-0-8018-3970-2.
178. Prestrud, P. Adaptations by the Arctic Fox (*Alopex lagopus*) to the Polar Winter. *Arctic* **1991**, *44*, 132–138. [CrossRef]
179. Dell'Arte, G.; Laaksonen, T.; Norrdahl, K.; Korpimäki, E. Variation in the Diet Composition of a Generalist Predator, the Red Fox, in Relation to Season and Density of Main Prey. *Acta Oecologica* **2007**, 276–281. [CrossRef]
180. Kapel, C.M.O. Diet of Arctic Foxes (*Alopex lagopus*) in Greenland. *Arctic* **1999**, *52*, 289–293. [CrossRef]
181. Elmhagen, B.; Tannerfeldt, M.; Verucci, P.; Angerbjörn, A. The Arctic Fox (*Alopex lagopus*): An Opportunistic Specialist. *J. Zool.* **2000**, *251*, 139–149. [CrossRef]
182. Angerbjorn, A.; Tannerfeldt, M.; Erlinge, S. Predator–Prey Relationships: Arctic Foxes and Lemmings. *J. Anim. Ecol.* **1999**, *68*, 34–49. [CrossRef]
183. Hersteinsson, P. Population Genetics and Ecology of Different Colour Morphs of Arctic Foxes *Alopex lagopus* in Iceland. *Finn. Game Res.* **1989**, *46*, 64–78.
184. Chesemore, D.L. Notes on the Food Habits of Arctic Foxes in Northern Alaska. *Can. J. Zool.* **1968**, *46*, 1127–1130. [CrossRef]
185. D'hondt, B.; Vansteenbrugge, L.; Van Den Berge, K.; Bastiaens, J.; Hoffmann, M. Scat Analysis Reveals a Wide Set of Plant Species to Be Potentially Dispersed by Foxes. *Plecevo* **2011**, *144*, 106–110. [CrossRef]
186. Savory, G.A. *Foxes and Food Subsidies: Anthropogenic Food Use by Red and Arctic Foxes, and Effects on Arctic Fox Survival, on the Arctic Coastal Plain of Alaska*; University of Alaska Fairbanks: Fairbanks, AK, USA, 2013.
187. Cavallini, P. Variation in the Social System of the Red Fox. *Ethol. Ecol. Evol.* **1996**, *8*, 323–342. [CrossRef]
188. Gravel, R.; Lai, S.; Berteaux, D. Long-Term Satellite Tracking Reveals Patterns of Long-Distance Dispersal in Juvenile and Adult Arctic Foxes (*Vulpes lagopus*). *R. Soc. Open Sci.* **2023**, *10*, 220729. [CrossRef] [PubMed]
189. Lai, S.; Bêty, J.; Berteaux, D. Spatio–Temporal Hotspots of Satellite–Tracked Arctic Foxes Reveal a Large Detection Range in a Mammalian Predator. *Mov. Ecol.* **2015**, *3*, 1–10. [CrossRef]
190. Lai, S.; Bêty, J.; Berteaux, D. Movement Tactics of a Mobile Predator in a Meta-Ecosystem with Fluctuating Resources: The Arctic Fox in the High Arctic. *Oikos* **2017**, *126*, 937–947. [CrossRef]
191. Walton, Z.; Samelius, G.; Odden, M.; Willebrand, T. Variation in Home Range Size of Red Foxes Vulpes Vulpes along a Gradient of Productivity and Human Landscape Alteration. *PLoS ONE* **2017**, *12*, e0175291. [CrossRef]
192. Fuglei, E.; Tarroux, A. Arctic Fox Dispersal from Svalbard to Canada: One Female's Long Run across Sea Ice. *Polar Res.* **2019**, *38*, 3512. [CrossRef]
193. Tarroux, A.; Berteaux, D.; Bêty, J. Northern Nomads: Ability for Extensive Movements in Adult Arctic Foxes. *Polar Biol.* **2010**, *33*, 1021–1026. [CrossRef]
194. Angerbjörn, A.; Hersteinsson, P.; Tannerfeldt, M. Arctic Foxes. Consequences of Resource Predictability in the Arctic Fox—Two Life History Strategies. In *The Biology and Conservation of Wild Canids*; Macdonald, D.W., Sillero-Zubiri, C., Eds.; Oxford University Press: Oxford, UK, 2004.
195. Frafjord, K. Reproductive Effort in the Arctic Fox Alopex Lagopus. *Nor. J. Agric. Sci.* **1993**, *7*, 301.
196. Strand, O.; Landa, A.; Linnell, J.D.; Zimmermann, B.; Skogland, T. Social Organization and Parental Behavior in the Arctic Fox. *J. Mammal.* **2000**, *81*, 223–233. [CrossRef]
197. Tannerfeldt, M.; Moehrenschlager, A.; Angerbjörn, A. Den Ecology of Swift, Kit and Arctic Foxes: A Review. In *The Swift Fox: Ecology and Conservation of Swift Foxes in a Changing World*; Sovada, M., Carbyn, L., Eds.; Canadian Plains Research Center: Regina, Saskatchewan, 2003; pp. 167–181.
198. Chevallier, C.; Lai, S.; Berteaux, D. Predation of Arctic Fox (*Vulpes lagopus*) Pups by Common Ravens (*Corvus corax*). *Polar Biol.* **2016**, *39*, 1335–1341. [CrossRef]
199. Tannerfeldt, M.; Angerbjörn, A. Life History Strategies in a Fluctuating Environment: Establishment and Reproductive Success in the Arctic Fox. *Ecography* **1996**, *19*, 209–220. [CrossRef]
200. Pamperin, N.J.; Follmann, E.H.; Petersen, B. Interspecific Killing of an Arctic Fox by a Red Fox at Prudhoe Bay, Alaska. *Arctic* **2006**, *59*, 361–364. [CrossRef]
201. Tannerfeldt, M.; Elmhagen, B.; Angerbjörn, A. Exclusion by Interference Competition? The Relationship between Red and Arctic Foxes. *Oecologia* **2002**, *132*, 213–220. [CrossRef]
202. D'Errico, F.; Zilhão, J.; Julien, M.; Baffier, D.; Pelegrin, J. Neanderthal Acculturation in Western Europe? A Critical Review of the Evidence and Its Interpretation. *Curr. Anthropol.* **1998**, *39*, S1–S44. [CrossRef]

203. Sáenz de Buruaga, A. *El Paleolítico Superior de La Cueva de Gatzarria (Zuberoa, País Vasco)*; Veleia (Anejos series maior); Instituto de Ciencias de la Antigüedad: Vitoria, Spain, 1991; Volume 6.
204. Sáenz de Buruaga Blázquez, A. Colgantes y Otras Manifestaciones Artísticas En Los Niveles Del Paleolítico Superior Inicial de La Cueva de Gatzarria (Zuberoa, País Vasco). *Veleia* **1989**, *6*, 21–48.
205. Guadelli, A. Kostni Artefakti Ot Paleolita v Bulgaria (Les Artefacts En Matière Dure Animale Du Paléolithique Bulgare). 2011. Available online: https://www.academia.edu/3132703/%D0%9A%D0%BE%D1%81%D1%82%D0%BD%D0%B8_%D0%B0%D1%80%D1%82%D0%B5%D1%84%D0%B0%D0%BA%D1%82%D0%B8_%D0%BE%D1%82_%D0%BF%D0%B0%D0%BB%D0%B5%D0%BE%D0%BB%D0%B8%D1%82%D0%B0_%D0%B2_%D0%91%D1%8A%D0%BB%D0%B3%D0%B0%D1%80%D0%B8%D1%8F_Les_artefacts_en_mati%C3%A8re_dure_animale_du_pal%C3%A9olithique_Bulgare_ (accessed on 15 August 2023).
206. Zwyns, N. *Laminar Technology and the Onset of the Upper Paleolithic in the Altai, Siberia*; Studies in human evolution; Leiden University Press: Leiden, The Netherlands, 2012; ISBN 978-90-8728-173-1.
207. Dinnis, R.; Bessudnov, A.; Reynolds, N.; Devièse, T.; Pate, A.; Sablin, M.; Sinitsyn, A.; Higham, T. New Data for the Early Upper Paleolithic of Kostenki (Russia). *J. Hum. Evol.* **2019**, *127*, 21–40. [CrossRef]
208. Sinitsyn, A.A. A Palaeolithic 'Pompeii' at Kostenki, Russia. *Antiquity* **2003**, *77*, 9–14. [CrossRef]
209. Soffer, O. *The Upper Paleolithic of the Central Russian Plain*; Academic: London, UK, 1985; ISBN 0-12-654270-8.
210. Fernández, E.Á.; Jöris, O. Personal Ornaments in the Early Upper Paleolithic of Western Eurasia: An Evaluation of the Record. *Eurasian Prehist.* **2008**, *5*, 29–42.
211. Kozlowski, J.K. The Balkans in the Middle and Upper Palaeolithic: The Gate to Europe or a Cul-de-Sac? In Proceedings of the Prehistoric Society; Cambridge University Press: Cambridge, UK, 1992; Volume 58, pp. 1–20.
212. Hahn, J. Aurignacian Signs, Pendants and Art Objects in Central and Eastern Europe. *World Archaeol.* **1972**, *3*, 252–266. [CrossRef]
213. Maureille, B.; Faivre, J.-P.; Gravina, B.; Rigaud, S.; Madelaine, S.; Bessou, M.; Couture-Veschambre, C. La grotte Castaigne (Commune de Torsac, Charente, France). Présentation d'un site méconnu riche en vestiges humains (dont des Néandertaliens). *PALEO. Rev. d'Archéol. Préhist.* **2020**, 196–221. [CrossRef]
214. Rigaud, S.; Roussel, M.; Rendu, W.; Primault, J.; Renou, S.; Hublin, J.-J.; Soressi, M. Les Pratiques Ornementales à l'Aurignacien Ancien Dans Le Centre-Ouest de La France: L'apport Des Fouilles Récentes Aux Cottés (Vienne). *Bulletin Soc. Préhist. Française* **2014**, *111*, 19–38. [CrossRef]
215. Michel, A.; d'Errico, F.; Lenoble, A.; Vanhaeren, M.; Grootes, P.; Renou, S.; Rendu, W.; Daulny, L.; Verna, C.; Roger, G. Nouvelles Fouilles Sur Le Site Aurignacien Chez Les Rois (Mouthiers-Sur-Boëme, Charente). In *Nouvelles fouilles dans le site aurignacien des Rois, Mouthiers-sur-Boëme (Charente)*; 2006; pp. 289–299. Available online: https://shs.hal.science/halshs-00451991/ (accessed on 15 August 2023).
216. Costamagno, S.; Laroulandie, V. *Mode de Vie Au Magdalénien: Apports de l'archéozoologie*; Archaeopress: Oxford, UK, 2003; ISBN 1-84171-516-6.
217. Dumitru, F. The Lost Art. An Overview on the Romanian Palaeolithic Art. *Annales D'Université "Valahia" Târgovişte. Sect. D'Archéologie d'Histoire* **2006**, *8*, 198–206. [CrossRef]
218. Akhmetgaleeva, N.B. The Worked Bone Artifacts from the Siuren I Rock-Shelter (Crimea): The 1990s Excavation. In *Siuren I rock-shelter. From Late Middle Paleolithic and Early Upper Paleolithic to Epi-Paleolithic in Crimea.*; Demidenko, Y.E., Otte, M., Noiret, P., Eds.; ERAUL: Liège, Belgium, 2012; Volume 129, pp. 79–90.
219. Vercoutère, C.; Giacobini, G.; Patou-Mathis, M. Une dent humaine perforée découverte en contexte Gravettien ancien à l'abri Pataud (Dordogne, France). *L'Anthropologie* **2008**, *112*, 273–283. [CrossRef]
220. Simon, U.; Händel, M.; Einwögerer, T.; Neugebauer-Maresch, C. The Archaeological Record of the Gravettian Open Air Site Krems-Wachtberg. *Quat. Int.* **2014**, *351*, 5–13. [CrossRef]
221. Teschler-Nicola, M.; Fernandes, D.; Händel, M.; Einwögerer, T.; Simon, U.; Neugebauer-Maresch, C.; Tangl, S.; Heimel, P.; Dobsak, T.; Retzmann, A.; et al. Ancient DNA Reveals Monozygotic Newborn Twins from the Upper Palaeolithic. *Commun. Biol.* **2020**, *3*, 1–11. [CrossRef]
222. Händel, M.; Einwögerer, T.; Simon, U. Krems-Wachtberg—A Gravettian Settlement Site in the Middle Danube Region. *Wiss. Mitteilungen Niederösterreichisches Landesmus.* **2008**, *19*, 91–107.
223. Svoboda, J.A.; Frouz, M. Symbolic Objects and Items of Decoration. In *Pavlov Excavations 2007–2011*; Institute of Archaeology at Brno, Academy of Sciences of the Czech Republic: Brno, Czech Republic, 2011.
224. Svoboda, J.A.; Bocheńnski, Z.M.; Čulíková, V.; Dohnalová, A.; Hladilová, S.; Hložek, M.; Horáček, I.; Ivanov, M.; Králík, M.; Novák, M.; et al. Paleolithic Hunting in a Southern Moravian Landscape: The Case of Milovice IV, Czech Republic. *Geoarchaeology* **2011**, *26*, 838–866. [CrossRef]
225. Lázničková-Galetová, M. Gravettian Ivory Ornaments in Central Europe, Moravia (Czech Republic). *L'Anthropologie* **2021**, *125*, 102870. [CrossRef]
226. Costamagno, S.; Laroulandie, V. L'exploitation Des Petits Vertébrés Dans Les Pyrénées Françaises Du Paléolithique Au Mésolithique: Un Inventaire Taphonomique et Archéozoologique. In *J.-P. Brugal, J. Desse (dir.), Petits Animaux et Sociétés Humaines. Du Complément Alimentaire aux Ressources Utilitaires. Actes des XXIVe Rencontres Internationales d'archéologie et d'histoire, Antibes, France, 23–25 October 2003*; APDCA: Antibes, France, 2004; pp. 369–382.
227. Beldiman, C.; Sztancs, D.-M. Art Mobilier Au Paleolithique Superieur En Transylvanie, Roumanie. *Rev. Bistrilei* **2006**, *20*, 9–20.

228. Cârciumaru, M.; Niţu, E.-C. *Symbolic Behaviour and Art on the Territory of Romania from the Middle Palaeolithic to the Mesolithic: (55,000–7,500 BP)*; Cetatea de Scaun: Suceava, Romania, 2018; ISBN 978-606-537-429-4.
229. Yanevich, A.; Péan, S.; Crépin, L.; Láznicková-Galetová, M.; Prat, S.; Prysyajnuk, V. Upper Palaeolithic Settlements in Buran-Kaya 3 (Crimea, Ukraine): New Interdisciplinary Researches of the Layers 5-2, 6-1 and 6-2. *Aoo aaa* **2009**, *20*, 187–202.
230. Taborin, Y. La Parure Des Solutréens et Des Magdaléniens Anciens Des Jamblancs. *Paléo Rev. d'Archéol. Préhist.* **1991**, *3*, 101–108. [CrossRef]
231. Peyrony, E. La Station Préhistorique Du Pech de La Boissière. *Bull. De La Soc. Préhist. De Fr.* **1934**, *31*, 194–213. [CrossRef]
232. Höck, C. *Das Magdalénien Der Kniegrotte: Ein Höhlenfundplatz Bei Döbritz, Saale-Orla-Kreis*; Weimarer Monographien zur Ur- und Frühgeschichte; Theiss-Verlag: Stuttgart, Germany, 2000; ISBN 978-3-8062-1595-3.
233. Müller, W.; Leesch, D. Einige Neubestimmungen Aus Der Magdalénien-Fundstelle Hollenberg-Höhle 3 Bei Arlesheim (Basel-Landschaft) Und Daraus Folgende Überlegungen Zur Nutzung Der Höhle. *Jahrb. Archäol. Schweiz* **2011**, *94*, 7–20.
234. Ladier, E. Nouvelles Recherches à La Grotte Du Cuzoulet (Bruniquel, Tarn-et-Garonne): Premiers Aperçus. *Bull. De La Soc. Archéol. De Tarn-Et-Garonne* **1998**, *123*, 7–20.
235. Rousseau, L. Le Magdalénien Dans La Vienne. Découverte et Fouille d'un Gisement Du Magdalénien à Angles-Sur-l'Anglin (Vienne). *Bull. De La Soc. Préhist. Française* **1933**, *30*, 239–256. [CrossRef]
236. Conard, N.J. Eiszeitlicher Schmuck Auf Der Schwäbischen Alb. In *Eiszeitschmuck. Status und Schönheit*; Urgeschichtliches Museum Blaubeuren: Blaubeuren, Germany, 2003; Volume Museumsheft 6.
237. Conard, N.J.; Janas, A. Fortsetzung Der Ausgrabungen Am Hohle Fels Und Die Entdeckung Einer Markierten Mammutrippe Aus Dem Gravettien. *Archäol. Ausgrabungen Baden-Württemberg* **2018**, *2017*, 52–55.
238. Conard, N.J.; Janas, A. Ausgrabungen Im Hohle Fels: Fundschichten Aus Dem Mittelpaläolithikum Und Neues Zur Jagdtechnik Der Neandertaler. *Archäol. Ausgrabungen Baden-Württemberg* **2021**, *2020*, 60–65.
239. Conard, N.J.; Malina, M. Neue Ergebnisse Zum Mittelpaläolithikum, Zum Aurignacien Und Zu Den Letzen Neandertalern Am Hohle Fels Bei Schelklingen, Alb-Donau-Kreis. *Archäol. Ausgrabungen Baden-Württemberg* **2006**, *2005*, 17–20.
240. Conard, N.J.; Malina, M. Die Ausgrabungen 2007 Im Hohle Fels Bei Schelklingen, Alb-Donau-Kreis, Und Neue Einblicke in Die Anfänge Des Jungpaläolithikums. *Archäol. Ausgrabungen Baden-Württemberg* **2007**, 17–20.
241. Conard, N.J.; Malina, M. Neue Belege Für Malerei Aus Dem Magdalénien Vom Hohle Fels. *Archäol. Ausgrabungen Baden-Württemberg* **2010**, *2009*, 52–56.
242. Conard, N.J.; Malina, M. Neue Eiszeitkunst Und Weitere Erkenntnisse Über Das Magdalénien Vom Hohle Fels Bei Schelklingen. *Archäol. Ausgrabungen Baden-Württemberg* **2011**, *2010*, 56–60.
243. Conard, N.J.; Langguth, K.; Uerpmann, H.-P. Die Ausgrabungen 2003 in Den Gravettien- Und Aurignacien-Schichten Des Hohle Fels Bei Schelklingen, Alb-Donau-Kreis, Und Die Kulturelle Entwicklung Im Frühen Jungpaläolithikum. *Archäol. Ausgrabungen Baden-Württemberg* **2004**, *2003*, 17–22.
244. Conard, N.J.; Lingnau, M.; Malina, M. Einmalige Funde Durch Die Nachgrabung Am Vogelherd Bei Niederstotzingen-Stetten Ob Lonetal, Kreis Heidenheim. *Archäol. Ausgrabungen Baden-Württemberg* **2007**, *2006*, 20–24.
245. Conard, N.J.; Malina, M.; Miller, C. Die Fortsetzung Der Nachgrabung Am Vogelherd Bei Niederstotzingen-Stetten Ob Lonetal, Kreis Heidenheim. *Archäol. Ausgrabungen Baden-Württemberg* **2008**, *2007*, 21–24.
246. Conard, N.J.; Malina, M.; Verrept, T. Weitere Belege Für Eiszeitliche Kunst Und Musik Aus Den Nachgrabungen 2008 Am Vogelherd Bei Niederstotzingen-Stetten Ob Lonetal, Kreis Heidenheim. *Archäol. Ausgrabungen Baden-Württemberg* **2009**, *2008*, 23–27.
247. Conard, N.J.; Malina, M.; Zeidi Kulehparcheh, M. Neue Kunst Und Erste Einblicke in Ungestörte Schichten Am Vogelherd. *Archäol. Ausgrabungen Baden-Württemberg* **2010**, *2009*, 57–61.
248. Conard, N.J.; Zeidi, M.; Bega, J. Die Letzte Kampagne Der Nachgrabungen Am Vogelherd. *Archäol. Ausgrabungen Baden-Württemberg* **2013**, *2012*, 84–88.

quaternary

Article

The Late Palaeolithic and Mesolithic Settlement of Northern Italy: Problems and Perspectives

Paolo Biagi

Department of Asian and North African Studies, Ca' Foscari University of Venice, Ca' Cappello, 30125 Venice, Italy; pavelius@unive.it

Abstract: This paper considers some problems of the Late Palaeolithic and the Mesolithic periods in Northern Italy. More precisely, it deals with chronology, settlement pattern, techno-typological characteristics of knapped stone assemblages, and climatic changes that have taken place in the region from the discovery of the first sites in the 1960s and the excavations that soon followed to the present state of research. The Italian Alps, the Piedmont, and the valleys that descend from the high massifs have yielded important traces of Late Palaeolithic (Final Epigravettian) and Mesolithic (Sauveterrian and Castelnovian) sites and findspots, some of which are rock shelters that were settled throughout several millennia. This paper describes and discusses the evidence available mainly from two regions of the western and eastern Alpine arc, which are characterised by very different landscapes and yielded a great variety of archaeological features.

Keywords: northern Italy; late palaeolithic; final epigravettian; mesolithic; alpine chain; settlement pattern; radiocarbon chronology

check for updates

Citation: Biagi, P. The Late Palaeolithic and Mesolithic Settlement of Northern Italy: Problems and Perspectives. *Quaternary* **2024**, *7*, 2. https://doi.org/10.3390/quat7010002

Academic Editor: Ran Barkai

Received: 28 November 2023
Revised: 21 December 2023
Accepted: 25 December 2023
Published: 4 January 2024

1. Introduction

Research and excavations carried out during the last fifty years have greatly improved our view of the Late Pleistocene and Early Holocene archaeology of Italy. Little was known of these two periods in the decades preceding the outbreak of the Second World War when our knowledge was based on the results of excavations at just a few sites [1,2]. In those years, debate regarding the terminology used to subdivide the different periods of Italian prehistory [3], and the existence or nonexistence of some of these periods, the Upper Palaeolithic for example, was quite intense [4].

The first summary paper regarding the Mesolithic in Italy was written in 1960 [5]. It was followed, a few years later, by the discovery of an important Mesolithic sequence in the Riparo Blanc, in the Circeo Promontory, ca 80 km southeast of Rome [6]. The terminology that we use nowadays to describe the Late Palaeolithic and Mesolithic cultures of Italy was introduced in the 1970s, when the archaeology of these two periods began to develop in some regions of northern Italy, thanks to the results achieved from new excavations and surveys organised by a few universities and local natural history museums.

The history of the Late Palaeolithic Epigravettian [7] is quite different from that of the Mesolithic Sauveterrian and Castelnovian [8]. Regarding the Epigravettian, previously called Tardigravettian, we must refer to the seminal works written by G. Laplace [9–11], which focused on the subdivision of the Italian Upper Palaeolithic as it was conceived [12], shaped [13], and soon after adopted by many Italian prehistorians with just a few changes [14]. From a chronological point of view, the Final Epigravettian is radiocarbon-dated to ca 14,000–10,000 uncal BP yrs (ca 15,000–9500 cal BC yrs). It is followed by the beginning of the Holocene, after the Younger Dryas cold event, which marks the end of the Pleistocene.

So far, only a few regions of northern Italy have yielded Final Epigravettian sites. Many have been discovered in Liguria and Veneto-Trentino-Friuli [15] (Figure 1). They are less represented in other regions, Piedmont for example, most probably because of the geomorphological characteristics of the territory and the scarcity of good-quality knappable chert sources. Their location and altitude are variable, as are the characteristics of the sites, among which are rock shelters, caves, and open-air stations.

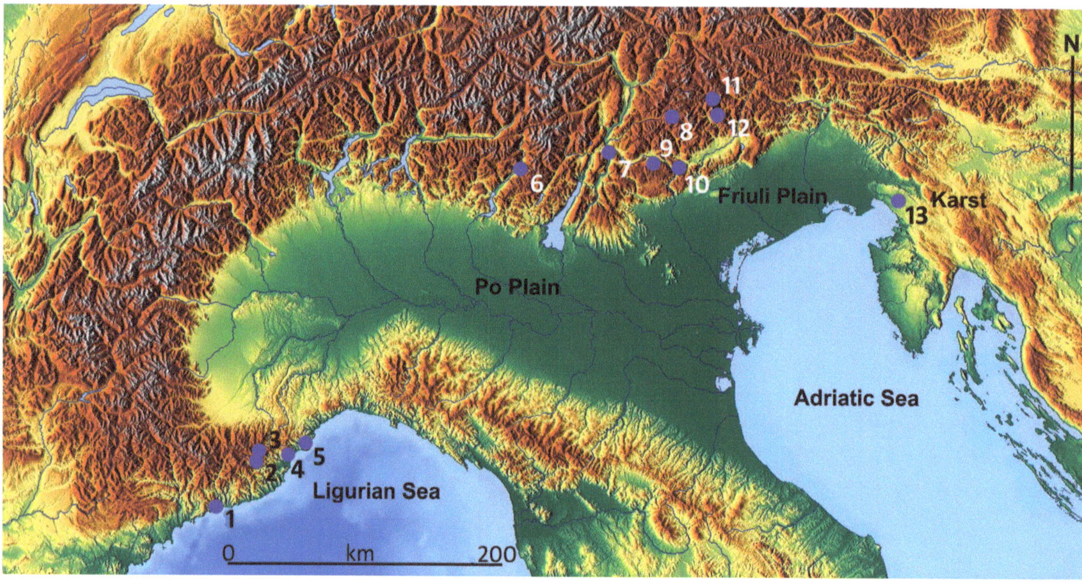

Figure 1. Distribution map of the most important sites reported in the text: Balzi Rossi (1), Bàsura (2), Val Pennavaira (3), Arma Veirana (4), Arene Candide (5), Cividate Camuno (6), Trento Basin (Vatte di Zambana, Romagnano, Pradestel), (7), Colbricon (8), Riparo Dalmeri (9), Val Cismon (10), Mondeval de Sora (11), Pian de la Lóra (12), and Trieste Karst (13) (drawing by P. Biagi).

The geography of northern Italy includes the Alpine arc, which is altogether very inhomogeneous, with a wide variety of landscapes. Among these are steep, narrow valleys incised between the high peaks of the western Alpine range, and the more walkable U-shaped valleys of the central and eastern chain, some of which lead to central and eastern Europe through high- and middle-altitude passes surrounded by glaciers, which nevertheless have always been crossed, at least since the Neolithic [16,17].

The central part is occupied by the Po Plain, a territory that has yielded very scarce evidence of Epigravettian occupation for reasons that are probably to be found in the recent Quaternary geology of the area [18] and intensive agricultural exploitation which lasted at least 4500 years [19]. However, this pattern contrasts with that available from other countries of Western Europe [20]. The Friuli Plain opens farther to the east. It is crossed by rivers which flow from the north. East of the plain rises the limestone Trieste and Slovene Karst Plateau. Many caves open in the Karst, though none has ever yielded evidence of Epigravettian occupation [21]. The western and eastern sides of this wide region are bounded by the Ligurian and Adriatic Seas. This description of northern Italy gives just an oversimplified idea of the complexity of a territory the extent of which has been greatly affected by the Final Pleistocene sea-level rise [7].

Regarding the Mesolithic, the late 1960s saw a dramatic improvement in the study of Early Holocene hunter-gatherers. This was due mainly to the results of the excavations carried out in three Adige Valley rock shelters whose long sequences, buried below metres of alluvial deposits, were uncovered in the Trento Basin: Vatte di Zambana, Romagnano, and Pradestel [22]. The results had a strong impact on the prehistorians of those years. However, even more unexpected was the discovery of high-altitude Mesolithic stations at the Colbricon Pass in the Trentino at around 2000 m of altitude [23,24]. This was followed a few years later by the discovery of other high-altitude camps in South Tyrol [25,26].

Until the 1970s, the Late Pleistocene and Early Holocene cultural sequence of most of Italy was based almost exclusively on the typological characteristics of knapped stone assemblages and their "evolution" [27,28]. In Italy, where Mesolithic sites were believed to be very few, or non-existent [5], many prehistorians studied the knapped stone artefacts following the taxonomic method of G. Laplace [8]. This approach changed only a few years later [22,29], when the lithic assemblages of the last hunter-gatherers of the Trentino began to be analysed according to the sites' locations, altitudes, functions, and structures, the provenance of the raw material employed for their manufacture [30–32], and the characteristics of the complementary activities carried out within different areas of the same site [33]. In addition, new anthropological and ethnographic approaches began to be applied to the study of the archaeological sites, especially the high-altitude stations [34,35] and the material culture remains.

Around the same time, pollen cores extracted from peat bogs located close to the Mesolithic sites began to be studied [36]. The scope was to interpret the altitude reached by the upper tree line during the progressive deglaciation of the Alpine chain between the Preboreal and Atlantic periods and the vegetation cover of the highland zones during the climatic oscillations of the end of the Last Glaciation [37,38]. The results conflicted with the ideas that were then in fashion in those years, according to which high altitudes were never settled before the Bronze and Iron Ages [39,40].

Moreover, while the anthropological–ethnographic approach to the study of prehistoric hunter-gatherers had been undertaken in several other countries from the 1960s onwards [41–44], in Italy, the new methods started to be applied a few years later [45,46]. However, high-altitude archaeology attracted only a few archaeologists despite its importance, novelty, and complexity [47,48].

The discoveries made in the Trentino were followed by several international round tables whose scope was to discuss the prehistoric peopling of the Alpine world and place Italian Alpine archaeology in its wider European context [49–52]. The important contributions of a few European colleagues [53–57], and the first systematic use of radiocarbon dating, led to the adoption of the term Mesolithic [58] to describe the Early Holocene cultural aspects which previously had been called Epipalaeolithic, following the French terminology which was widely employed in those years in Italy [59].

2. Key Sites and Sequences

2.1. The Ligurian Sites

This paper considers, almost exclusively, two regions of northern Italy (Liguria and Veneto-Trentino-Friuli) that have provided important data regarding the Early Holocene and the first 1300 years of the Northgrippian stage [60], during which the last hunter-gatherers became extinct.

Liguria is squeezed between high mountains and the Ligurian (part of the Tyrrhenian) Sea. The north-westernmost Apennines meet the south-easternmost fringes of the Alps at the Colle di Cadibona (459 m a.s.l.) in western Liguria (Liguria di Ponente). The only plain is the small Piana di Albenga in the Savona Province, which extends over ca 10 sq km. It is formed by the alluvial sediments of the Arroscia, Neva, and Pennavaira torrents (Figure 1).

In 1983 A. Palma di Cesnola wrote an important paper in which he described and discussed the knapped stone assemblages retrieved from Final Epigravettian sites excavated in the region and suggested their preliminary seriation according to the techno-typological characteristics of the artefacts [61]. A first glance at the subsistence and environmental aspects of the Ligurian sites of this period was provided only a few years later [15,61].

The excavations carried out in the narrow Pennavaira Valley (Figure 2) between the 1950s and the 1970s led to the discovery of two important Final Epigravettian sequences in the Nasino (246 m a.s.l.) and Stefanin rock shelters (440 m a.s.l.) [62–64]. A few more cave sites occupied during other periods of prehistory were discovered along the sides of the same valley. The knapped stone assemblages from the lower layers of the Nasino and Stefanin rock shelters have been attributed to the Final Epigravettian on purely techno-typological grounds. However, one of the Final Epigravettian layers of the Stefanin sequence has been dated from charcoal to 12,700 ± 300 BP, 14,029–12,111 cal BC (HAR-6915) [61] (Figure 2). No radiocarbon dates are available for Nasino, the assemblages from which have been attributed to the very end of the Final Epigravettian. Following the study of the bone remains, Stefanin has been interpreted as a winter camp exploited by ibex hunters *"who spent most of the year at lower elevations"* [61].

Figure 2. The Pennavaira Valley (Liguria) near the village of Alto (ca 630 m a.s.l.) where Arma dello Stefanin opens (white dot) (**top**). Arma dello Stefanin: profile of the Epigravettian–Neolithic sequence excavated in 1982 with the location of the Epigravettian radiocarbon date (white rectangle) (**bottom**) (photograph by P. Biagi, 1986 and 1982).

Two burials excavated in the outer part of the Nasino rock shelter have been radiocarbon-dated to the Boreal Mesolithic [65]. The bones of a child discovered in the Arma (cave) Veirana, which opens along the neighbouring Neva River Valley, have been dated to the same period [66]. These discoveries show that Mesolithic Sauveterrian groups moved along the narrow valleys of western Liguria during the Boreal. Most probably they aimed to reach the watershed to enter the Po Plain following an easy route. Mesolithic sites are almost unknown in western Liguria [67]. This fact contrasts with that of eastern Liguria (Liguria di Levante), where several middle- and high-altitude knapped stone concentrations of different Mesolithic periods were discovered mainly in the 1970s [68]. However, Final Epigravettian findspots are still unrecorded from this region. All the Mesolithic assemblages so far discovered are from the surface. None of them has ever been radiocarbon-dated because of the absence of organic material, bones, and charcoal.

Returning to the Final Epigravettian, two important Palaeolithic sequences are known from the Balzi Rossi caves and rock shelters that open close to the French border [69]. The coastal area is very rich in caves and rock shelters. Two of them have yielded evidence of Final Epigravettian occupation: the Mochi rock shelter and the Fanciulli cave. In both cases, the Final Epigravettian horizon has been found partly eroded at the very top of the sequences. No radiocarbon dates are available from the topmost layers B and A of the Mochi rock shelter [63] (Figure 3). The Fanciulli layer B has been dated to 12,200 ± 400 BP, 13,614–11,394 cal BC (MC-402) from marine shells, while burial GE1 has been dated directly to 11,130 ± 100 BP, 11,287–10,836 cal BC (GifA-94197) [70]. Besides these, two radiocarbon dates (12,310 ± 60 BP, 12,873–12,125 cal BC (GrA-69598) and 12,370 ± 60 BP, 12,902–12,199 BC (GrA-69597)) tell us that the well-known Bàsura cave, which opens in western Liguria, was visited around this time. Although the cave did not yield any knapped stone artefact, the footprints left by five individuals tell us that some corridors of the cave were explored by members of a Final Epigravettian community [71].

Figure 3. Riparo Mochi at the Balzi Rossi (Liguria): Profile of the old sequence (photograph by P. Biagi, 1990).

More impressive Younger Dryas discoveries have been made in the Arene Candide cave (Savona Province) [72] (Figure 4). The excavations carried out in 1940–1942 revealed a "cemetery" at the top of the Palaeolithic sequence with evidence of complex, unique funerary rituals (Figure 5) [73]. According to the radiocarbon results the "cemetery" was in use between ca 10,820 ± 40 BP, 10,876–10,776 cal BC (GrM-16978) and 10,245 ± 40 BP, 10,479–9807 cal BC (GrM-22252) [74].

Figure 4. Arene Candide (Liguria): cave entrance in the 1980s (photograph by P. Biagi, 1986).

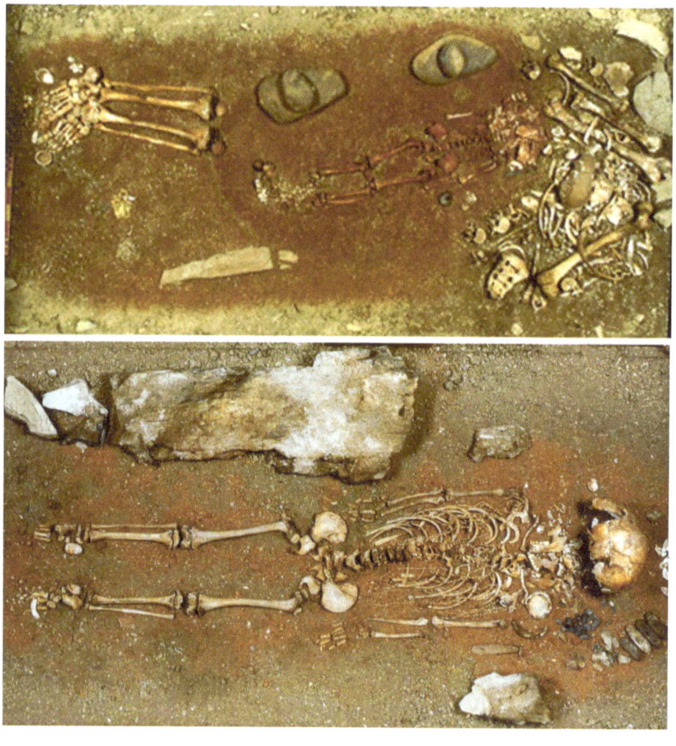

Figure 5. Arene Candide (Liguria) burials, grave goods, and ochre staining from the Final Epigravettian cemetery (photographs by P. Biagi, 2000).

To sum up, in Liguria, our knowledge of this crucial period of prehistory is fragmentary and highly differentiated between the two zones (western/Ponente and eastern/Levante) into which the region is commonly subdivided. This may be due to landscape characteristics, coastal uncontrolled urban development and road construction, the limited number of surveys, alluvial deposits, and dense forestation in some areas. Most discoveries are from caves, some of which have been known for at least a century [63,70,74]. However, Mesolithic Sauveterrian and Castelnovian findspots have been discovered at different altitudes, mainly in eastern Liguria [68].

2.2. The Veneto-Trentino-Friuli Sites

The problems related to the end of the Late Palaeolithic in north-eastern Italy have been summarised in a few papers in which many aspects of the Late Glacial period have been discussed [75–77]. Surveys and excavations carried out during recent decades have greatly improved our knowledge regarding settlement patterns, raw material supply and circulation [75,78–80], vegetation history, and exploitation of faunal resources [81–83]. Moreover, a good series of radiocarbon dates have helped us interpret the events that took place in the region from the 14th to the beginning of the 10th millennium BP (ca 15,000–9500 cal BC) [74].

An important aspect is the location of the archaeological sites [75,84–86] and the chronology of the earliest peopling of this part of the Alpine chain. Undoubtedly, the sites' distribution has been affected by the Late Pleistocene, Adriatic sea-level rise which dramatically reduced the extent of the territory available to the last Palaeolithic hunters [87,88], changed the sites' distribution pattern [7], and led to the present situation most probably around the beginning of the Atlantic or soon after [89].

The impressive amount of data collected in recent decades also includes mobile art [90]. The Riparo Dalmeri has yielded the remains of a hut structure ca 5.5 m in diameter and inside which many cobbles with red ochre painted animal, shamanic, and other naturalistic representations were recovered. The shelter opens at ca 1240 m above present sea level (a.s.l.) in a narrow valley, which descends from the northernmost edge of the Marcesina Plateau (Trento) [87,88]. According to the available eight radiocarbon dates the site was repeatedly occupied between 11,550 ± 70 BP, 11,634–11,309 cal BC (UtC-5040) and 10,800 ± 110 BP, 11,107–10,556 cal BC (R-425) [91,92].

Another important rock shelter, Riparo Villabruna, was settled a few centuries earlier than Riparo Dalmeri. Riparo Villabruna opens at ca 500 m a.s.l. at the bottom of the Cismon Valley between the Veneto and Trentino regions (Figure 6). The sequence has been radiocarbon-dated between 12,040 ± 150 BP, 12,846–11,536 cal BC (R-2023) and 11,910 ± 120 BP, 12,103–11,562 cal BC (UtC-1979). A male burial was found at the bottom of the sequence. The skeleton was covered with stone blocks, some of which were decorated with red ochre naturalistic patterns and one with a dancing shaman [93].

An important, still-debated subject concerns the "continuity" or "discontinuity" between the Younger Dryas and Preboreal Mesolithic. The problem was already being addressed in the 1980s following analysis of the knapped stone assemblages of the two periods. According to some authors, they show some clear differences in the techno-typological characteristics of the cores, tool types, presence of hypermicrolithic geometric armatures, and debitage dimension [94] (p. 672). Other authors suggest a "continuity" between the two periods, which is reinforced by a small set of new radiocarbon results [95–98]. However, the number of radiocarbon dates currently available for the south alpine Preboreal sites is limited [26,99–101].

Another problem concerns the distribution of the sites relative to the deglaciation of the region and the practicability of the valley bottoms and watersheds. The data at our disposal show that the distribution of the valley bottom sites is conditioned by several natural and anthropic factors, among which are slope stability and vegetation cover, which have hidden the sites below metres of alluvium. Consequently, the discovery of valley bottom sites is "*often the result of road construction and quarrying at the base of the limestone cliffs in the valleys*" [12] (p. 61). This is the case in the Adige Valley, where the most important Holocene sequences, radiocarbon-dated from the Preboreal onwards, have been excavated in the Trento Basin [102]. However, the discovery of an important Late Epigravettian site at Cividate Camuno, in neighbouring central Valcamonica, ca 45 km from the northernmost edge of the Po Plain [103], radiocarbon-dated to 13,805 ± 440 BP, 16,117–13,627 cal BC (GX-17273), shows that some of the alpine valley bottoms were already accessible many centuries before the end of the Late Glacial.

Figure 6. Val Cismon (Veneto/Trentino): the Villabruna rock shelter along the left side of the valley, where industrial quarrying has removed the alluvial fan (photograph by P. Biagi, 1989).

The first Mesolithic valley bottom site of the south Alpine arc was discovered by chance in February 1968 during industrial quarrying at Vatte di Zambana, along the right bank of the Adige Valley, ca 10 km north of Trento. The excavations that followed brought to light an important Mesolithic sequence, at the bottom of which the burial of a mature woman was found just below layer 10 [104], radiocarbon-dated on charcoal to 8000 ± 110 BP, 7309–6601 cal BC (R-491) and 7740 ± 150 BP, 7047–6266 cal BC (R-491α) [105]. The dates attribute the burial to the end of the Boreal Sauveterrian Mesolithic and confirm the results of typological analysis of the lithic assemblages [102]. The skeleton was placed supine, without grave goods, inside a burial pit that was later covered with limestone blocks, a ritual similar to that of the Final Epigravettian burial of Riparo Villabruna [93].

The discovery of two more important rock shelters in the Trento Basin, Romagnano III and Pradestel (Figure 7), followed a few months later. The three sites yielded long Mesolithic sequences. Their study led to a preliminary assessment of the Mesolithic cultures of northern Italy [22]. These were subdivided into Preboreal and Boreal Sauveterrian (previously called Sauveterroid) and early Atlantic Castelnovian (previously called Tardenoid) and were compared with those of the Trieste Karst caves [21,102]. In broad terms, the analysis of the knapped stone assemblages showed that the three periods mentioned above were characterised by assemblages with well-defined geometric hypermicrolithic or microlithic tools of triangular or trapezoidal shape, which changed throughout the three periods: hypermicrolithic isosceles triangles in the Preboreal, scalene triangles in the Boreal, and microlithic trapezes of different shapes in the Early Atlantic [57,102]. Radiocarbon dating of the three Trentino sequences confirmed the reliability of the subdivision suggested by the analysis of the knapped stone assemblages [99].

Figure 7. The rock shelter sequences of Romagnano III (**bottom**) and Pradestel (**top**) (Trentino) when they were discovered in 1969 during industrial quarrying. The red square shows the location of the Mesolithic sequence under excavation (photographs by P. Biagi, 1969, 1973).

In 1970, occasional research led to the discovery of the first high-altitude Mesolithic sites at the Passo Colbricon in the southern Alps of the Trentino (1925 m a.s.l.) (Figure 8) [23,24,106]. Hundreds of other sites of this period were discovered in the following years [85]. They showed that many high-altitude stations had been settled during the Boreal, and less so during the early Atlantic [8]. However, the precise chronology of the latter is often uncertain. The available distribution pattern would suggest a variable dislocation of the sites of the three different periods. They broadly reaffirm the impression that the Early Holocene peopling of the southern Alps took place gradually from the Preboreal, through the Boreal to the early Atlantic [8,85], though recently this view has been called into question [107].

Figure 8. One of the two Colbricon lakes (Trentino) where the first high-altitude Mesolithic sites were discovered (photograph by P. Biagi, 1983).

A unique discovery was made in the Veneto Dolomites in the 1980s. The burial of a Late Mesolithic Castelnovian hunter was discovered beneath a huge erratic boulder at 2150 m a.s.l. during the excavation of the Boreal site of Mondeval de Sora. Direct dating of the supine skeleton yielded a date of 7425 ± 55 BP, 6425–6095 cal BC (OxA-7468). The discovery is unique because all the bones were perfectly preserved despite the acidity of the mountain soil. Moreover, the grave goods have provided us important information about the possible routes followed by the Mesolithic hunter to reach the Dolomites, the artefacts he used to repair his hunting weapons, and the symbolic meaning of certain items found in association with the skeleton [108]. The presence of Castelnovian hunters in the Dolomites has been confirmed by the discovery of a Late Mesolithic camp at Pian de la Lóra in the Civetta Valley (1930 m a.s.l.), radiocarbon-dated on charcoal from a fireplace to 7920 ± 50 BP, 7035–6651 cal BC (GrN-31265) [109]. However, Mondeval de Sora is not an isolate burial case. Another Late Mesolithic supine individual was found in 1991 in the Adige Valley bottom near Mezzocorona, ca 25 km north of Trento [108].

Moving east, the Trieste Karst is the region of northern Italy that has yielded the first evidence of Mesolithic occupation. All the sites known so far are caves, the investigation of which started in the early 1960s [110]. The importance of the Karst caves in the definition of the Italian Mesolithic was stressed by A.M. Radmilli from the beginning [5], but emphatically revoked many years later [111]. The excavations carried out at the Edera Cave near Aurisina, close to the border with Slovenia, have yielded a long sequence with superposed Mesolithic layers, which were radiocarbon-dated, respectively, to the Preboreal, Boreal, and Early Atlantic [112] (Figure 9). All the Trieste Karst Mesolithic sequences are discontinuous and show long gaps between the aforementioned three periods [113]. For example, the Edera sequence shows a gap of ca 1500 years between the Preboreal (GrN-14108: 9810 ± 70 BP, 9647–8932 cal BC) and the Boreal (GrN-25139: 8350 ± 120 BP, 7588–7078 cal BC) (Figure 10).

Figure 9. Edera (Trieste Karst): the cave from the air (**left**) and the sequence brought to light during the 1992–1999 excavations with the radiocarbon-dated Mesolithic layers from 3d to 3a (**right**) (photographs by G. Marzolini, 1964 and P. Biagi, 1999).

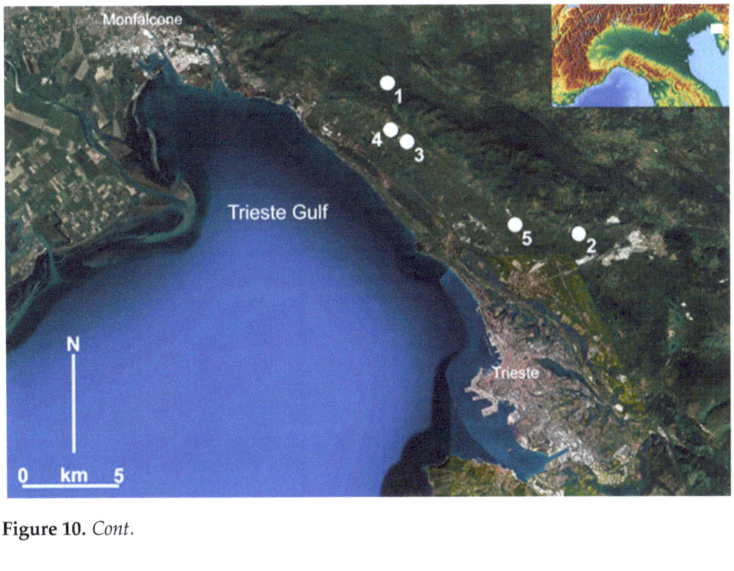

Figure 10. *Cont.*

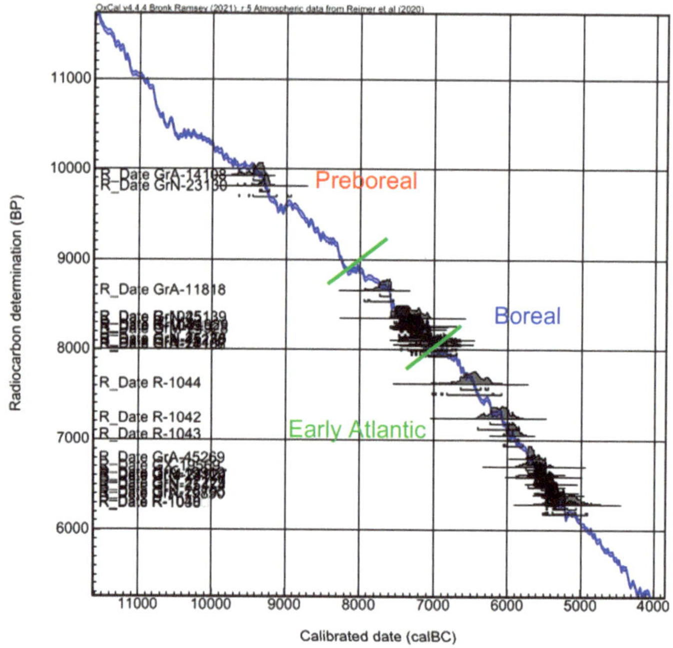

Figure 10. Distribution map of the Trieste Karst caves with radiocarbon-dated Mesolithic sequences: Benussi (1), Ciclami (2), Azzurra di Samatorza (3), Edera (4), and Zingari (5) (**top**). Plot of the available radiocarbon dates from the five cave sequences (**bottom**) (after Biagi, Spataro, 2024, Figure 1 [113]).

3. Discussion

The problems presented in the preceding section show how complex and fragmentary our knowledge of Final Pleistocene–Early Holocene archaeology is in the two regions considered in this paper. This is due mainly to their different landscapes, geologic and climatic characteristics, and research history.

Our knowledge of the Final Epigravettian in Liguria is restricted to a few cave sites, most of which were already known in the 19th century when the earliest pioneering research started to be conducted in the region [114]. Study of the Pennavaira Valley, Balzi Rossi, Bàsura, and Arene Candide caves has shed new light on this period of Ligurian prehistory and has improved our knowledge of the burial rituals and chronology of the Final Epigravettian and Mesolithic sites of the Ponente. At present, we know that Mesolithic hunter-gatherers also lived in the western part of the region. This evidence contrasts with that of the Levante, where Boreal and Early Atlantic middle- and high-altitude lithic concentrations have been found in the Apennines, close to good raw material outcrops. They are particularly numerous in the interior of the Tigullio Gulf [67].

However, despite the presence of Final Epigravettian sites and burials, so far we have no evidence for Preboreal stations. The Arene Candide cave sequence described by L. Cardini [73] tells us that the Final Epigravettian cemetery, which he imprecisely attributed to the Mesolithic, is sealed by a stalagmite layer above which are the Neolithic deposits [67].

For Veneto-Trentino-Friuli, our knowledge has greatly improved during recent decades thanks to the discovery of hundreds of high-altitude lithic concentrations and sites [85]. The first high-altitude Final Epigravettian sites were excavated in the Trentino in the early 1970s. Theories regarding the earliest peopling of this part of the Alpine chain were suggested soon after the excavations carried out at Andalo, Terlago, and Bondone [115,116]. They were followed by preliminary considerations of the lithic raw material supply and circulation. This subject was later developed in greater detail [33,75] as were territoriality [32] and mobility [117,118], settlement location at the upper limit of the tree line, and seasonality [119], comparing the complexity of spatial and altitude patterning observed in other zones of the Alps [120,121]. However, the Mesolithic peopling in this region [122] can be compared to that known in the Swiss and French Jura Alps [123], where the number of Boreal stations is greater than that of the Preboreal, while the Late Mesolithic ones are even fewer. The latter is still, nowadays, the lesser-known Mesolithic period [8], despite the great number of characteristic trapezoidal geometric armatures found isolated all over northern Italy [8]. We know almost nothing of the internal chronology of the Castelnovian in northeastern Italy, the way the knapped stone assemblages developed or changed, or how and when it disappeared. Only one high-altitude site excavated in the central Alps has yielded information regarding these aspects [46].

Regarding the Trieste Karst, chronological discontinuities are evident in the five Mesolithic radiocarbon-dated sequences (Figure 10 bottom). This fact has been reported from other European countries, where long Final Palaeolithic and Mesolithic stratigraphies have been systematically dated [124,125]. This consideration opens up new perspectives on the interpretation of the Mesolithic sequences and, consequently, on the study of the knapped stone assemblages.

The discovery of a high-altitude Late Mesolithic burial in the Dolomites is unique. It does not fit into the general pattern, according to which caves and rock shelters are preferred places for burying Final Epigravettian and Mesolithic hunters not only in the study regions but in Europe in general [126].

4. Conclusions

Many questions are still to be answered to achieve discrete knowledge of the Final Epigravettian and Mesolithic periods in northern Italy. Our current view is very fragmentary, though it has greatly improved in recent decades thanks to new surveys, improved excavation methods, and new series of radiocarbon dates. However, we still know very little of the characteristics of the sites' features. Just a few high- and middle-altitude Alpine sites excavated in South Tyrol and Lombardy have yielded different types of structural remains [8,45,46,101,119], which are in most cases difficult to interpret. This is also due to the mountain soil that, with very few exceptions, does not favor the preservation of bone remains. The Riparo Dalmeri has yielded bone remains of several mammal species, though the most important hunted for meat is ibex, followed by red deer [127]. The Sauveterrian site of Mondeval de Sora in the Dolomites was settled during the summer months. This is shown by the faunal assemblage, among which there are red deer and ibex juvenile bones [128]. Overall, the few data at our disposal from the two study regions show that the economic strategy of the Final Epigravettian and Early Mesolithic hunters was based on ibex [82].

Regarding the techno-typological characteristics of the lithic assemblages, we still know very little about the reasons for the hypermicrolithisation of the knapped stone artefacts that reached its apex during the Preboreal, similar to the pattern observed in other countries of Europe [129,130]. We know about differences between the Preboreal and Boreal assemblages, especially as regards the techniques employed in manufacturing triangular micro-armatures [131]. Another important problem concerns the dramatic climatic changes that took place at the end of the Boreal and beginning of the Atlantic, when the lithic assemblage characterised by trapezoidal geometrics took over [8,132,133] and slightly varied throughout ca one millennium, roughly from the beginning of the eighth to the beginning of the seventh millennium BP (ca 7000–6000 cal BC). Did hunting methods change not only in northern Italy, but throughout most of Europe, as the varieties of microlithic geometric tools would suggest? [134–136]. The Mesolithic sequences excavated in the Crimea [137] and radiocarbon-dated pollen cores from the same region [138] are indicative in this respect, though a North African origin for the Late Mesolithic blade and trapeze complexes has also been proposed [139].

Mobility is another topic of major importance in hunter-gatherer studies [120], as are the locations and functions of individual campsites, which can help us interpret mobility patterns, foraging radius and raw material exploitation [140]. The discovery of burials and cemeteries of different periods with varied ritual characteristics can help us in the interpretation of mobility. This is the case not only for the Late Mesolithic hunters of Mondeval de Sora but also for the Sauveterrian Val Pennavaira and Arma Veirana burials. Moreover, the impressive Final Epigravettian cemetery of the Arene Candide cave shows that social complexity was already emerging in that period, as the variety of rituals and grave goods would suggest [141].

To conclude, despite the progress made during recent decades, we can still direct the debate towards what has recently been published regarding Central European hunter-gatherer archaeology: "*Although hundreds of scientific publication deal with the Mesolithic ...., there are still only few high-quality excavations of Mesolithic sites with good preservation conditions. .... This starts with the still incomplete chronology, continues with the poor preservation of organic finds and ends with the rarity of secure archaeological contexts*" [126]. This also describes very well the situation in northern Italy.

Funding: This research paper has not been supported by any special funding.

Data Availability Statement: All data used during the study appear in the article.

Acknowledgments: The author is very grateful to E. Starnini (University of Pisa, I) and C. Franco (Ca' Foscari University of Venice) for the critical review of the text and to two anonymous reviewers whose recommendations have helped to improve the paper. Special thanks are due to C. Bonsall (Edinburgh University, UK) for the revision of the original English text. Throughout the text, radiocarbon calibrations have been made according to OxCal v4.4.4 [142,143].

Conflicts of Interest: The author declares no conflict of interest.

References

1. Blanc, G.L.; Romanelli, G., II. Dati ecologici e paletnologici. In *Atti della Prima Riunione dell'Istituto Italiano di Paleontologia Umana (21–24 Aprile 1927)*; Cardini, L., Ed.; Archivio per l'Antropologia e la Etnologia; Società Italiana di Antropologia e Etnologia: Firenze, Italy, 1927; Volume LVIII, pp. 365–522.
2. Cardini, L. Gli strati mesolitici e paleolitici della caverna delle Arene Candide. *Riv. St. Liguri* **1946**, *12*, 3–11.
3. Mochi, A. Schema di proposta sulla nomenclatura paletnologica italiana. In *Atti della Prima Riunione dell'Istituto Italiano di Paleontologia Umana (21–24 Aprile 1927)*; Cardini, L., Ed.; Archivio per l'Antropologia e la Etnologia; Società Italiana di Antropologia e Etnologia: Firenze, Italy, 1927; Volume LVIII, pp. 31–36.
4. Cipriani, L. Il primo assertore del Paleolitico Superiore in Italia: Ettore Regalia. In *Atti della Prima Riunione dell'Istituto Italiano di Paleontologia Umana (21–24 Aprile 1927)*; Cardini, L., Ed.; Archivio per l'Antropologia e la Etnologia; Società Italiana di Antropologia e Etnologia: Firenze, Italy, 1927; Volume LVIII, pp. 118–121.
5. Radmilli, A.M. Considerazioni sul Mesolitico Italiano. *Ann. Univ. Ferrara* **1960**, *XV*, 29–48.
6. Taschini, M. Il livello mesolitico del Riparo Blanc al Monte Circeo. *Bull. Paletnologia Ital.* **1964**, *73*, 65–88.
7. Ruiz-Redondo, A.; Vukosavljević, N.; Tomasso, A.; Peresani, M.; Davies, W.; Vander Linden, M. Mid and Late Upper Palaeolithic in the Adriatic Basin: Chronology, transitions and human adaptations to a changing landscape. *Quat. Sci. Rev.* **2022**, *276*, 107319. [CrossRef]
8. Franco, C. *La Fine del Mesolitico in Italia. Identità Culturale e Distribuzione Territoriale Degli Ultimi Cacciatori-Raccoglitori*; Società per la Preistoria e Protostoria della Regione Friuli-Venezia Giulia: Trieste, Italy, 2011; Quaderno 13.
9. Laplace, G. Essai de typologie systématique. *Ann. Univ. Ferrara* **1964**, *XV*, 1–86.
10. Laplace, G. Les subdivisions du Leptolithique Italien. Etude de typologie analytique. *Bull. Paletnologia Ital.* **1964**, *73*, 25–63.
11. Laplace, G. *Recherches sur l'Origine et l'Évolution des Complexes Leptolithiques. Ecole Française de Rome*; Mélanges d'Archéologie et d'Histoire, 4; De Boccard: Paris, France, 1966.
12. Bietti, A. The Late Upper Paleolithic in Italy: An Overview. *J. World Prehist.* **1990**, *4*, 95–155. [CrossRef]
13. Bartolomei, G.; Broglio, A.; Palma di Cesnola, A. Chronostratigraphie et ecologie de l'Epigravettien en Italie. In *La Fin des Temps Glaciaires en Europe*; de Sonneville-Bordes, D., Ed.; Editions du CNRS: Paris, France, 1979; pp. 297–324.
14. Palma di Cesnola, A. (Ed.) Actes du Colloque International La position taxonomique et chronologique des industries à pointes à dos autour de ta Méditerranée européenne. *Riv. Sci. Preist.* **1983**, *38*, 7–426.
15. Mussi, M.; Peresani, M. Human settlement of Italy during the Younger Dryas. *Quat. Int.* **2011**, *242*, 360–370. [CrossRef]
16. Barfield, L.H. Patterns of North Italian Trade. In *Archaeology and Italian Society. Prehistoric, Roman and Medieval Studies*; Barker, G., Hodges, R., Eds.; Papers in Italian Archaeology II; BAR Publishing: Oxford, UK, 1981; Volume 102, pp. 27–51.

17. Hafner, A. Schnidejoch und Lötschenpass. In *Archäologische Forschungen in der Bernen Alpe*; Archäologischer Dienst des Kanton Bern: Bern, Switzerland, 2015; Volume 2.
18. Cremaschi, M. *Paleosols and Vetusols in the Central Po Plain (Northern Italy). A Study in Quaternary Geology and Soil Development*; UNICOPLI: Milano, Italy, 1987.
19. Cremaschi, M.; Nicosia, C.; Salvioni, M. L'uso del suolo nell'Eneolitico e nel Bronzo antico, nuovi dati dalla Pianura Padana centrale. In *Atti della XLIII Riunione Scientifica dell'Istituto Italiano di Preistoria e Protostoria, L'età del Rame in Italia*; Istituto Italiano di Preistoria e Protostoria: Firenze, Italy, 2012; pp. 225–231.
20. Boemke, B.; Maier, A.; Schmidt, I.; Römer, W.; Lehmkuhl, F. Testing the representativity of Palaeolithic site distribution: The role of sampling bias in the European Upper and Final Palaeolithic record. *Quat. Sci. Rev.* **2023**, *316*, 108220. [CrossRef]
21. Cremonesi, G.; Pitti, C.; Radmilli, A.M. Considerazioni sul Mesolitico del Carso Triestino. In *Il Mesolitico sul Carso Triestino*; Società per la Preistoria e Protostoria della Regione Friuli-Venezia Giulia: Trieste, Italy, 1984; Quaderno 5; pp. 229–240.
22. Clark, R. *The Mesolithic Hunters of the Trentino. A Case Study in Hunter-Gatherer Settlement and Subsistence from Northern Italy*; Hadrian Books; BAR Publishing: Oxford, UK, 2000; Volume 832.
23. Bagolini, B. Primi risultati delle ricerche sugli insediamenti epipaleolitici del Colbricon (Dolomiti). *Preist. Alp.* **1972**, *8*, 107–149.
24. Bagolini, B.; Dalmeri, G. I siti mesolitici di Colbricon (Trentino). Analisi spaziale e fruizione del territorio. *Preist. Alp.* **1987**, *23*, 7–188.
25. Lunz, R. *Vor- und Frühgeschichte Südtirols, Band 1 Steinzeit*; Arti Grafiche Vallagarina: Calliano, Italy, 1986.
26. Grimaldi, S. Radiocarbon dating of the early Mesolithic Colbricon site (north-eastern Italian Alps). *J. Anthr. Sci.* **2006**, *84*, 137–145.
27. Bietti, A. The Mesolithic Cultures in Italy: New Activities in Connection with Upper Palaeolithic Cultural Traditions. In *Veröffentlichungen des Museums für Ur- und Frühgeschichte Potsdam*; VEB Deutscher Verlag der Wissenschaften: Berlin, Germany, 1981; Volume 14–15, pp. 33–50.
28. Bietti, A. Normal Science and Paradigmatic Biases in Italian Hunter-Gatherer Prehistory. In *Perspectives on the Past. Theoretical Biases in Mediterranean Hunter-Gatherer Research*; Clark, G.A., Ed.; University of Pennsylvania Press: Philadelphia, PA, USA, 1991; pp. 258–281.
29. Sackett, J.R. Straight Archaeology French Style: The Phylogenetic Paradigm in Historic Perspective. In *Perspectives on the Past. Theoretical Biases in Mediterranean Hunter-Gatherer Research*; Clark, G.A., Ed.; University of Pennsylvania Press: Philadelphia, PA, USA, 1991; pp. 109–139.
30. Dalmeri, P.; Grimaldi, S.; Lanzinger, M. Il Paleolitico e il Mesolitico. In *Storia del Trentino, la Preistoria e la Protostoria*; Il Mulino: Bologna, Italy, 2001; Volume 1, pp. 15–117.
31. Grimaldi, S. Un tentativo di definire un modello di territorio e mobilità dei cacciatori raccoglitori Sauveterriani dell'Italia nord-orientale. *Preist. Alp.* **2006**, *41*, 73–88.
32. Kompatscher, K.; Hrozny Kompatscher, N.M. Dove piantare il campo: Modelli insediativi e di mobilità nel Mesolitico in ambiente alpino. *Preist. Alp.* **2007**, *42*, 137–162.
33. Kompatscher, K.; Hrozny Kompatscher, N.M.; Bassetti, M. Intra-site spatial organization of a Mesolithic hunting camp: The open air site SA 42 Cresta di Siusi/Seiser Alm auf der Schneide (Dolomiti-Italy). *Preist. Alp.* **2022**, *52*, 63–107.
34. Séara, F. Approche de l'organisation spatiale de campements de chasseurs-cueilleurs mésolithiques: Le cas du Ruff ey-sur-Seille dans le Jura (France). In *MESO '97. Actes de la Table Ronde «Epipaléolithique et Mésolithique» Lausanne, 21–23 novembre 1997. Cahiers d'Archéologie Romande*; Crotti, P., Ed.; Musée Cantonal d'Archéologie et d'Histoire: Lausanne, Switzerland, 2000; Volume 81, pp. 139–154.
35. Pignat, G.; Winiger, A. Les occupations Mésolithiques de l'Abri du Mllendruz. Abri Freymond commune de Mont-la-Ville (VD, Suisse). In *Cahiers d'Archéologie Romande*; Musée Cantonal d'Archéologie et d'Histoire: Lausanne, Switzerland, 1998; Volume 72, pp. 3–210.
36. Oeggl, K. The palynological record of human impact on highland zone ecosystem. In *Highland Zone Exploitation in Southern Europe. Monografie di Natura Bresciana*; Biagi, P., Nandris, J., Eds.; Museo Civico di Scienze Naturali: Brescia, Italy, 1994; Volume 21, pp. 107–122.
37. Pini, R.; Ravazzi, C.; Aceti, A.; Castellano, L.; Perego, R.; Quirino, T.; Valle, F. Ecological changes and human interaction in Valcamonica, the rock art valley, since the last deglaciation. *Alp. Mediterr. Quat.* **2016**, *29*, 19–34.
38. Pini, R.; Ravazzi, C.; Raiteri, L.; Guerreschi, A.; Castellano, L.; Comolli, R. From pristine forests to high-altitude pastures: An ecological approach to prehistoric human impact on vegetation and landscapes in the western Italian Alps. *J. Ecol.* **2017**, *105*, 1580–1597. [CrossRef]
39. Ravagni, G. *Profilo Preistorico del Trentino: Età della Pietra e del Bronzo*; Tridentum: Trento, Italy, 1930.
40. Pauli, L. *The Alps: Archaeology and Early History*; Thames and Hudson: London, UK, 1983.
41. Binford, L.R. Archaeology as Anthropology. *Am. Antiq.* **1962**, *28*, 217–225. [CrossRef]
42. Binford, L.R. Methodological Considerations of the Archaeological Use of Ethnographic Data. In *Man the Hunter the First Intensive Survey of a Single, Crucial Stage of Human Development—Man's Once Universal Way of Life*; Lee, R.B., DeVore, I., Eds.; Aldine: New York, NY, USA, 1968; pp. 268–273.
43. Binford, L.R. Working at Archaeology: The Debate, Arguments of Relevance, and the Generation Gap. In *Working at Archaeology*; Binford, L.R., Ed.; Academic Press: New York, NY, USA, 1983; pp. 157–167.

44. Clarke, D. Mesolithic Europe: The Economic Basis. In *Problems in Economic and Social Archaeology*; Sieveking, G.d.G., Longworth, I.H., Wilson, K.E., Eds.; Duckworth: London, UK, 1976; pp. 449–482.
45. Accorsi, C.A.; Bandini Mazzanti, M.; Leoni, L.; Pavarani, M. Il sito mesolitico sopra Fienile Rossino sull'Altipiano di Cariadeghe (Serle–Brescia): Aspetti pedostratigrafici, archeologici, antracologici e palinologici. *Nat. Brescia.* **1986**, *23*, 239–292.
46. Baroni, C.; Biagi, P. (Eds.) *Excavations at the High Altitude Mesolithic Site of Laghetti del Crestoso (Bovegno, Brescia–Northern Italy)*; Ateneo di Brescia, Accademia di Scienze Lettere ed Arti: Brescia, Italy, 1997; pp. 5–109.
47. Jochim, M.A. *Hunter-Gatherer Subsistence and Settlement. A Predictive Model*; Academic Press: London, UK, 1976.
48. Bettinger, R. Aboriginal Occupation at High Altitude: Alpine Villages in the White Mountains of Eastern California. *Am. Anthropol.* **1991**, *93*, 656–679. Available online: https://www.jstor.org/stable/679941 (accessed on 25 November 2023). [CrossRef]
49. Bagolini, B. Atti della Tavola Rotonda Internazionale Il Popolamento delle Alpi in Età Mesolitica VIII–V millennio a.c. *Preist. Alp.* **1983**, *19*, 1–294.
50. Bagolini, B.; Biagi, P.; Broglio, A.; Kozlowski, J.K.; Kozlowski, S.K.; Lanzinger, M. Human Adaptations to the Mountain Environment in the Upper Palaeolithic and Mesolithic. *Preist. Alp.* **1990**, *28*, 1–2.
51. Biagi, P. (Ed.) *The Neolithisation of the Alpine Region*; Monografie di Natura Bresciana; Museo Civico di Scienze Naturali: Brescia, Italy, 1990; Volume 13, pp. 5–235.
52. Biagi, P.; Nandris, J. (Eds.) *Highland Zone Exploitation in Southern Europe*; Monografie di Natura Bresciana; Museo Civico di Scienze Naturali: Brescia, Italy, 1994; Volume 20, pp. 7–338.
53. Kozłowski, S.K. *Cultural Differentiation of Europe from 10th to 5th Millennium B.C*; University Press: Warsaw, Poland, 1975.
54. Kozłowski, J.K.; Kozłowski, S.K. *Upper Palaeolithic and Mesolithic in Europe. Taxonomy and Palaeohistory*; Prace Komisji Archeologicznej; Polska Akademia Nauk, Oddział w Krakowie: Kraków, Poland, 1979; Volume 18.
55. Kozłowski, J.K.; Kozłowski, S.K. (Eds.) *Chipped Stone Industries of the Early Farming Cultures in Europe*; Archaeologia Interregionalis, Warsaw University: Warsaw, Poland; Jagiellonian University Cracow: Kraków, Poland, 1987.
56. Kozłowski, J.K.; Machnik, J. *Problèmes de la Néolithisation dans Certain Régions de l'Europe*; Prace Komisji Archeologicznej; Polska Akademia Nauk, Oddział w Krakowie: Kraków, Poland, 1980; Volume 21.
57. Broglio, A.; Kozłowski, S.K. Tipologia ed evoluzione delle industrie mesolitiche di Romagnano III. *Preist. Alp.* **1987**, *19*, 93–148.
58. Kozłowski, S.K. Introduction to the history of Europe in Early Holocene. In *The Mesolithic in Europe*; Kozłowski, S.K., Ed.; University Press: Warsaw, Poland, 1973; pp. 331–366.
59. Rozoy, J.G. The Franco-Belgian Epipalaeolithic. Current Problems. In *The Mesolithic in Europe*; Kozłowski, S.K., Ed.; University Press: Warsaw, Poland, 1973; pp. 503–530.
60. Walker, M.; Gibbard, P.; Head, M.J.; Berkelhammer, M.; Björck, S.; Cheng, H.; Cwynar, L.C.; Fisher, D.; Gkinis, V.; Long, A.; et al. Formal Subdivision of the Holocene Series/Epoch: A Summary. *J. Geol. Soc. India* **2019**, *93*, 135–141. [CrossRef]
61. Palma di Cesnola, A. L'Épigravettien évolué et final de la région haute-tyrrhénienne. *Riv. Sci. Preist.* **1983**, *XXXVIII*, 301–318.
62. Barker, G.; Biagi, P.; Clark, G.; Maggi, R.; Nisbet, R. From hunting to herding in the Val Pennavaira (Liguria–northern Italy). In *The Neolithisation of the Alpine Region*; Biagi, P., Ed.; Monografie di Natura Bresciana; Museo Civico di Scienze Naturali: Brescia, Italy, 1990; Volume 13, pp. 99–121.
63. Leale Anfossi, M. II giacimento dell'Arma dello Stefanin (Val Pennavaira–Albenga). Scavi 1952–1962. *Riv. Sci. Preist.* **1972**, *XXVII*, 249–321.
64. Palma di Cesnola, A. Arma di Nasino: L'industria dei livelli epipaleolitici. *Atti XVI Riun. Scient. Ist. Ital. Preist. Protost.* **1974**, 97–109.
65. Sparacello, V.S.; Goude, G.; Varalli, A.; Dori, I.; Gravel-Miguel, C.; Riel-Salvatore, J.; Palstra, S.W.L.; Moggi-Cecchi, J.; Negrino, F.; Starnini, E. Human remains from Arma di Nasino (Liguria) provide novel insights into the paleoecology of early Holocene foragers in northwestern Italy. *Sci. Rep.* **2023**, *13*, 16415. [CrossRef]
66. Hodgkins, J.; Orr, C.M.; Gravel-Miguel, C.; Riel-Salvatore, J.; Miller, C.E.; Bondioli, L.; Nava, A.; Lugli, F.; Talamo, S.; Hajdinjak, M.; et al. An infant burial from Arma Veirana in northwestern Italy provides insights into funerary practices and female personhood in early Mesolithic Europe. *Sci. Rep.* **2021**, *11*, 23735. [CrossRef]
67. Biagi, P.; Maggi, R.; Nisbet, R. Liguria 11,000–7000 BP. In *The Mesolithic in Europe. Papers Presented at the Third International Symposium*; Bonsall, C., Ed.; John Donald: Edinburgh, UK, 1985; pp. 533–540.
68. Biagi, P.; Maggi, R. Aspects of the Mesolithic age in Liguria. *Preist. Alp.* **1984**, *19*, 159–168.
69. Mussi, M. Le 15 figurine paleolitiche scoperte da Louis Alexandre Jullien ai Balzi Rossi. *Origini* **2004**, *XXVI*, 7–63.
70. Alciati, G.; Pesce Delfino, V.; Vacca, E. (Eds.) *Catalogue of Italian Fossil Human Remains from the Palaeolithic to the Mesolithic*; IStituto Italiano di Antropologia: Rome, Italy, 2005; No. 83.
71. Avanzini, M.; Salvador, I.; Starnini, E.; Arobba, D.; Caramiello, R.; Romano, M.; Citton, P.; Rellini, I.; Firpo, M.; Zunino, M.; et al. Following the Father Step in the Bowels of the Earth: The Ichnological Record from the Bàsura Cave (Upper Palaeolithic, Italy). In *Reading Prehistoric Human Tracks. Methods and Material*; Pastoors, A., Lennsen-Erz, T., Eds.; Springer Nature: Cham, Switzerland, 2021; pp. 251–276. [CrossRef]
72. Sparacello, V.S.; Rossi, S.; Pettitt, P.; Roberts, C.; Riel-Salvatore, J.; Formicola, V. New insights on Final Epigravettian funerary behaviour at Arene Candide Cave (Western Liguria, Italy). *J. Anthr. Sci.* **2018**, *96*, 161–184. [CrossRef]
73. Cardini, L. La necropoli Mesolitica delle Arene Candide (Liguria). *Mem. Ist. Ital. Paleont. Um.* **1980**, *3*, 9–31.

74. Sparacello, V.S.; Dori, I.; Rossi, S.; Varalli, A.; Riel-Salvatore, J.; Gravel-Miguel, C.; Riga, A.; Seghi, F.; Goude, G.; Palstra, S.W.L.; et al. New human remains from the Late Epigravettian necropolis of Arene Candide (Liguria, northwestern Italy): Direct radiocarbon evidence and inferences on the funerary use of the cave during the Younger Dryas. *Quat. Sci. Rev.* **2021**, *268*, 101731. [CrossRef]
75. Bertola, S.; Broglio, A.; Cassoli, P.F.; Cilli, C.; Cusinato, A.; Dalmeri, G.; de Stefani, M.; Fiore, I.; Fontana, F.; Giacobini, G.; et al. L'Epigravettiano recente nell'area prealpina e alpina orientale. In *L'Italia tra 15.000 e 10.000 Anni fa. Cosmopolitismo e Regionalità nel Tardoglaciale. Millenni Studi di Archeologia Preistorica*; Martini, F., Ed.; Museo Fiorentino di Preistoria: Firenze, Italy, 2007; Volume 5, pp. 39–94.
76. Peresani, M.; Ravazzi, C. (Eds.) *Le Foreste dei Cacciatori Paleolitici*; Società Naturalisti Silvia Zenari: Pordenone, Italy, 2009.
77. Ravazzi, C.; Peresani, M.; Pini, R.; Vescovi, E. Il Tardoglaciale nelle Alpi Italiane e in Pianura Padana. Evoluzione stratigrafica, storia della vegetazione e del popolamento antropico. *Ital. J. Quat. Sci.* **2007**, *20*, 163–184.
78. Bertola, S.; Fontana, F.; Visentin, D. Lithic raw material circulation and settlement dynamics in the Upper Palaeolithcic of the Venetian Alps. A key role for palaeoclimatic and landscape changes across the LGM? In *Palaeolithic Italy. Advanced Studies on Early Human Adaptations in the Apennine Peninsula*; Borgia, V., Cristiani, E., Eds.; Sidestone Press: Leiden, The Netherlands, 2018; pp. 219–246.
79. Peresani, M. Flint exploitation at Epigravettian and Mesolithic sites on the Asiago Plateau (Veneto Prealps). In *Highland Zone Exploitation in Southern Europe*; Biagi, P., Nandris, J., Eds.; Monografie di Natura Bresciana; Museo Civico di Scienze Naturali: Brescia, Italy, 1994; Volume 21, pp. 221–234.
80. Visentin, D. *The Early Mesolithic in Northern Italy and Southern France. An Investigation into Sauveterrian Lithic Technical Systems*; Archaeopress: Oxford, UK, 2018.
81. Cattani, L.; Ricciardi, S. Gli ambienti dei cacciatori paleolitici tra 15.000 e 10.000 anni fa. Aspetti floristici. In *L'Italia tra 15.000 e 10.000 Anni fa. Cosmopolitismo e Regionalità nel Tardoglaciale. Millenni Studi di Archeologia Preistorica*; Martini, F., Ed.; Museo Fiorentino di Preistoria: Firenze, Italy, 2007; Volume 5, pp. 11–19.
82. Sala, B. Mammalofaune tardoglaciali dell'Italia continentale. In *L'Italia tra 15.000 e 10.000 Anni fa. Cosmopolitismo e Regionalità nel Tardoglaciale. Millenni Studi di Archeologia Preistorica*; Martini, F., Ed.; Museo Fiorentino di Preistoria: Firenze, Italy, 2007; Volume 5, pp. 21–38.
83. Tagliacozzo, A.; Fiore, I. Hunting strategies in a mountain environment during the Late Glacial in north eastern Italy. *Preist. Alp.* **2009**, *44*, 79–94.
84. Angelucci, D.E.; Bassetti, M. Humans and their landscape from the Alpine Last Glacial Maximum to the Middle Holocene in Trentino: Geoarchaeological considerations. *Preist. Alp.* **2009**, *44*, 59–78.
85. Broglio, A.; Lanzinger, M. Considerazioni sulla distribuzione dei siti tra la fine del Paleolitico superiore e l'inizio del Neolitico nell'Italia nord-orientale. In *The Neolithisation of the Alpine Region*; Biagi, P., Ed.; Monografie di Natura Bresciana; Museo Civico di Scienze Naturali: Brescia, Italy, 1990; Volume 132, pp. 53–69.
86. Dalmeri, G.; Pedrotti, A. Distribuzione topografica dei siti del Paleolitico Superiore finale e Mesolitico in Trentino Alto-Adige e nelle Dolomiti Venete (Italia). *Preist. Alp.* **1992**, *28*, 247–267.
87. Brommer, M.B.; Weltje, G.J.; Trincardi, F. Reconstruction of sediment supply from mass accumulation rates in the Northern Adriatic Basin (Italy) over the past 19,000 years. *J. Geophys. Res.* **2009**, *114*, F02008. [CrossRef]
88. Lambeck, K.; Antonioli, F.; Purcell, A.; Silenzi, S. Sea-level change along the Italian coast for the past 10,000yr. *Quat. Sci. Rev.* **2004**, *23*, 1567–1598. [CrossRef]
89. Bortolini, E.; Pagani, L.; Oxilia, G.; Posth, C.; Fontana, F.; Badino, F.; Saupe, T.; Montinaro, F.; Margaritora, D.; Romandini, M.; et al. Early Alpine occupation backdates westward human migration in Late Glacial Europe. *Curr. Biol.* **2021**, *31*, 2484–2493.E7. [CrossRef]
90. Broglio, A.; Montoya, C. Les transformations techniques dans l'industrie lithique et l'art mobilier de l'Epigravettien recent des Préalpes de Vénetie. In *Pitture Paleolitiche nelle Prealpi Venete. Grotta di Fumane e Riparo Dalmeri*; Broglio, A., Dalmeri, G., Eds.; Memorie del Museo Civico di Storia Naturale di Verona, 2 Serie, Sezione Scienze Uomo; Museo di Storia Naturale: Verona, Italy, 2005; Volume 9, pp. 184–187.
91. Montoya, C. Apport de l'analyse technique à la compréhension de l'évolution des groupes humaines épigravettiens d'Italie Nord Orientale: La production lithique de l'US15a-65 du Riparo Dalmeri. *Preist. Alp.* **2008**, *43*, 191–208.
92. Dalmeri, G.; Neri, S.; Bassetti, M.; Cusinato, A.; Kompatscher, K.; Kompatscher, M.; Duches, R.; Flor, E.; Legrand, E.; Girardi, S.; et al. Riparo Dalmeri: Le pietre dipinte dell'area rituale. *Preist. Alp.* **2011**, *45*, 67–117.
93. Broglio, A. Considerazioni sulla produzione artistica dell'Epigravettiano recente del Veneto e del Trentino. *Riv. Sci. Preist.* **1998**, *XLIX*, 103–121.
94. Bagolini, B.; Broglio, A. Il ruolo delle Alpi nei tempi preistorici (dal Paleolitico al Calcolitico). In *Studi di Paletnologia in Onore di Salvatore M. Puglisi*; Liverani, M., Palmieri, A., Peroni, R., Eds.; Università La Sapienza: Roma, Italy, 1985; pp. 663–705.
95. Guerreschi, A. L'Epigravettiano di Piancavallo (Pordenone). *Preist. Alp.* **1975**, *11*, 255–293.
96. Bassetti, M.; Cusinato, A.; Dalmeri, G.; Kompatscher, M.H.; Kompatscher, K.; Wierer, U. Updating the Final Palaeolithic-Mesolithic transition in Trentino (NE Italy). *Preist. Alp.* **2009**, *44*, 121–135.
97. Dalmeri, G. (Ed.) Studi sul Riparo Cogola (Carbonare di Folgaria–Trento). Frequentazione Umana e Paleoambiente. *Preist. Alp.* **2004**, *40*, 89–200.

98. Cusinato, A.; Dalmeri, G.; Kompatscher, K.; Hrozhy Kompatscher, M. Gli insiemi litici della sequenza preistorica di Riparo Cogola e la problematica relativa alla transizione tra Epigravettiano e Mesolitico in Area Alpina. *Preist. Alp.* **2004**, *40*, 125–154.
99. Alessio, M.; Allegri, L.; Bella, F.; Broglio, A.; Calderoni, G.; Cortesi, C.; Improta, S.; Martinez, M.P.; Petrone, V.; Turi, B. ^{14}C datings of three Mesolithic series of Trento Basin in the Adige Valley (Vatte di Zambana, Pradestel, Romagnano) and comparison with Mesolithic series of other regions. *Preist. Alp.* **1983**, *19*, 245–254.
100. Alessio, M.; Angelucci, D.E.; Broglio, A.; Improta, S. New data for the chronology of the Mesolithic in the Dolomites. The radiocarbon dates from Plan de Frea (Selva Val Gardena, Italy). *Preist. Alp.* **1994**, *30*, 145–154.
101. Dalmeri, G.; Kompatscher, K.; Hrozny Kompatscher, M.; Bassetti, M.; Cusinato, A.; Piazzi, O. Dinamiche comportamentali degli ultimi cacciatori raccoglitori in area alpina. Il caso di studio del sito LR3 del Laghetto delle Regole (Castelfondo, Trento). *Preist. Alp.* **2004**, *40*, 5–26.
102. Broglio, A. Risultati preliminari delle Ricerche sui complessi epipaleolitici della Valle dell'Adige. *Preist. Alp.* **1971**, *7*, 135–241.
103. Lo Vetro, D.; Bertola, S.; Poggiani Keller, R.; Martini, F. Le industrie litiche paleolitiche e mesolitiche di Cividate Camuno–Via Palazzo (Valle Camonica, Brescia): Sistemi tecnici e materie prime. *Riv. Sci. Preist.* **2022**, *LXXII*, 837–844.
104. Corrain, C.; Graziati, G.; Leonardi, P. La sepoltura epipaleolitica nel riparo di Vatte di Zambana (Trento). *Preist. Alp.* **1976**, *12*, 175–212.
105. Alessio, M.; Bella, F.; Cortesi, V.C.; Turi, B. University of Rome Carbon-14 dates VII. *Radiocarbon* **1969**, *11*, 482–498. [CrossRef]
106. Bagolini, B.; Barbacovi, F.; Castelletti, L.; Lanzinger, M. Colbricon (scavi 1973–1974). *Preist. Alp.* **1975**, *11*, 201–235.
107. Fontana, F.; Cristiani, E.; Bertola, S.; Briois, F.; Guerreschi, A.; Ziggiotti, S. A snapshot of Late Mesolithic life through death: An appraisal of the lithic and osseous grave goods from the Castelnovian burial of Mondeval de Sora (Dolomites, Italy). *PLoS ONE* **2020**, *15*, e0237573. [CrossRef]
108. Dalmeri, G.; Mottes, E.; Nicolis, F. The Mesolithic burial of Mezzocorona-Borgonuovo (Trento): Some preliminary comments. *Preist. Alp.* **2001**, *34*, 129–138.
109. Franco, C. Excavations at the high altitude Mesolithic site of Pian de la Lóra (Val Civetta–Venetian Dolomites, Italy). *Preist. Alp.* **2016**, *48*, 83–87.
110. Cannarella, D. La storia delle ricerche del Mesolitico sul Carso. In *Il Mesolitico sul Carso Triestino*; Società per la Preistoria e Protostoria della Regione Friuli-Venezia Giulia: Trieste, Italy, 1984; Quaderno 5; pp. 13–20.
111. Pluciennik, M. Reconsidering Radmilli's Mesolithic. In *Studi sul Paleolitico, Mesolitico e Neolitico del Bacino dell'Adriatico in Ricordo di Antonio M. Radmilli*; Biagi, P., Ed.; Società per la Preistoria e Protostoria della Regione Friuli-Venezia Giulia: Trieste, Italy, 2000; Quaderno 8; pp. 171–183.
112. Biagi, P.; Starnini, E.; Voytek, B. The Mesolithic-Neolithic transition in the Trieste Karst (north-eastern Italy) as seen from the excavations at the Edera Cave. In *The Iron Gates in Prehistory. New Perspectives*; Bonsall, C., Boroneanţ, V., Radovanović, I., Eds.; BAR Publishing: Oxford, UK, 2008; Volume 1893, pp. 251–260.
113. Biagi, P.; Spataro, M. The Trieste Karst and its caves: Discontinuities and continuities in the Early Holocene. In *Papers Presented in Honour of T. Težak-Gregl*; Karavanić, I., Karavanić, I., Potrebica, H., Burić, M., Šošić Klindžić, R., Vukosavljević, N., Mavrović Mokos, J., Eds.; Zagreb, Croatia, 2024; *in press*.
114. De Pascale, A.; Negrino, F. La formazione delle collezioni preistoriche in Liguria. In *Colligite Fragmenta 2. Aspetti e Tendenze del Collezionismo Archeologico in Liguria "Un Altro Modo di Fare l'Italia"*; De Pascale, A., Gandolfi, D., Eds.; Istituto Internazionale di Studi Liguri: Bordighera, Italy, 2017; pp. 83–96.
115. Guerreschi, A. Il sito Epigravettiano di Andalo (Trento) ed alcune considerazioni sull'Epigravettiano finale nel nord Italia. *Preist. Alp.* **1984**, *20*, 15–38.
116. Cavulli, F.; Grimaldi, S. Raw material and settlement strategies at the Pleistocene/Holocene boundary in Trentino (north-eastern Italian Alps): A GIS approach. In *Mesolithic Horizons. Papers Presented at the Seventh International Conference on the Mesolithic in Europe, Belfast 2005*; McCartan, S., Schulting, R., Warren, C., Warren, P., Eds.; Oxbow Books: Oxford, UK, 2009; pp. 96–101.
117. Kompatscher, K.; Hrozny Kompatscher, N.M.; Bassetti, M.; Castiglioni, E.; Rottoli, M.; Wierer, U. Mesolithic settlement and mobility patterns at high altitudes. The site of Staller Sattel STS 4A (South Tyrol, Italy). *Quat. Int.* **2016**, *423*, 23–48. [CrossRef]
118. Preston, P.R.; Kador, T. Approaches to Interpreting Mesolithic Mobility and Settlement in Britain and Ireland. *J. World Prehistory* **2018**, *31*, 321–345. [CrossRef]
119. Fontana, F.; Guerreschi, A. Highland occupation in the southern Alps during the Early Holocene. In *Mesolithic on the Move*; Larsson, L., Kindgren, H., Knutsson, K., Loeffler, D., Åkerlund, A., Eds.; Oxbow Books: Oxford, UK, 2003; pp. 97–102.
120. Eriksen, B.V. Cultural Change or Stability in Prehistoric Hunter-Gatherer Societies. A Case Study from the Late Palaeolithic–Early Mesolithic in Southwestern Germany. In *Contributions to the Mesolithic in Europe. Papers Presented at the Fourth International Symposium 'The Mesolithic in Europe', Leuven*; Vermeersch, P.M., Ven Peer, P., Eds.; Leuven University Press: Leuven, Belgium, 1990; Volume 1990, pp. 193–202.
121. Gehlen, B.; Schön, W. The Mesolithic of Temperate Europe. In *Reference Module in Social Sciences*; Elsevier: Amsterdam, The Netherlands, 2023. [CrossRef]
122. Fontana, F.; Visentin, D.; Bertola, S.; Cristiani, E.; Dipino, N.; Flor, E.; Fontana, A. Investigating the Early-to-Late Mesolithic Transition in Northeastern Italy: A Multifaceted Regional Perspective. *Open Archaeol.* **2023**, *9*, 20220284. [CrossRef]
123. Cupillard, C.; Perrenoud-Cupillard, N. The Mesolithic of the Swiss and French Jura and its margins: 10,150–6000 BP. In *Mesolithic on the Move*; Larsson, L., Kindgren, H., Knutsson, K., Loeffler, D., Åkerlund, A., Eds.; Oxbow Books: Oxford, UK, 2003; pp. 82–95.

124. Biagi, P.; Spataro, M. Plotting the evidence: Some aspects of the radiocarbon chronology of the Mesolithic-Neolithic transition in the Mediterranean basin. *Atti Soc. Preist. Protohist. Friuli-VG* **2000**, *XII*, 15–54.
125. Biagi, P. The Last Hunter-Gatherers of the Northern Coast of the Black Sea and their Role in the Mesolithic of Europe. A View from Crimea. In *Southeast Europe before Neolithisation*; Krauss, R., Floss, H., Eds.; RessourcenKulturen; Tübingen University Press: Tübingen, Germany, 2016; Volume 1, pp. 113–129.
126. Orschiedt, J. The Late Upper Palaeolithic and earliest Mesolithic evidence of burials in Europe. *Philos. Trans. R. Soc. B* **2018**, *373*, 20170264. [CrossRef]
127. Fiore, I.; Tagliacozzo, A. Oltre lo stambecco: Gli altri mammiferi della struttura abitativa dell'US 26c a Riparo Dalmeri (Trento). *Preist. Alp.* **2008**, *43*, 209–236.
128. Alciati, G.; Cattani, L.; Fontana, F.; Gerhardinger, E.; Guerreschi, A.; Milliken, S.; Mozzi, P.; Rowley-Conwy, P. Mondeval de Sora: A high altitude Mesolithic campsite in the Italian Dolomites. *Preist. Alp.* **1994**, *28*, 351–366.
129. Nuzhnyi, D. Development of Microlithic Projectile Weapons in the Stone Age. *Anthropol. Préhistoire* **2000**, *11*, 95–101.
130. Biagi, P.; Starnini, E. The microlithisation of the chipped stones in the Old World: Case-studies from the Crimea and northern Italy. In *Multas per Gentes and Multa per Saecula*; Valde-Nowak, P., Sobczyk, K., Nowak, M., Źrałka, J., Eds.; Jagiellonian University: Kraków, Poland, 2018; pp. 229–241.
131. Chesnaux, L. Variability in the Manufacturing of Triangular Geometric Microliths during the Early Mesolithic: Toward a Simplification of Barb Manufacturing? A Comparative Techno-functional Analysis of Microlithic Assemblages from Saint-Lizier at Creysse (24) and La Grande Rivoire at Sassenage (28). *Palethnologie* **2014**, *6*, 54–64. [CrossRef]
132. Biagi, P.; Starnini, E. The origin and spread of the Late Mesolithic Blade and Trapeze Industries in Europe: Reconsidering J.G.D. Clark's hypothesis fifty years after. In *Interactions, Changes and Meanings. Essays in Honour of Igor Manzura on the Occasion of His 60th Birthday*; Ţerna, S., Govedarica, B., Eds.; High Anthropological School: Kishinev, Moldova, 2016; pp. 33–45.
133. Gronenborn, D. Migrations before the Neolithic? The Late Mesolithic blade-and-trapeze horizon in central Europe and beyond. In *Migration and Integration from Prehistory to the Middle Ages*; Meller, H., Daim, F., Krause, J., Risch, R., Eds.; Tagungen des Landesmuseums für Vorgeschichte Halle; Landesamt für Denkmalpflege und Archäologie Sachsen-Anhalt: Halle, Germany, 2017; Volume 17, pp. 113–128.
134. Binder, D. Mesolithic and Neolithic interaction in southern France and northern Italy: New data and current hypotheses. In *Europe's First Farmers*; Price, T.D., Ed.; Cambridge University Press: Cambridge, UK, 2000; pp. 117–143.
135. Kozłowski, S.K. *Thinking Mesolithic*; Oxbow Books: Oxford, UK, 2009.
136. Cziesla, E. *Grenzen im Wald. Stabilität und Kontinuität Während des Mesolithikums in der Mitte Europas*; Berliner Archäologische Forschungen; Verlag Marie Leidorf GmbH: Rahden, Germany, 2015; Volume 15.
137. Bibikov, S.N.; Stanko, V.N.; Koen, V.Y. *The Final Palaeolithic and Mesolithic of the Crimean Mountains*; I.I. Mechnikov University: Odessa, Ukraine, 1994.
138. Cordova, C.E.; Lehman, P.H. Holocene Environmental Change in Southwestern Crimea (Ukraine) in Pollen and Soil Records. *Holocene* **2005**, *15*, 263–277. [CrossRef]
139. Marchand, G.; Perrin, T. Why this revolution? Explaining the major technical shift in Southwestern Europe during the 7th millennium cal. BC. *Quat. Int.* **2017**, *428*, 73–85. [CrossRef]
140. Binford, L.R. The Archaeology of Place. *J. Anthropol. Archaeol.* **1982**, *1*, 5–31. [CrossRef]
141. Rowley-Conwy, P. Time, change and the archaeology of hunter-gatherers: How original is the 'Original Affluent Society'? In *Hunter-Gatherers. An Interdisciplinary Perspective*; Panter-Brick, C., Layton, R.H., Rowley-Conwy, P., Eds.; Cambridge University Press: Cambridge, UK, 2001; pp. 39–72.
142. Reimer, P.J.; Austin, W.E.N.; Bard, E.; Bayliss, A.; Blackwell, P.G.; Bronk Ramsey, C.; Butzin, M.; Cheng, H.; Edwards, R.L.; Friedrich, M.; et al. The IntCal20 Northern Hemisphere Radiocarbon Age Calibration Curve (0–55 Cal kBP). *Radiocarbon* **2020**, *62*, 725–757. [CrossRef]
143. Bronk Ramsey, C. *OxCal, Version 4.4*; University of Oxford: Oxford, UK, 2021.

MDPI AG
Grosspeteranlage 5
4052 Basel
Switzerland
Tel.: +41 61 683 77 34

Quaternary Editorial Office
E-mail: quaternary@mdpi.com
www.mdpi.com/journal/quaternary

www.ingramcontent.com/pod-product-compliance
Lightning Source LLC
Chambersburg PA
CBHW041405210926
51850CB00013B/3183
* 9 7 8 3 7 2 5 8 3 1 6 2 3 *